Amniotic Membrane

Ana Catarina Mamede • Maria Filomena Botelho
Editors

Amniotic Membrane

Origin, Characterization and Medical Applications

Springer

Editors
Ana Catarina Mamede
Health Sciences Research Center
CICS-UBI
Covilhã
Portugal

Maria Filomena Botelho
Biophysics Unit
Faculty of Medicine, University of Coimbra
Coimbra
Portugal

ISBN 978-94-024-0287-2 ISBN 978-94-017-9975-1 (eBook)
DOI 10.1007/978-94-017-9975-1

Springer Dordrecht Heidelberg New York London

Softcover reprint of the hardcover 1st edition 2015

Printed on acid-free paper

Springer Science+Business Media B.V. Dordrecht is part of Springer Science+Business Media (www.springer.com)

Foreword

The amniotic membrane remains one of the most mysterious fetal structures, with properties that are poorly understood by most clinicians. Nevertheless, amniotic membrane diseases are one of the most important and prevalent aspects of clinical obstetrics, and the structure is used for treatment in several areas of clinical medicine.

This book, edited by Maria Filomena Botelho and Ana Catarina Mamede, is an important contribution to the existing knowledge in this field. It includes topics ranging from embryology to the biochemical and biophysical properties of the amniotic membrane.

Another section reviews the use of this tissue in different areas of clinical medicine, such as stem cell treatment, ophthalmology, treatment of burns, cancer, oral medicine, and gynecology. A final chapter touches on the ethical aspects associated with research and treatment.

The book constitutes interesting reading for obstetricians, basic science researchers, and healthcare professionals interested in amniotic membrane therapy. It is also an excellent review of the topic for all those who, for different reasons, are developing a special interest in this fetal structure.

Diogo Ayres-de-Campos
Associate Professor – Medical School,
University of Porto
Treasurer – FSPOG Federation of Portuguese
Societies of Obstetrics and Gynecology
Safe Motherhood and Newborn Health Committee – FIGO International
Federation of Gynecology and Obstetrics

Contents

Part I Origin and Characterization of Amniotic Membrane

1 **Embryology and Anatomy of Placental Membranes** 3
José Joaquim de Sousa Barros

2 **Biochemical Properties of Amniotic Membrane** 19
Sandra Catarina Moreira Rocha and Cláudio Jorge Maia Baptista

3 **Biophysical Properties of Amniotic Membrane** 41
Ana Margarida Abrantes, João Casalta-Lopes, and Maria Filomena Botelho

4 **Stem Properties of Amniotic Membrane-Derived Cells** 57
Maddalena Caruso, Anna Cargnoni, and Ornella Parolini

5 **Amniotic Membrane in Health and Disease: An Obstetrical Perspective** 77
Ana Luísa Areia and Paulo Moura

Part II Medical Applications of Amniotic Membrane

6 **Amniotic Membrane in Ophthalmology** 105
Esmeralda Costa and Joaquim Neto Murta

7 **Amniotic Membrane in the Treatment of Burns** 123
Ahmad Sukari Halim, Ehfa Bujang-Safawi, and Arman Zaharil Mat Saad

8 **Amniotic Membrane in Cancer** 139
Ana Catarina Mamede, Ana Salomé Pires, and Ana Filipa Brito

9 **Amniotic Membrane in Oral Medicine** 153
Manuel Marques Ferreira and Eunice Virgínia Palmeirão Carrilho

10 Amniotic Membrane in Gynaecology 165
Maria João Carvalho

11 Amniotic Membrane and the Controlled Drug Release 177
Miklós D. Resch, László Marsovszky, Erzsébet Csányi, Mária Budai-Szűcs, and János Németh

Part III Technical Issues About Amniotic Membrane

12 Isolation and Characterization of Mesenchymal Stem Cells from Amniotic Membrane 195
Ana Lopes, Mário Grãos, and Artur Paiva

13 Preservation of Amniotic Membrane 209
Mafalda Laranjo

14 Ethical and Legal Concerns of the Use of Amniotic Membrane 231
Guilherme de Oliveira and André Pereira

Index 245

Editors

Editor 1

Ana Catarina Mamede (✉)
CICS-UBI, Health Sciences Research Center,
University of Beira Interior, Covilhã, Portugal

Biophysics Unit,
Institute for Biomedical Imaging and Life Sciences (IBILI),
Center of Investigation in Environment, Genetics and Oncobiology (CIMAGO),
Faculty of Medicine, University of Coimbra, Coimbra, Portugal
Email: ana_mamede@hotmail.com

Editor 2

Maria Filomena Botelho
Biophysics Unit,
Institute for Biomedical Imaging and Life Sciences (IBILI),
Center of Investigation in Environment, Genetics and Oncobiology (CIMAGO),
Faculty of Medicine, University of Coimbra, Coimbra, Portugal
Email: mfbotelho@fmed.uc.pt

Part I
Origin and Characterization of Amniotic Membrane

Chapter 1
Embryology and Anatomy of Placental Membranes

José Joaquim de Sousa Barros

Abstract All vertebrates have extraembryonic tissues known as placental membranes. In humans they are composed of the so-called reflected membranes and those of the chorionic plate on the placental disc. Although the membranes are usually seen as a placental appendage, they are distinct from the chorion frondosum, which is the actual hemochorionic placental mass. Their origin along with their structural and functional complexity has recently been receiving considerable attention mainly because of an interest in the composition of the various extracellular connective tissue components and due to the stem cell properties of their epithelial and stromal cells. Although molecular mechanisms regulating trophectoderm and inner cell mass differentiation are beginning to be known, our understanding of cell fate decisions during human embryonic development still remains limited. In addition to the chorion and amnion, there is a third component of the membranes – the mesodermal derivatives, which form the middle aspect of the placental membranes.

Keywords Epiblast • Placenta • Amnion • Chorion • Placental membranes

J.J. de Sousa Barros
Faculty of Medicine, University of Coimbra, Coimbra, Portugal

Department of Obstetrics and Gynecology,
Coimbra Hospital and University Center, Coimbra, Portugal
e-mail: josebarros@huc.min-saude.pt

A.C. Mamede, M.F. Botelho (eds.), *Amniotic Membrane: Origin, Characterization and Medical Applications*,
DOI 10.1007/978-94-017-9975-1_1

1.1 General Considerations

All vertebrates have extraembryonic tissues known as placental or fetal membranes, also referred to as the chorioamniotic membranes. In humans they are composed of the so-called reflected membranes and those of the chorionic plate on the placental disc. Normally, membranes insert on the edge of the placental disc, run on the chorionic plate, and represent a fluid-filled membranous sac, also known as bag of waters that encloses the fetus throughout pregnancy (Fig. 1.1). Although the membranes are usually seen as a placental appendage, they are distinct from the chorion frondosum, which is the actual hemochorionic placental mass and forms a thickened villous tissue. Both are specialized areas of important maternal-fetal interactions, which are essential for the normal pregnancy evolution. Several distinct layers are present in placental membranes. Their origin along with their structural and functional complexity has recently been receiving considerable attention mainly because of an interest in the composition of the various extracellular connective tissue components and due to the stem cell properties of their epithelial and stromal cells. With the unlimited availability of placentas, large numbers of cells can be isolated from the placental membranes and their immunomodulatory properties make them highly attractive for cell based reparative and regenerative medicine. Hoping to use these cells for clinical applications, promising results have been achieved, in particular using fibroblast like cells isolated from the amnion and chorion. Indeed, these fetal membranes are no longer seen as waste tissues but as a source of cells and bioactive molecules for therapeutic applications.

1.2 Origin of the Placental Membranes

1.2.1 Yearly Placenta

The close relationship between the embryo-fetus and the mother is one of the most important traits of human development. In order to develop and grow during intrauterine life, the fetus must establish and maintain an intimate relation with the

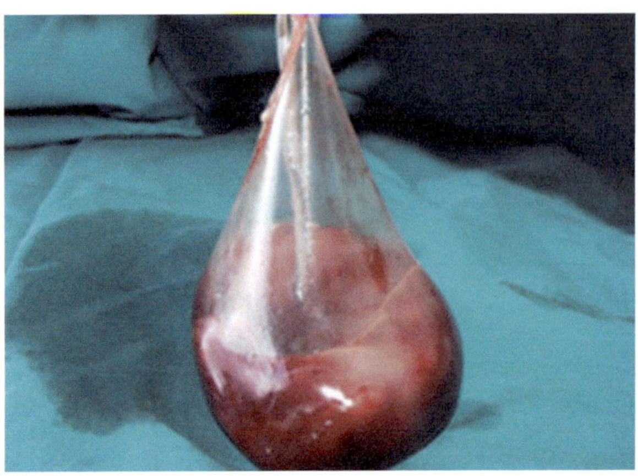

Fig. 1.1 Chorioamniotic membranes also known as "bag of waters". When observed from the maternal surface, are clear and have a bluish hue and are devoid of vasculature

mother's body for acquiring nutrients, oxygen and eliminating waste products. After conception, a series of cell divisions create a mass of totipotent cells, the morula. The first differentiation event occurs after compaction of the morula with formation of a vesicle called blastocyst by the end of day 5. Up to the time of implantation the blastocyst consists of a spherical structure with a central cavity surrounded outwardly by trophoblastic cells that lie on the outside of the morula. At one pole, within the cavity of the blastocyst, a thickened small group of large cells form the inner cell mass or embryoblast. Both cellular components of the blastocyst, the embryoblast and the trophoblastic cells also called trophoectoderm are precursors of other tissues that appear in subsequent stages of embryo development. The trophoblast is the forerunner of the placenta and chorionic villi while the inner cell mass will form the embryo/fetus, yolk sac, amnion and umbilical cord.

Although molecular mechanisms regulating trophectoderm and inner cell mass differentiation are beginning to be known, our understanding of cell fate decisions during human embryonic development still remains limited. During the initial stages of embryo development a differentiation event begins at the compacted morula stage where the outer layer of cells segregates to become the trophectoderm that gives rise to the embryonic part of the placenta. Within the first 2 days of development, the embryo undergoes successive cleavage divisions to produce an eight-cell embryo and at this stage cell-cell contact increase to produce a compacted morula. Subsequent divisions increase the positional complexity of the morula cells with some cells taking a position on its interior being fully enclosed by the outermost cells. Alternatively, cells can be stored in contact with the exterior, forming the outer lining wall of the morula and later of the blastocyst. Lineage tracing experiments suggested that trophoblastic cells (trophectoderm) are derived mainly from outside cells whereas inside cells contribute to the inner cell mass, which will produce the embryonic lineages cells. The differentiation of trophectoderm may be regarded as the hallmark event in mammalian preimplantation development, as it is the first tissue that becomes differentiated during embryogenesis being morphologically distinct from the inner cell mass. The trophectoderm, which will form the placenta, begins as an epithelial sheet enclosing the inner cell mass, the future embryo. Molecularly, trophectoderm and inner cell mass lineages can be distinguished by the expression patterns of several lineage-specific transcription factors. Interactions among these factors during blastocyst formation are thought to reinforce trophectoderm and inner cell mass fates. The mechanisms behind the establishment of these cell fate decisions include a combination of interactions between transcriptional factors, signalling, cell polarity and cell position within the morula. At cellular level, the segregation results in spatially and molecularly distinct cell populations. While the trophoectoderm is the first tissue that becomes differentiated during embryogenesis due to its biological function, the inner cell mass needs to retain its pluripotency status much longer, since the establishment of the other germ cell layers.

By the end of the first week following fecundation, the implantation of the blastocyst takes the first steps. Successful implantation and development of the blastocyst depends on a series of complex and coordinated cellular and molecular interactions between trophoblast and endometrial tissues that is facilitated by the trophoblast that start a differentiation process into cell lines with migratory capacity.

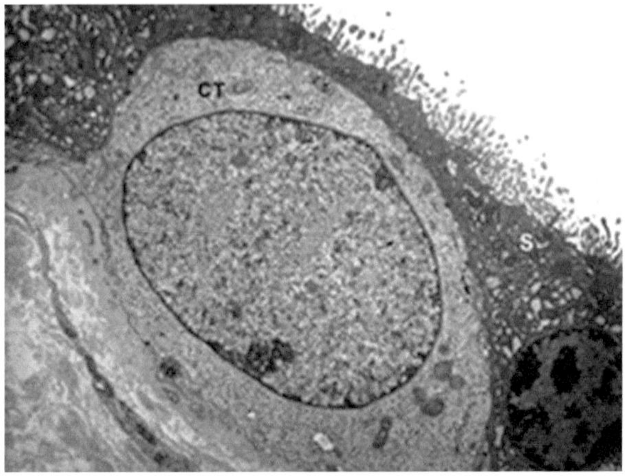

Fig. 1.2 Transmission electron micrograph of second trimester placenta. An undifferentiated cytotrophoblast (*CT*) can be clearly distinguished from the overlying syncytium (*S*). During all stages of gestation, most of the villous cytotrophoblast belongs to this type of proliferating stem cell, ×9,000

On the 7 day, the trophoblast embeds deeper into the endometrium and undergoes two different pathways of differentiation that leads to the development of villous and extravillous trophoblast. In the following days the trophoblastic cells proliferate and the embryoblast is separated from maternal tissues and blood by an inner layer of cytotrophoblast cells and an outer layer of syncytiotrophoblast – trophoblastic shell. The cytotrophoblast is an undifferentiated, mononucleated stem cell from which the other forms of trophoblast derived (Fig. 1.2). The syncytiotrophoblast facing the maternal tissue is transformed by fusion of the underlying cytotrophoblastic cells. It consists of a continuous multinucleated system without intercellular spaces, individual cells or syncytial units. Within the first 2 weeks of development, a labyrinth of lacunae appears within the syncytiotrophoblast forming the intervillous space and, soon after their formation, the invading front of trophoblast will have penetrated the endometrial blood vessels, allowing the maternal blood to enter the lacunae from branches of the spiral arteries. At the same time, the trophoblastic shell continues to grow and develop until the earliest evidence of fetal circulation and of the chorionic villi formation can be identified, by the third week after fecundation. Twenty eight days following fecundation, a deep interstitial blastocyst implantation within the endometrium with differentiation of invasive trophoblast into the decidualized stroma occurs. At 12 weeks of gestation, true maternal blood flow enters the intervillous space when uterine spiral arteries form the uteroplacental circulatory system.

Simultaneously with the overall development of the placental architecture cytotrophoblast differentiation continues in several pathways functioning as the precursor cells for all the other trophoblast types including the extravillous trophoblast located in the nonvillous structures of the placenta.

In the meanwhile, at the implantation pole of the blastocyst, the trophoblastic cells start to proliferate to form the syncytiotrophoblast by fusion of neighboring cytotrophoblastic cells, making the trophoblast thicker. This area of preferential growth, with the appearance of extraembryonic mesoderm, is later transformed into

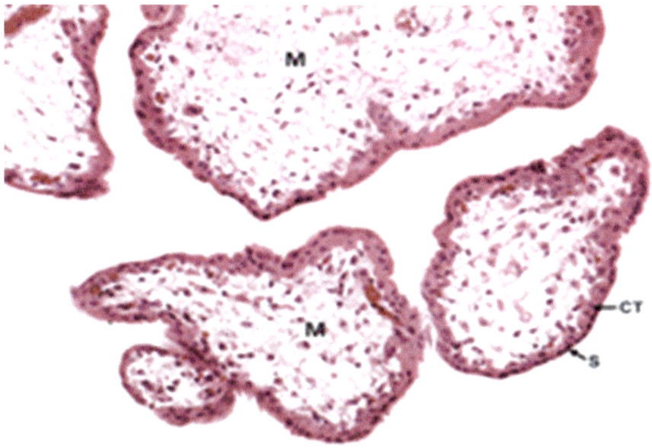

Fig. 1.3 Light microscopic appearance of a first trimester placental villi. Note the trophoblastic covering consisting of relatively thick syncytiotrophoblast (*S*) and well-defined cytotrophoblastic layer (*CT*). Villi still belong to the mesenchymal type (*M*). Paraffin section, Hematoxylin and Eosin stain, ×200

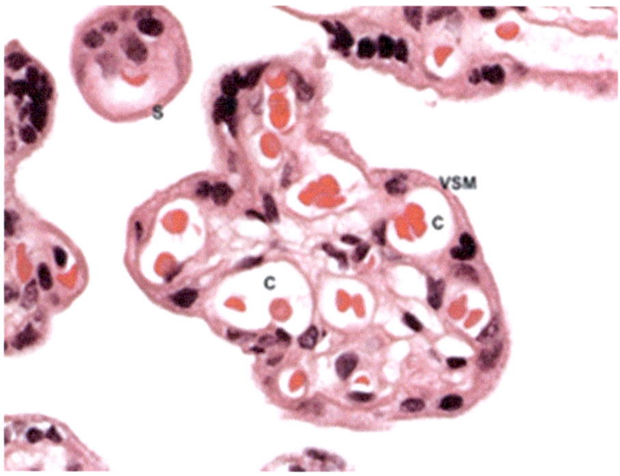

Fig. 1.4 Light microscopic appearance of third trimester placental villi. Note the vasculosyncytial (*VSM*) membranes formed by thinned anucleate syncytiotrophoblast (*S*) and fetal capillaries (*C*). Paraffin section, Hematoxylin and Eosin stain, ×100

a mass of chorionic tissue, which is identified as the placental villous mass, the so-called chorion frondosum. Endodermal and mesodermal components of the placenta and of the amnion–chorion membrane are later derivatives of the inner cellular mass, so mesenchyme and blood vessels then form the villi, which branch and grow along with the chorion frondosum.

The main functional units of this placental mass are the chorionic villi, a finger-like structure mediating nutrient absorption, oxygen transport, waste elimination and generating the bulk of the hormones produced by the placenta during gestation. Cross sectioning a chorionic villous reveals the basic components of this part of the placenta. A cross section of a first trimester chorionic villous shows a central mesenchymal core with embedded fetal capillaries surrounded by an inner layer composed of cytotrophoblast and an outer layer of syncytiotrophoblast (Fig. 1.3). A term chorionic villous exhibits the same basic structure with an increased number of fetal capillaries. Some of these capillaries are close enough to the outer edge of the villous to facilitate nutrient exchange, which is why these areas are known as vasculosyncytial membranes (Figs. 1.4 and 1.5).

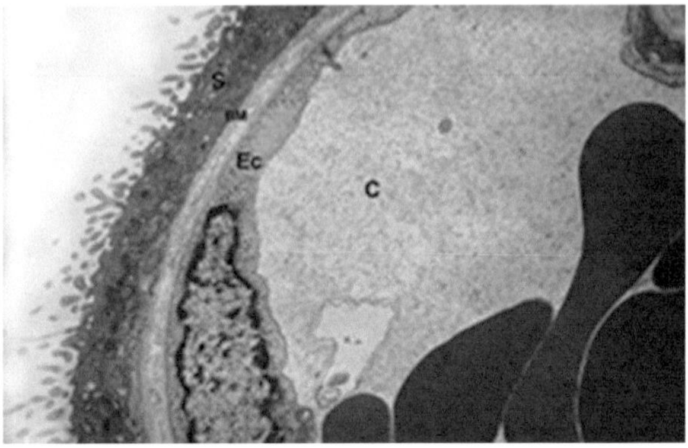

Fig. 1.5 Transmission electron micrograph of term villus showing the vasculosyncytial membrane formed by the thinned syncytiotrophoblast (*S*), basement membranes (*BM*), endothelial cytoplasm (*Ec*) and the fetal capillary (C). The capillary basement membrane and the trophoblast basement membrane may come into such close proximity that they fuse, ×8,000

Although the human oocyte is implanted interstitially and the chorionic sac is completely covered with villi at the beginning, the growth of the chorionic villi at the abembryonic pole is still limited. As early as 8 weeks of development, these capsular chorionic villi adjacent to the endometrial cavity and farthest away from the maternal blood supply are slowly pushed into the uterine cavity (decidua capsularis) by the expanding amniotic sac, which surrounds the embryo. These villi degenerate, atrophy and form the chorionic layer of the external membranes, the so-called smooth chorion or chorion laeve. At around 20 weeks of gestation, the combined amnion–chorion membrane makes contact with the opposite side of the uterus, where it fuses with the decidualized maternal endometrium (decidua parietal), forming the complete external membrane consisting of amnion, nonvillous chorion and decidua layers.

1.2.2 Amnion and Chorion

At an early stage in pregnancy, the human intrauterine implantation site is represented by differentiation of the endoderm and ectoderm, being surrounded by the extraembryonic mesenchyme and trophoblast. Indeed, just before the embryo implants into the endometrium, a differentiation event that includes the segregation of two cellular lineages within the inner cell mass occurs, leading up to the formation of the basic body plan, the bilaminar germ disc. Ultimately, these changes lead to the rearrange of the embryoblast into an epithelial configuration. Twenty-four hours after blastocyst formation, the second lineage decision leads to the establishment of two morphologically distinct cellular populations from the inner cell mass, the primitive ectoderm or epiblast, which is the embryonic lineage, and the primitive endoderm or hipoblast. The primitive endoderm forms a monolayer of small, cuboidal cells on the surface of the inner cell mass directly facing the blastocyst cavity that will give rise to the extraembryonic endoderm layer of the visceral yolk sac. The primitive ectoderm, which is the main upper layer of high columnar cells and the embryonic lineage, lies between the primitive endoderm and the trophoectoderm.

The classical model of epiblast – hipoblast lineage formation proposed that embryoblast cells of the early blastocyst are a homogeneous population of bipotential cells, each one with the ability to become either primitive ectoderm or primitive endoderm. Cell fate would be determined by the position of the cells within the inner cell mass. Surface cells would differentiate as primitive endoderm while enclosed cells would differentiate as primitive ectoderm. This model is consistent with the ultimate arrangement of primitive ectoderm and primitive endoderm cells within the late blastocyst being supported by data provided by in vitro studies using relatively homogeneous cell populations. Recently, this original assumption of cell segregation has been clarified by the identification of the mechanisms involved in the process of epiblast and hipoblast formation and maintenance of these two cell lines. Indeed, several studies suggest that cells of the inner cell mass may not be a homogeneous population but a heterogeneous mixture of progenitors cells that will sort out into the appropriate layer. The cells of the inner cell mass are arranged in a mosaic pattern, according to their lineage-specific transcription factors: some cells represent the precursors of the epiblast, and others the precursors of the hypoblast.

The basis for the differentiation of these two distinct precursor cell lines is not completely understood, but according to some hypothesis, those cells that enter the inner cell mass earliest are biased to perpetuate their pluripotency and assume an epithelial configuration as they form the epiblast. Possibly because of the influence of growth factors, secreted by these first arrivals to the inner cell mass, later incomers are then biased to produce molecules that increase their adhesive properties, as well as their mobility, making their way to the lower surface of the inner cell mass to form the hypoblast. Primitive endoderm precursors exhibit an apparently random distribution within the inner cell mass of the early blastocyst and then segregate to their final position lining the cavity by the late blastocyst. Some studies using early expressed proteins that are markers of cell lineage have shown that before blastocyst formation, lineage-specific factors are expressed in an overlapping manner. Subsequently, a gradual progression towards a mutually exclusive expression of primitive endoderm and epiblast-specific markers occurs. An analysis of lineage specific factors revealed a series of sequential and distinct phases in the process of primitive endoderm formation, in which the early overlapping expression of transcription factors precedes the maturation of inhibitory regulatory pathways and lineage restricted expression. Cell sorting then occurs via a combination of behaviors, including cell movement, adhesion and selective apoptosis. Studies of the events leading to formation of the primitive endoderm in the preimplantation mouse embryo indicate that, during early stages of development, cells co-express markers for different lineages. Subsequently, primitive endoderm fate results from a progressive restriction of expression of primitive endoderm-specific markers followed by cell sorting. Along with tracing studies, this has led to a model for lineage specification in which epiblast and primitive endoderm precursors are specified in a possibly random manner within the inner cell mass and later segregate to their respective layers. This is in contrast to the previously assumed model that cell position with respect to the blastocyst cavity is the primary determinant of primitive endoderm fate. After the hypoblast has become a well-defined cellular layer and the epiblast has taken on an epithelial configuration, the former inner cell mass is transformed into a bilaminar disk, which is a two layer structure with the epiblast on the dorsal surface and the hypoblast on the ventral surface lining the blastocyst cavity.

The epiblast contains the cells that produce the embryo itself, but also contribute to the extraembryonic tissues of the conceptus. By the eighth day following fecundation, the next layer to appear after the hypoblast is the amnion, a layer of extraembryonic ectoderm, at which time its margin is attached around the free edge of the embryonic disc that later loosely encloses the entire embryo in a fluid-filled chamber called the amniotic sac. During implantation, a cavity is formed between the embryoblast and basal trophoblast, developing the placental mass or chorion frondosum. The formation of the human amnion and amniotic cavity are not completely understood, but some studies on primate embryos indicate that the amniotic cavity appears in the dorsal region of the embryonic knot with the beginning of implantation or immediately afterwards. A small primordial amniotic cavity first arises by cavitation, that is, intercellular spaces appear in the inner cells mass, within the pre-epithelial epiblast and coalesce to form the amniotic cavity, which is covered by cells derived from the inner cell mass. According to some investigators, in the case of amniogenesis by cavitation, a cavity comes into existence through dehiscence in the region of the embryonic pole of the blastocyst. This primordial amniotic cavity develops within the dorsal embryonic knot through cavitation and enlarges peripherally in connection with the formation and growth of the primitive ectoderm. The roof of the primordial amniotic cavity gets thinner until finally opening up, exposing the cavity to the overlying trophoblast. This cavity is formed at the bottom by the embryoblast and at the roof by the trophoblast. Soon thereafter the edges of the epiblast fold up dorsally and grow towards each other along the trophoblastic membrane, forming the definitive amniotic cavity with a solid roof of original amniotic epithelium. Soon, in the further course of embryologic development a new cellular form arises, the denominated mesodermal cells. These, in combination with amniotic epithelium (ectoderm) will produce a thin extraembryonic membrane called amnion. In the further course of development mesodermal cells soon follow and the combination of amniotic epithelium (ectoderm) and mesoderm produces a thin extraembryonic membrane called amnion. The fibroblast layer of the amnion originates from the extraembryonic mesenchyme that develops from the hypoblast and expands to cover the surface of the amniotic epithelium and becomes the amniotic mesoderm. After 7–10 weeks of gestation, the amniotic mesoderm joins to the chorionic mesoderm, starting at the umbilical cord insertion site, at the chorionic plate. However, amnion and chorion never becomes closely attached and thus, the two membranes can easily slide against each other. Embryoblast derived mesenchyme also contribute to the formation of extraembryonic mesoderm.

According to this concept the amniotic cavity has its origin in the embryonic knot and the amniotic membrane is made up of the embryoblast material. Since the amniotic epithelium develops prior to gastrulation, a period during which the three embryonic germ layers form from the epiblast, it has been suggested that pluripotent stem cells derived from the epiblast are retained in the amnion even at term pregnancy.

Although the hypoblast plays a fundamental role in the early stages of development, its ultimate fate appears to be extra-embryogenesis. In fact, about 9 days after fecundation, the early embryo is still sunk in the endometrium, hypoblast cells begin to migrate in the direction of the abembryonic pole of the blastocyst and coat the inner surface of the trophoblast with a continuous layer of extraembryonic endoderm called parietal endoderm. When the endodermal spreading is completed, a new extraembryonic cavity is formed ventrally in front of the embryonic disc called primary yolk sac. The gap between the primary yolk sac and the trophoblast enlarges due to the extension of the blastocyst but is further on interconnected with cells and their extensions in a network fashion. Later on, this interspace becomes the chorionic cavity that will be filled, in the third week of development, by extra-embryonic mesoderm.

In addition to the chorion and amnion, there is a third component of the membranes – the mesodermal derivatives, which form the middle aspect of the placental membranes. These mesodermal investments are the mesenchymal core of the chorionic villi and the connective tissue layer of the inner surface of the chorion, in the chorionic plate, where they house fetal blood vessels, and of the chorion leave adjacent to the amnion.

The precise embryologic derivation of the extraembryonic mesoderm is unknown, but it is believed that the first cells arise from a transformation of parietal endodermal cells and not from trophoblast delamination. These first mesodermal cells are later joined by extraembryonic mesodermal cells that have originated from the primitive streak and thus from the epiblast. Some mesoderm cells split off and migrate to form the extraembryonic mesoderm, which also surrounds the amniotic cavity as previously described.

The chorionic cavity is lined by extraembryonic mesoderm and covered in trophoblast cells forming chorionic villi. In the human embryo, during the third week, this space forms outside the yolk sac and surrounds the amniotic sac. The extraembryonic mesoderm becomes the tissue that supports the epithelium of the amnion, the yolk sac and the chorionic villi, which arise from the trophoblastic tissues.

The cellular structure of the extraembryonic mesoderm takes on the character of a slack connecting tissue in the chorionic cavity where it provides a lining of the chorion and a further covering of the amnion and of the yolk sac. It becomes increasingly filled with fluid while its cell quantity decreases. This substance without a well-defined structure is sometimes called magma reticulare. It forms a transient space, the extraembryonic coelom that is lost by expansion of the amniotic sac, which fuses to the chorionic membrane. When the amniotic cavity starts extending at the cost of the chorionic cavity, the remnants of the mesodermal cells are integrated into the chorionic and the amniotic mesoderm until the chorionic cavity is obliterated, by the end of the third month.

As described, the amnion fuses with the chorion during expansion of the amniotic cavity. The inner surface of the chorion is bordered by the outer layer of the amnion and the outer layer of the chorion in conjunction with trophoblastic villi that sprout from its surface is in contact with the decidua. The decidua capsular is above the chorion laeve and fuses with the parietal decidua of the uterine wall when the chorionic sac grows to occupy the entire endometrial cavity. Therefore, the relationship of the chorion with the placenta has two contiguous aspects. One is related with the chorion laeve or the reflected membranes, sandwiched between amnion and decidua while the other is related with the surface of the placental mass or chorionic plate, which it is also covered by amnion but bordered beneath by trophoblast associated with the chorion frondosum or parenchyma of the placenta.

As previously mentioned, the entire surface of the early conceptus is covered by chorionic villi and, with the appearance of the first chorionic villi the trophoblast, the forerunner of the placenta, at the implantation pole, becomes the chorion frondosum. In the opposite implantation pole, the capsular chorion frondosum initially undergoes a similar development. However, as the gestational sac enlarges, regression of formed villi and obliteration of the relative intervillous space begins, a process that spreads laterally over about two thirds of the surface of the chorionic sac and the chorion, the obliterated intervillous space, the villous remnants, and the trophoblastic shell fuse, forming the nonvillous chorion. So, this smooth chorion consists of several layers of undifferentiated trophoblast and stromal layers. With the enlargement and outward expansion of the amniotic cavity, the amnion and chorion fuse to form the chorioamniotic membrane obliterating the chorionic cavity. After delivery, the chorion frondosum and the relatively avascular smooth chorion can be observed (Fig. 1.6).

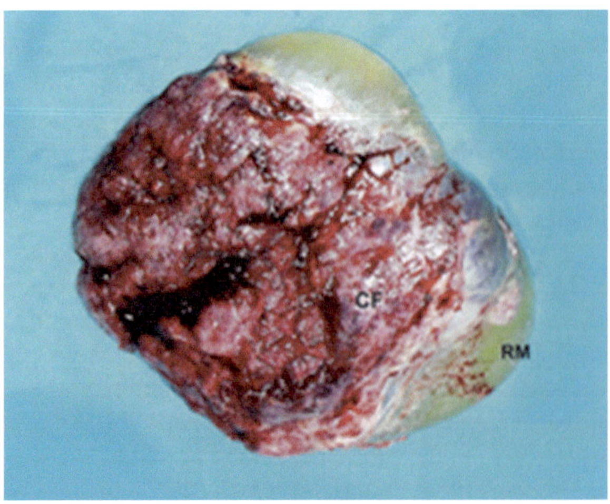

Fig. 1.6 Intact placenta and fetus at 23 weeks of pregnancy. Macroscopic appearance of normal membranes. Note the chorion frondosum (*CF*) and the reflected membranes of the fetal sac composed of amnion and chorion leave (*RM*). The free membranes meet the placental surface at the margin of the vascular or chorionic plate

1.3 Anatomy of the Placental Membranes

1.3.1 Amnion and Chorion

Before microscopic observation of term fetal membranes, it is important to know the basic elements of a singleton fetus gestation placenta. We can say that full-term normal placentas are composed of the so-called placental disc, membranes and the umbilical cord. Rarely do the membranes retain their normal anatomic relations during the delivery of the placenta. Most commonly, the fetal sac is attached to the rim of the placenta and may have inverted over the maternal surface of the placenta, so that the fetal surface or inner aspect of the amniotic cavity is demonstrated. Then, the fetal cavity can be reconstructed by manually flipping back the membranes to their normal anatomical in vivo position (Fig. 1.7). The location of the point of rupture through which the fetus was delivered and the distance between the opening in the membranes and the placental disc margin can be identified. The membranes are composed of the reflected membranes of the fetal sac and those of the chorionic plate or fetal surface on the placental disc. Both are in continuity and composed of two distinct layers, the amnion and chorion. These layers can be separated along their natural cleavage plane by grasping them at the point of rupture with toothed pickups and the amnion trimmed from the fetal surface of the placental disc. As observed by gross examination, the amnion is translucent with a bluish sheen and the chorion is relatively more opaque (Fig. 1.8). The chorion and the amnion are distinct layers and should be understood as they relate to chorion frondosum at the chorionic plate and to the chorion laeve of the so-called free or reflected membranes. Therefore, there are two contiguous aspects of the chorion that relate to the placenta: one associated with the chorion leave or the reflected membranes and the other linked to the surface of the placental disc or chorionic plate, here also covered

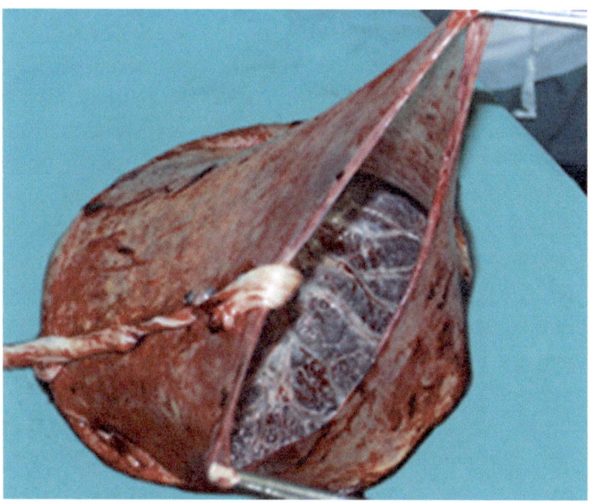

Fig. 1.7 Reconstruction of the chorioamniotic cavity. Note the opening for the exit of the fetus

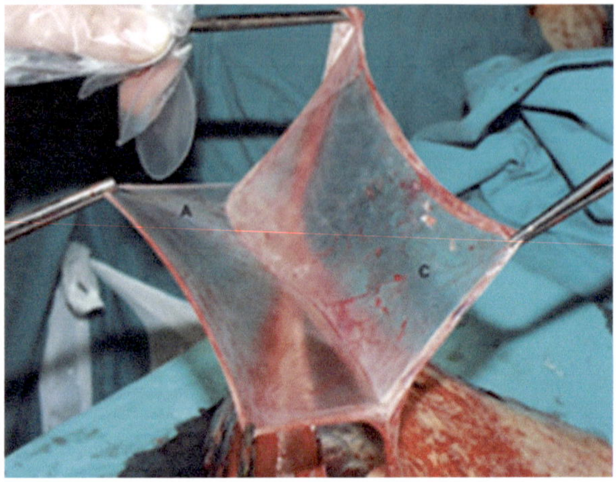

Fig. 1.8 General appearance of the fetal membranes. From the border of the opening of the sac it is easy to strip the amniotic layer from the chorion. The amnion (*A*) is a thin translucent layer of tissue loosely attached to the chorion (*C*)

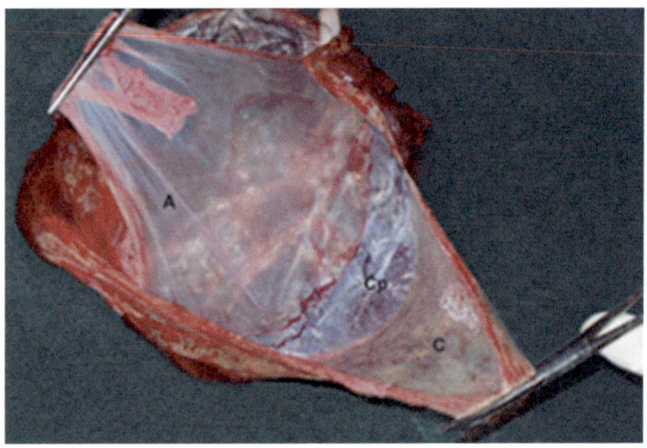

Fig. 1.9 The amnion (*A*) and the chorion (*C*) are separated. When the amniotic membrane arrives at its insertion on the placental surface, continuing to strip the amnion away clearly leaves the chorionic plate underneath (*Cp*)

by amnion but bordered beneath by the chorion frondosum or parenchyma of the placenta (Figs. 1.9 and 1.10). By grasping the membranes at the border and lifting, the chorioamniotic cavity may be reconstructed. The thickness after separation from the uterine wall is about 200–300 μm, in average. Using one's fingers or atraumatic forceps, it is easy to dissect the membranes from the top to the bottom of the amniotic sac and identify both the amnion and the chorion.

The amnion is a complex thin avascular membrane of only about 35–60 μm in greatest thickness that is probably nourished by mesenchymal-epithelial cells present between the chorion and amnion, the amniotic fluid, fetal surface vessels and by glands of the decidua, through diffusion mechanisms. The early amniotic membrane consists of an inner single layer of extraembryonic cuboidal ectodermal cells lined by a nonvascular and nerveless mesothelium-like layer of extraembryonic mesoderm. However, as the conceptus ages, distinct intermediate layers appear, originating a human amnion composed of five layers starting from the innermost layer: epithelium, basement membrane, compact layer, fibroblast layer and spongy layer.

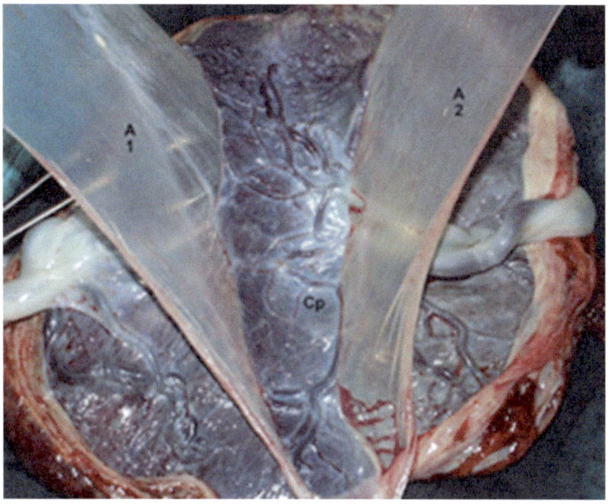

Fig. 1.10 Two layers of very thin translucent amniotic tissue (*A1* and *A2*) from the dividing membrane of a monochorionic diamniotic twin placenta. At their basis it is easy to peel them off the underlying chorionic surface leaving the chorionic plate (*Cp*)

It is currently believed that only a single cytological element should be present in the amnion. However, this cellular form may take on several appearances because of its localization, degrees of degeneration and other artifactual aberrancies. The epithelium is composed of a single layer of flat to cuboidal cells. More columnar cells can be seen at the margin of the chorionic plate, and flatter cells are generally present at greater distances radially from the center of the placenta. The free surfaces of these cells have brush borders and tend to be rounded. The amniotic epithelium lies over a delicate reticular basal lamina or basement membrane, which is connected to a thin connective tissue layer by filamentous strands. The connective tissue consists of a thicker more compact stromal layer (the strongest layer of the amnion) and a lower fibroblastic zone. The compact stromal layer contains bundles of collagen, scattered elastic fibers and sometimes histocytes. The fibroblast layer consists of a network of fibroblasts and a few macrophages. Although the amniotic connective tissue layer is divided into a compact stromal layer and a fibroblast layer, these are difficult to identify on histologic sections. Beneath this and directly overlying the chorion, there is a spongy layer with relative absence of fibroblastic cells. The spongy layer results from the tissue of the extraembryonic coelom with bundles of reticular fibers and mucin, render routine staining difficult. This is the zone of fusion between the mesodermal layer of the amnion and that of the chorion with a few isolated fibroblasts, presenting a highly variable thickness (Figs. 1.11 and 1.12).

Cytochemical and histochemical studies of the human amnion show a bipolar nuclear cytoplasmic cyclic activity. Both scanning and electron transmission microscopy confirm that both apical and basal surfaces of the amniotic epithelium are active in transport of solutes and water. In the apical surface of the amniotic epithelial cells, there are numerous microvilli while the basal border bears podocytic invaginations. Microvilli, rough endoplasmic reticulum, glycogen, pinocytotic vesicles and lipid can be found in the amniotic cells by performing an electron microscopy observation. Ultrastructural studies also demonstrate the presence of numerous pinocytic vesicles in the cytoplasm. Indeed, the amnion is well adapted to

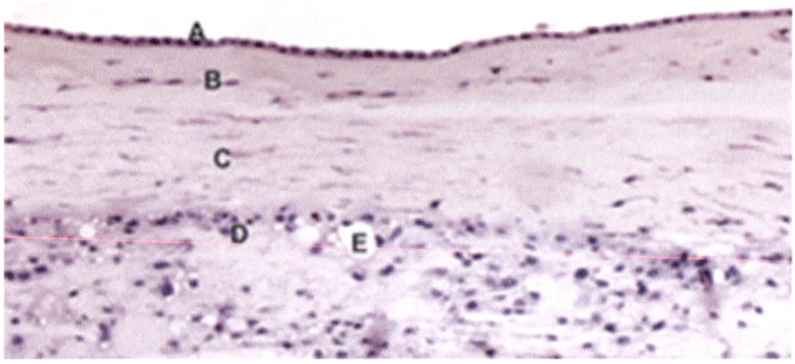

Fig. 1.11 Microscopic appearance of free or reflected membranes in the third trimester. The amniotic epithelium (*A*) and the connective tissue layer of the extraembryonic coelom (*B*) compose de amnion. The connective tissue layer of the amnion continues without sharp demarcation into the chorionic mesoderm (*C*). The chorion presents variable amounts of connective tissue and a large amount of trophoblast (*D*) within which are atrophic villi of the chorion laeve (*E*). Paraffin section, Hematoxylin and Eosin stain, ×100

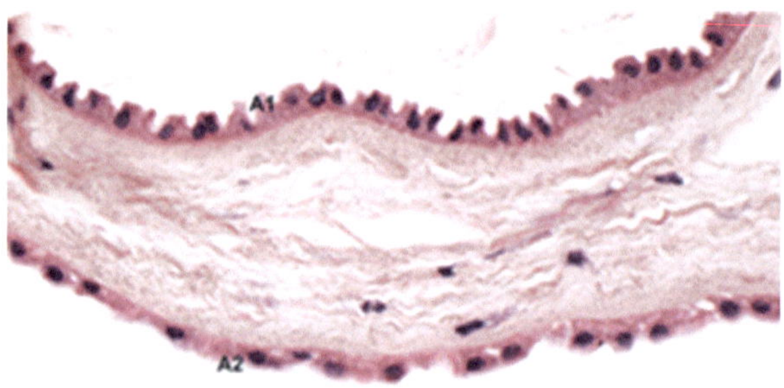

Fig. 1.12 Part of the membrane of a diamniotic septum of twin placenta with only two layers of amnion (*A1* and *A2*) back to back without interposed chorion

the transfer of many solutes through paracellular and transcellular pathways. The amniotic epithelium contains not only many cell organelles and inclusions, but also a large number of lipid inclusions, which are more numerous near term. This fact may be related to absorption of the surfactant synthesized in the fetal lungs and found in the amniotic fluid. In humans, large cells that look like macrophages are found in the wall of the amniotic sac. Those are analogous to the Hofbauer cells present in the stromal core connective tissue of the chorionic villi.

The nonvillous chorion is composed of several layers of polygonal cells and is up to 0.4 μm thick. Due to the existence of the spongy layer, which is the result of incomplete fusion of the amniotic and chorionic mesoderm and is thus between the two layers, it is easily separated from the amnion. The spongy layer of the amnion continues without sharp demarcation into the chorionic mesoderm (Fig. 1.11). The composition of the chorionic mesoderm is similar to the fibroblast layer of the amnion as it consists of an inner cellular layer of fibroblasts, myofibroblasts and macrophages and a reticular layer of collagen fibers. Finally, there is a pseudo-basement membrane that is highly variable in thickness and structure and an outer trophoblastic layer. When observed by transmission electron microscopy the

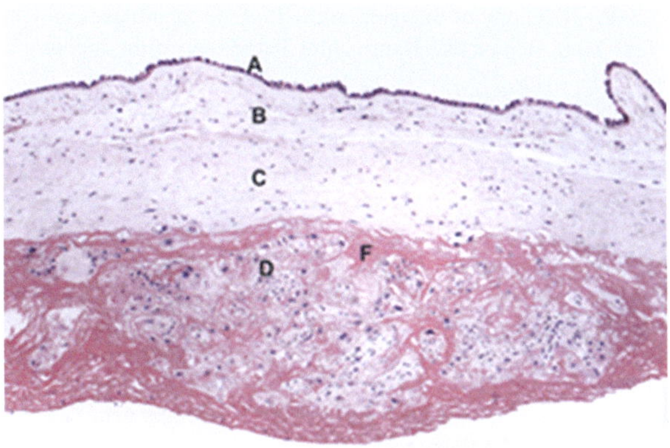

Fig. 1.13 Microscopic layers of the attached membranes of the chorionic plate from term placenta are shown. Amniotic epithelium (*A*) and the connective tissue of the extraembryonic coelom (*B*) are considered as amnion. The chorion is formed primarily by fibrous connective tissue (*C*), separable from amniotic connective tissue in an ill-defined plane, and clusters of extravillous trophoblast cells (*D*) typically incorporated into fibrinoid (*F*). Paraffin section, Hematoxylin and Eosin stain, ×200

ultrastructure of amnion and chorion stromal cells differ considerably. Human amnion stromal cells show features of mesenchymal and epithelial cells, a finding that suggested a multi-potentiality and/or an ability to undergo epithelial to mesenchymal transitions. On the other hand, the ultrastructure of human chorion stromal cells is similar to haematopoietic progenitor cells and therefore a more differentiated phenotype in stem cell hierarchy.

The trophoblastic layer is the deepest of the chorion and consists of two to ten layers of trophoblast cells in contact with the decidua capsular on their deeper aspect.

The layering of the chorionic plate is consistent with that described for reflected membranes. The layers of the amnion are structurally similar to the corresponding layer of the free membranes. However, in the chorionic mesoderm of the chorionic plate or fetal surface of the placenta there are distinct layers of connective tissue. Abundant scattered or aggregated extravillous trophoblast cells and occasional remnants of atrophied chorionic villi can be seen in the fibrinoid layer (Fig. 1.13).

Acknowledgement The author would like to thank the Institute of Histology and Embryology, Faculty of Medicine, University of Coimbra, the facilities granted in obtaining some of the images for this chapter [1–26].

References

1. Alviano F, Fossati V, Marchionni C et al (2007) Term amniotic membrane is a high throughput source for multipotent mesenchymal stem cells with the ability to differentiate into endothelial cells in vitro. BMC Dev Biol 7:1–14

2. Baergen RN (2006) Histology of the chorionic villi, fetal membranes and umbilical cord. In: Baergen RN (ed) Man. Benirschke Kaufmann's Pathology of the human placenta. Springer, New York, pp 80–94
3. Baergen RN (2005) Development and histology of the nonvillous portions of the placenta. In: Baergen RN (ed) Man. Benirschke Kaufmann's Pathology of the human placenta. Springer, New York, pp 107–134
4. Barros J (2007) Postpartum placental examination. In: Blickstein I, Keith LG (eds) Prenatal assessment of multiple pregnancy. Informa Healthcare, London, pp 15–26
5. Benirschke K, Kaufman P, Baergen RN (2006) Early development of the human placenta. Pathology of the human placenta, 5th edn. Springer, Berlin, pp 42–48
6. Benirschke K, Kaufman P, Baergen R (2006) Anatomy and pathology of the placental membranes. Pathology of the human placenta, 5th edn. Springer, Berlin/Heidelberg/New York, pp 321–366
7. Plusa B, Piliszek A, Frankenberg S et al (2008) Distinct sequential cell behaviours direct primitive endoderm formation in the mouse blastocyst. Development 135:3081–3091
8. Bourne G (1962) The foetal membranes: a review of the anatomy of normal amnion and chorion and some aspects of their function. Postgrad Med J 38:193–201
9. Bourne G (1960) The microscopic anatomy of the human amnion and chorion. Am J Obstet Gynecol 79:1070–1073
10. Bryant-Greenwood G (1998) The extracellular matrix of the human fetal membranes: structure and function. Placenta 19:1–11
11. Chopra A, Thomas BS (2013) Amnitic membrane: a novel material for regeneration and repair. Biomim Biomater Tissue Eng 18:1–8
12. Carlson BM (2014) Formation of germ layers and early derivatives. In: Carlson BM (ed) Human embryology and developmental biology, 5th edn. Elsevier-Saunders, Philadelphia, pp 75–85
13. Carlson BM (2014) Placenta and extraembryonic membranes. In: Carlson BM (ed) Human embryology and developmental biology., 5th edn. Elsevier-Saunders, Philadelphia, pp 117–130
14. Cross J (2006) Trophoblast cell fate specification. In: Moffett A, Loke C, McLaren A (eds) Biology and pathology of trophoblast. Cambridge University Press, New York, pp 3–14
15. Frank H, Kaufmann P (2006) Nonvillous parts and trophoblast invasion. Pathology of the human placenta, 5th edn. Springer, New York, pp 191–309
16. Guibourdenche J, Fournier T, Malassiné A, Evain-Brion D (2009) Development and hormonal functions of the human placenta. Folia Histochem Cytobiol 47:S35–S42
17. Hargitai B, Marton T, Cox P (2004) Examination of the human placenta. J Clin Pathol 57:785–792
18. Ilancheran S, Moodley Y, Manuelpillai U (2009) Human fetal membranes: a source of stem cells for tissue regeneration and repair? Placenta 30:2–10
19. Lewis S, Gilbert-Barness (1999) Placental membranes. In: Lewis SH, Perrin E (eds) Pathology of the human placenta, 2nd edn. Churchill Livingstone, New York, pp 137–160
20. Madeja ZE, Sosnowski J, Hryniewicz K et al (2013) Changes in sub-cellular localisation of trophoblast and inner cell mass specific transcription factors during bovine preimplantation development. BMC Dev Biol 13:32
21. Pereira PN, Dobreva MP, Graham L et al (2011) Amnion formation in the mouse embryo: the single amniochorionic fold model. BMC Dev Biol 11:48
22. Toda A, Okabe M, Yoshida T, Nikaido T (2007) The potential of amniotic membrane/amnion-derived cells for regeneration of various tissues. J Pharmacol Sci 105:215–228
23. Vogler H (1987) Chorion and yolk sac. In: Lierse W (ed) Human blastogenesis. Formation of the extraembryonic cavities. Bibl. Anat. Kargar, Hamburg, pp 51–101
24. Vogler H (1987) Amnion. In: Lierse W (ed) Human blastogenesis. Formation of the extraembryonic cavities. Bibl. Anat. Kargar, Hamburg, pp 102–121
25. Whittle W, Gibb W, Challis J (2000) The characterization of human amnion epithelial and mesenchymal cells: the cellular expression, activity and glucocorticoid regulation of prostaglandin output. Placenta 21:394–401
26. Yamanaka Y, Ralston A, Stephenson RO, Rossant J (2006) Cell and molecular regulation of the mouse blastocyst. Dev Dyn 235:2301–2314

Chapter 2
Biochemical Properties of Amniotic Membrane

Sandra Catarina Moreira Rocha and Cláudio Jorge Maia Baptista

Abstract The amniotic membrane, also known as amnion, has been recently characterized as a promising source for tissue transplantation. It has several properties that become an attractive tool in several applications. Among the several properties, it could be highlighted their ability to reduce the inflammation status and to secrete pro-apoptotic factors. The cells of amniotic membrane are divided into two subpopulations, amniotic epithelial cells and amniotic mesenchymal cells, which present similar characteristics to stem cells. These cells express pluripotent markers, present high expansion in vitro and can be differentiated into all three germ layer. In this chapter, it will be reviewed the main characteristics of amniotic membrane and its biochemical components.

Keywords Amniotic membrane • Extracellular matrix • Transcriptome • Proteome • Secretome

2.1 Introduction

The fetal membrane or chorioamniotic membrane is a thin membrane that surrounds the developing fetus and forms the amniotic cavity. The fetal membrane is derived from fetal tissue and it is composed of two layers: chorion (outer layer) and the amnion (inner layer). The chorion is a more opaque membrane that exists between the developing fetus and maternal tissue. The amnion or amniotic membrane is a translucent structure adjacent to the amniotic fluid, which provides nutrients to the

S.C.M. Rocha • C.J. Maia Baptista (✉)
CICS-UBI, Health Sciences Research Center, University of Beira Interior, Covilhã, Portugal
e-mail: cmaia@fcsaude.ubi.pt

A.C. Mamede, M.F. Botelho (eds.), *Amniotic Membrane: Origin, Characterization and Medical Applications*,
DOI 10.1007/978-94-017-9975-1_2

amniotic membrane cells [1]. Based on the morphology and biochemical properties, human amniotic membrane has been found to have a number of characteristics that make it uniquely suited for several clinical applications. These characteristics of the amniotic membrane are: (1) anti-inflammatory effects through the production of a great variety of anti-inflammatory factors, such as hyaluronic acid (HA) [2]; (2) suppression the pro-inflammatory cytokines interleukin 1 alpha (IL-1α) and 1 beta (IL-1β) [3, 4]; (3) production of natural metalloproteinases (MMP's) inhibitors [3, 5]; (4) anti-bacterial properties because express natural antimicrobial molecules such as β-defensins and elafin [6]; (5) non-immunogenic and low antigenicity due to amniotic epithelial cells do not express on their surfaces human leukocyte antigens A, B, C, and DR antigens, or β2-microglobulin [7]; (6) it provides an extracellular matrix (ECM) for cellular migration and proliferation [8]; (7) It promotes an increased healing and enhancement of the wound healing process [5]; (8) anti-scarring and anti-adhesive activity because reduces proteases activity via the secretion of tissue inhibitors of metalloproteinases (TIMP's) [5] and down-regulates transforming growth factor beta (TGF-β) which is responsible for the activation of fibroblasts and prevent the adhesion reducing the risk of fibrosis [9, 10]; (9) it has both angiogenic and anti-angiogenic properties, which is surface dependent, i.e., promotes angiogenesis when applied to ischemic organs and inhibits angiogenesis in pathological conditions such as cancer [11]; (10) it secretes certain substances which act primarily as pro-apoptotic agents [12] and therefore, the amniotic membrane could be considered a promising candidate for anticancer agents [13]; and (11) it provides a natural biological barrier [14–16]. Due to these properties, in recent years, amniotic membrane has been used for clinical applications. The amniotic membrane is used by its high concentrations of cytokines and growth factors. Usually, the membrane is placed with the epithelial side down in contact with the wound surface in order to efficiently release the growth factors to wound site. The use of the amniotic membrane to cover inflamed or exposed areas favorably influences the wound healing process, as well as, reducing the levels of pain and discomfort of the patient [17]. The wide research done is to allow the application of this membrane in treatments associated with the genitor-urinary tract, oral cavity, stomach, cartilage and brain [18–21]. However, so far, the best known and most promising application of amniotic membrane are ocular surface reconstruction, skin application and tissue engineering [16, 22, 23]. The special properties, availability, low-cost and its non-invasive recover of amniotic membrane make it an ideal candidate for several clinical applications.

2.2 Histology of the Amniotic Membrane

Amniotic membrane develops from extra-embryonic tissue and consists of a fetal component, the chorionic plate and a maternal component, the deciduas. It is a translucent biological structure that has no nerves, muscles or lymph vessels which represents the innermost layer of the sac that encloses the fetus. The amniotic

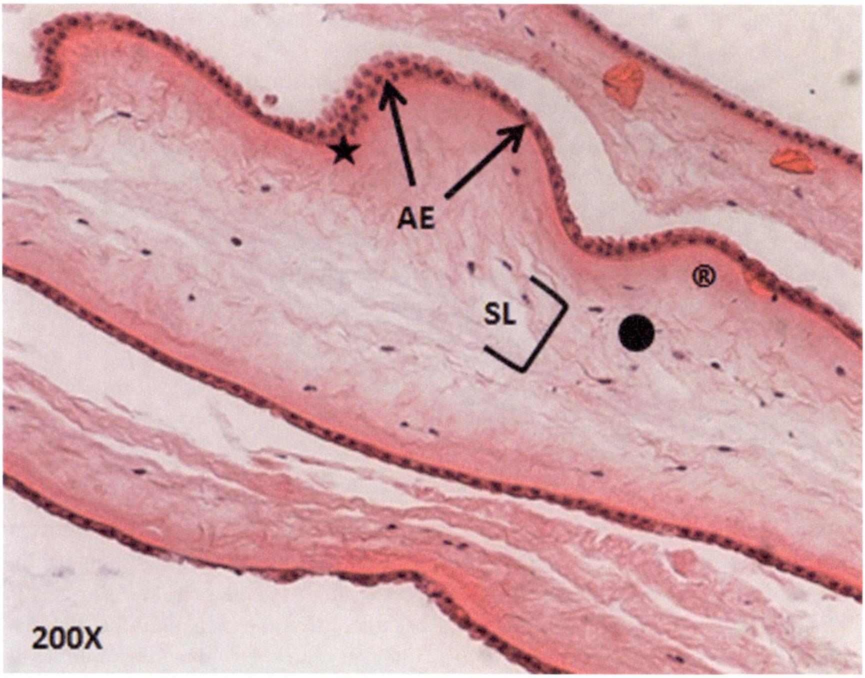

Fig. 2.1 Structure of the human amniotic membrane stained with hematoxylin and eosin. *AE* amniotic epithelium; ★: basement membrane; ®: compact layer; ●: fibroblastic layer and *SL* spongy layer

membrane (Fig. 2.1) is a thin (up to 2 mm), elastic and semi-permeable fetal membrane attached to the chorionic membrane. Both the amnion and chorion constitute the amniotic sac filled with amniotic fluid, providing and protecting the fetal environment. The outer layer, the chorion, consists of trophoblastic chorionic and mesenchymal tissues. The inner layer, the amniotic membrane, consists of a single layer of ectoderm-derived epithelium uniformly arranged on the basement membrane, which is one of the thickest membranes found in any human tissue, and a collagen-rich mesenchymal layer [24]. This mesenchymal layer can be subdivided into the compact layer forming the main fibrous skeleton of the amniotic membrane, the fibroblast layer and an intermediate layer, which is also called the spongy layer or zona spongiosa [16, 19].

2.2.1 Amniotic Epithelium

The innermost layer, nearest to the fetus and in directly contact with the amniotic fluid, is called the amniotic epithelium and is constituted by a single homogeneous layer of simple cuboidal epithelial cells [16, 25]. Amniotic epithelial cells have many microvilli at their apical surface and probably have an active secretory function and intra- and trans-cellular transport functions [26]. These cells have a large irregular nucleus with a large homogeneous nucleolus and many intracytoplasmic organelles and pinocytic vesicles [25]. The cells of amniotic epithelium have several characteristics that make it a great source of stem cells. Recent studies have pointed out the amniotic epithelium with stem cell-like characteristics because these cells express the surface markers associated with embryonic stem cells, such as stage-specific embryonic antigen 3 and 4 (SSEA-3 and -4), TRA-1-60 and TRA-1-81. In addition, these cells also express pluripotent stem cells-specific transcription factors, such as octamer-binding transcription factor 4 (Oct-4) and Nanog, suggesting that amniotic epithelium cells have a pluripotent potential to differentiate [27, 28].

2.2.2 Basement Membrane

It has been shown that molecules that made up the basement membrane affect the differentiation and survival of cells [29]. The main components that compose the basement membrane include laminins, type IV and VII collagen, and fibronectin [30]. Laminin is the major component of basement that interacts with the underlying cells via cell surface receptors, such as integrins and dystroglycan complex. The laminins regulate gene expression and influence cell fate. Furthermore, laminins critically contribute to cell differentiation, cell shape and movement, the maintenance of tissue phenotypes, and the promotion of tissue survival [31, 32]. Fibronectin and collagen are glycoproteins that constitute the ECM. The fibronectin is capable to mediate the cell adhesion and collagen performs structural functions enabling the maintenance of membrane integrity [33].

2.2.3 Avascular Stroma

The stromal layer is composed by a compact layer adjacent to the basement membrane, fibroblast layer and spongy layer, as can be seen in Fig. 2.1. The compact layer of stromal matrix forms the main fibrous skeleton of the amniotic membrane. The main constituent of the stromal matrix is the collagen [33]. The types I, III, V and VI collagens of the compact layer are secreted by mesenchymal cells located in the fibroblast layer. Type V and VI collagen are considered to be a minor fibrillar collagen and performs an anchoring function [34]. This two types of collagen gives rise to heterotypic fibrils with Interstitial collagens (types I and III) [33], in order to maintain the mechanical integrity of amniotic membrane [16]. The spongy layer of the stromal matrix sits adjacent to the chorionic membrane. Beyond of collagen, it is also abundant in proteoglycans and glycoproteins that produce a spongy appearance in histological preparations and it contains a nonfibrillar meshwork of mostly type III collagen [35].

2.3 Components of Amniotic Membrane

The major components of the amniotic membrane are cells and the ECM. The former are responsible for the synthesis, degradation and turnover of the latter. The ECM in turn influences the functions of the cellular components throughout pregnancy. The major tensile strength of the amniotic membrane is provided by the collagens of the compact layer beneath the amniotic epithelium. However, this arrangement of interstitial collagens beneath and connected to the collagens of the amniotic basement membrane provides extra strength in tissue exposed to mechanical forces [33]. ECM materials form the structural components of the architecture of the amniotic membrane and contain a variety of specialized proteins [36]. Collagen and proteoglycan molecules are, along with elastin, fibronectin and laminin, the major components of the amniotic membrane ECM, accounting for the tissue integrity and mechanical properties [33].

Collagen is composed of a triple helix, which generally consists of two identical chains ($\alpha 1$) and an additional chain that differs slightly in its chemical composition ($\alpha 2$). The amino-acid composition of collagen is atypical for proteins, particularly with respect to its high hydroxyproline content. The most common motifs in the amino-acid sequence of collagen are glycine-proline-X and glycine-X-hydroxyproline, where X is any amino-acid other than glycine, proline or hydroxyproline [37]. Collagen biosynthesis and assembly follows the normal pathway for a secreted protein. The collagen chains are synthesized as longer inactive precursors called pro-collagens; the growing peptide chains are co-translationally transported into the lumen of the rough endoplasmic reticulum (ER). In the ER, the pro-collagen chain undergoes a series of processing reactions including hydroxylation, glycosylation, and disulfide-bond formation. Interchain disulfide bonds between the C-terminal pro-peptides of three pro-collagens align the chains in register and initiate formation of the triple helix [38]. Genetically distinct types of collagen have been identified so far. All types of collagens are triple helices and they differ not only in their amino-acid sequences but also in their spatial conformations, conferring to them specific functional properties [39]. Collagen types I and III, together with smaller amounts of types IV, V and VI, are thought to be the main collagen components of amniotic membrane. Collagens I, III and V belong to the same family, which aggregate in quarter stagger arrangement to form higher order structures or fibers. They are presumed to be the primary components of the collagen network-like structure dominating the mechanical responses of the tissue. Type IV collagens molecules form structural sheets within the basal lamina of the amniotic epithelium and basement membranes of the amniotic membrane. Finally, type VI collagen, whose functional properties remain uncertain, is thought to provide anchoring filaments at the interface of the amniotic membrane [40].

Proteoglycans are special glycoproteins composed of a core protein to which sulfated polysaccharides or glycosaminoglycans (i.e. repeating disaccharide units) are covalently attached. Proteoglycans interact with collagen molecules to promote network cross-linking and to regulate collagen fibril formation. The smaller proteoglycans like decorin or biglycan have been identified in the ECM of amniotic membrane [41], and decorin constitute two thirds of the total amount of proteoglycans, with the rest consisting of biglycan. A non-sulfate glycosaminoglycans, HA, is also present in the tissue. HA form a complex network and does not attach a core protein. Essential contributors to the viscoelastic response of the connective tissue [42], proteoglycans also play a structural role in maintaining the tissue integrity [43]. In addition to structural roles, proteoglycans are important for cell proliferation and differentiation and perform, through binding of growth factors, essential functions in remodeling processes [44].

Fibronectins are glycoproteins synthesized by a wide variety of cell types which also direct their subsequent organization into fibrils of the ECM. This complex protein family is a product of a single gene, which can, by alternative splicing, give rise to 20 different forms of human fibronectin subunits [45]. Fetal fibronectin is a protein that acts as a "glue" during pregnancy attaching the amniotic sac, due to their multiple binding domains for cells, as well as other matrix components, which stabilize the whole system of cells and matrix [46]. This is accomplished by specialized domains, of at least six peptide sites capable of mediating cell adhesion. Most cells can adhere to fibronectin through its centrally located cell-binding domain, the Arg-Gly-Asp (RGD) sequence is crucial for this. There are six different cell surface receptors or integrin receptors which recognize this site, but the conformation of the fibronectin is also important for the specificity and affinity of this cell-fibronectin recognition [45]. Integrin receptors can mediate cell adhesion and migration on fibronectin and other extracellular molecules [47].

Laminins are a major component of basement membranes and are formed of several subunits linked together by disulphide bonds, forming a cross-like structure. Unlike fibronectin, laminins are the product of several genes and thereby originating seven different isoforms. The cells also interact with laminins via specific recognition sequences and integrin receptors [45]. Ensuring the "anchoring" of the epithelial cells to the underlying stroma via the basement membrane, laminins are thought to perform a significant strengthening function in the amniotic membrane [48].

Finally, elastin confers the property of elastic recoil to elastic fibres which are assembled from a family of tropoelastin precursors. These are covalently cross-linked to form insoluble elastin by formation of desmosine and isodesmosine, catalyzed by the enzyme lysyl oxidase [49]. Elastin molecules, which correspond to complex insoluble proteins cross-linked to fibrillin-based microfibrils, are suspected to be abundant in the reticular and compact layers of the amniotic membrane [50]. Having mechanical properties analogous to those of pure rubber, they may confer to the amniotic membrane part of its intrinsic elasticity [33].

2.4 Biochemical Functions of Amniotic Membrane

Amniotic membrane is not only a simple avascular structure, but it also has multiple metabolic functions such as the transport of water, soluble materials, the ionic transfer and the production of bioactive factors, including peptides, growth factors and cytokines [51, 52]. The main pathway for membrane water transport is through water-selective channels – aquaporins (AQPs). AQPs are transmembrane proteins that are organized in membrane as homotetramers with approximately 28 kDa in size for monomers; they are channels that regulate the input or output of water and small neutral solutes [53, 54]. All three subgroups of the AQPs family are expressed in the amniotic membrane: the classic AQPs with AQP1 and AQP8, the aquaglyceroporins with AQP3 and AQP9, and super-AQPs with AQP11 [55–57]. This expression pattern suggests an interesting cytoplasmic (AQP11) and transmembranous (AQPs 1, 3, 8 and 9) network enabling water transport for all expressed AQPs, as well as for other molecules (urea, glycerol, ammonia, anion as nitrate, and neutral solutes) for the aquaglyceroporins [58]. They act as a major contributor to amniotic fluid volume and homeostasis, and also facilitate rapid changes of cell volume to migrate easily into the ECM [54].

The amniotic membrane produces a great variety of soluble factors, such as HA, TIMPs and MMPs, IL, migration-inhibitory factor and prostaglandins [59]. HA is present at high levels in the amniotic membrane, which in amniotic membrane stroma, mediates the entrapment of inflammatory cells, including lymphocytes, through the binding to cluster of differentiation 44 (CD44) expressed on inflammatory cells [2]. Additionally, HA covalently linked to heavy chain (HC) of inter-α-inhibitor and form HC*HA complex, is the active component in amniotic membrane responsible in part for clinically observed anti-inflammatory and anti-scarring actions [60].

The amniotic membrane contains various tissue inhibitors (TIMP-1, -2, -3, -4) of MMPs [61]. They regulate many crucial processes in inflammation and fibrotic processes including chemotactic migration of inflammatory cells, mitosis of fibroblasts, and synthesis and degradation of extracellular components [16]. IL-1 receptor agonist and IL-10 are potent anti-inflammatory cytokines that are expressed by amniotic membrane [3]. IL-10 counteracts the action of different pro-inflammatory cytokines, such as IL-6, IL-1, IL-8 and tumor necrosis factor (TNF) [62]. In particular, the stroma of the amniotic membrane suppresses the expression of potent pro-inflammatory cytokines IL-1α and IL-1ß. This fact may explain the effect of amniotic membrane transplantation in reducing inflammation [4]. The epithelial cells of amniotic membrane secrete the macrophage migration-inhibitory factor (MIF) [63], which inhibits migration of macrophages and natural killer (NK) cell-mediated lytic activity [64].

The amniotic membrane also plays an important role during parturition. In the initiation and maintenance of uterine contraction, prostaglandins expressed by amniotic membrane play a pivotal role. The cells of amniotic epithelium beyond expressed prostaglandin E2 (PGE2) [65], also expresses prostaglandin-biosynthesis enzymes such as phospholipase, prostaglandin synthase and cyclooxygenase [66]. Moreover, these enzymes are regulated by human chorionic gonadotropin (hCG), and its receptor is expressed on the amniotic epithelium [67]. PGE2 regulates the maturation and antigen presentation of dendritic cells and inhibits T cell proliferation and proinflammatory cytokine production [68]. Amniotic epithelium is highly metabolically active throughout gestation, and it is also responsible for regulating the pH of the amniotic fluid, keeping it constant at about 7.10 [20].

2.5 Transcriptomics and Proteomics of Amniotic Membrane

From amniotic membrane it is possible to isolate principally two cell types, which have been definitely defined in a recent report as amniotic epithelial cells (AEC) and amniotic mesenchymal stem cells (AMSC) derived from the amniotic epithelium and amniotic stromal layer, respectively [69–71]. These two cell types have a different embryological origin: both cell types are originated during the pregastrulation stages of the developing embryo, before the delineation of the three primary germ layers, but whereas AEC are derived from the embryonic ectoderm, AMSC originate from the extraembryonic mesoderm [69, 72, 73]. Several protocols have been established for AECs and AMSCs isolation and culture [28, 69, 74], however, significant contamination of AMCs with AECs and vice versa frequently occurs. Barbati and theirs collaborators describes an efficient and rapid method to separate, mechanically, amniotic mesoderm from amniotic epithelium in order to obtain, after subsequent enzymatic digestions, purified population of AMSCs and AECs [75]. Starting in purified populations becomes easier to study the transcriptome and proteome of cellular constituents of amniotic membrane.

Transcriptomic analysis using microarrays has been reported for AMSCs (Table 2.1) [76]. These experimental data provided information on the AMSC gene expression pattern compared to gene expression profiles of the other three MSC groups (amniotic fluid, cord blood, and bone marrow-MSCs). Several up-regulated genes in AMSCs are involved in regulation of the immune adaptation regulation between the maternoplacental interface. Among others, spondin 2 (SPON2), interferon, alpha inducible protein 27 (IFI27), bradykinin receptor B1 (BDKRB1), small inducible cytokine subfamily B member 5 and 6 (SCYB5, SCYB6), and Yamaguchi sarcoma viral related oncogene homolog (LYN) were found to be upregulated [76]. In addition, other genes with increased expression in AMSCs compared to other three MSC groups, included transcription factors, such as forkhead box F1 (FOXF1), heart and neural crest derivatives expressed 2 (HAND2), and transcription factor 21 (TCF21) and metabolic enzymes, such as dipeptidyl-peptidase 6 (DPP6), tryptophan 2,3-dioxygenase (TDO2), and ST6-N-acetylgalactosaminide α-2,6-sialytransferase 5 (ST6GALNAC5) [76].

A detailed approach for studying amniotic membrane proteins using high-resolution two-dimensional gel electrophoresis (2-DE), core technology for arraying complex protein mixtures, was described by Hopkinson and theirs collaborators [77, 78]. These authors performed a proteomic analysis of amniotic membrane samples that were prepared for human transplantation. The wash media from the amniotic membrane samples were also examined and the secreted proteins were identified. These proteins were mostly soluble cytoplasmic proteins and were categorized according to their subcellular localization and physiological function [78]. A spectrum of proteins identified for this group, the most abundant are thrombospondin (THBS1), mimecan (also named osteoglycin – OGN), βIG-H3 (also known TGFβI) and integrin α6. THBS1 participates in cell-to-cell and cell-to-matrix communication [79], and has been implicated in the mediation of cellular adhesion, proliferation,

Table 2.1 Genes identified in human amniotic membrane

Gene	Symbol	Function (UniProtKB/Swiss-Prot)
Annexin A3	ANXA3	Inhibitor of phospholipase A2, also possesses anti-coagulant properties
Bradykinin receptor B1	BDKRB1	Factor in chronic pain and inflammation
CD55 molecule	CD55	Involved in the complement activation, classical pathway
Chemokine (C-X-C motif) ligand 1, 5 and 6	CXCL1, CXCL5 and CXCL6	Chemotactic activity for neutrophils
Desmocollin 3	DSC3	Component of intercellular desmosome junctions. Involved in the interaction of plaque proteins and intermediate filaments mediating cell-cell adhesion
Dipeptidyl-peptidase 4 (CD26)	DPP4	Cell surface glycoprotein receptor involved in the T-cell stimulation. Involved in the migration and invasion of endothelial cells into the ECM
Forkhead box F1	FOXF1	Transcription activator
Heart and neural crest derivatives expressed 2	HAND2	Essential for cardiac morphogenesis, particularly for the formation of the right ventricle and of the aorticarch arteries
Homeobox B6	HOXB6	Sequence-specific transcription factor
Interferon, α-inducible protein 27	IFI27	Promotes cell death
Interleukin 1, α and β	IL1A and IL1B	Involved in the inflammatory response
Interleukin 13 receptor, α2	IL13RA2	Binds as a monomer with high affinity to interleukin-13
Interleukin 33	IL33	Acts as a chemoattractant tor Th2 cells, and may function as an "alarmin", that amplifies immune responses during tissue injury
Membrane metallo-endopeptidase (CD10)	MME	Important role in the proteolysis
Niban protein	FAM129A	Regulates phosphorylation of a number of proteins involved in translation regulation
Paired-like homeodomain transcription factor 2	PITX2	Controls cell proliferation in a tissue-specific manner and is involved in morphogenesis
S100 calcium binding protein A4	S100A4	Involved in the regulation of cell cycle progression, modulating intercellular adhesion, and invasive and metastatic properties of cancer cells
Spondin 2, extracellular matrix protein	SPON2	Promotes adhesion cellular. Essential in the initiation of the innate immune
ST6-N-acetylgalactosaminide α-2,6-sialyltransferase 5	ST6GALNAC5	Involved in the biosynthesis of ganglioside GD1a from GM1b. It exhibits higher activity with glycolipids than with glycoproteins
Transcription factor 21	TCF21	Play a role in the specification or differentiation of one or more subsets of epicardial cell types

(continued)

Table 2.1 (continued)

Gene	Symbol	Function (UniProtKB/Swiss-Prot)
Tryptophan 2,3-dioxygenase	TDO2	Involved in the tryptophan catabolic process to acetyl-CoA and kynurenine
v-yes-1 Yamaguchi sarcoma viral related oncogene homolog	LYN	Non-receptor tyrosine-protein kinase that transmits signals from cell surface receptors and plays an important role in the regulation of innate and adaptive immune responses, hematopoiesis, responses to growth factors and cytokines, integrin signaling, but also responses to DNA damage and genotoxic agents. Plays an important role in the regulation of B-cell differentiation, proliferation, survival and apoptosis, and is important for immune self-tolerance

differentiation, migration and apoptosis [80]. More importantly, THBS1 is reported to control a number of physiological processes such as wound repair, inflammatory response, and angiogenesis [79, 81]. Mimecan represents a small leucine-rich proteoglycan, found in the ECM of connective tissue and is reported to maintain the tensile strength and hydration nature of the tissue [78, 82–84]. βIG-H3 is an ECM adhesive molecule acting as a membrane-associated growth factor during cell differentiation and wound healing [78, 85], and integrin α6 is the key hemidesmossomal protein involved in cell-to-matrix attachments and plays an important role in mediating cell adhesion and wound repair signaling pathways [78, 86].

Another important study carried out by Baharvand and theirs collaborators was focused on the analysis of epithelium-denuded human amniotic membrane [87]. They investigated the proteome of the human amniotic membrane epithelium, and detected 515 spots in all the 2-DE gels, but only 43 proteins were identified using MALDI TOF/TOF MS. The most abundant proteins identified were different isoforms of lumican (LUM) and OGN, both members of the proteoglycan family. In particular, OGN might play role in many biological processes including cell growth, angiogenesis, and inflammation [84]. Other proteins detected includes collagen VI α-1/α-2, fibrinogen β chain (FGB), transglutaminase 2 isoform A (TGM2A), β-actin variant (ACTB), periplakin (PPL), heat shock 70 kD protein 5 (HSPA5), nidogen 2 (NID2), integrin α6, βIG-H3, and tubulointerstitial nephritis (TIN) [88].

The more ample study on the analysis of the human amniotic membrane proteome using 2-DE followed by peptide mass fingerprinting (PMF) by MALDI-TOF MS was performed by Park and theirs collaborators [89]. They separated the soluble and membrane fraction of the amniotic membrane and observed less spots from the membrane fraction than from the soluble fraction. Theses spots identified 19 membrane proteins and 92 soluble proteins. The soluble fraction has based on agreement from three separate analyses and several different spots from the same gel corresponded to the same protein. Thirty-two proteins were from cytoplasm, seven from both cytoplasm and nucleus, four from endoplasmic reticulum lumen, and one from mitochondrial matrix. The subcellular locations of these proteins were consistent with the hydrophilic character of the soluble fraction [89].

The identification of 111 amniotic membrane proteins (92 soluble proteins (Table 2.2) and 19 membrane proteins (Table 2.3)) from term delivery patients represents the first extensive proteome analysis obtained from human amniotic membrane tissue.

Table 2.2 Proteins identified from soluble fraction of human amniotic membrane

Protein	Symbol	Function (UniProtKB/Swiss-Prot)
14-3-3 protein sigma	SFN	Adapter protein implicated in the regulation of a large spectrum of both general and specialized signaling pathways
14-3-3 protein zeta/delta	YWHAZ	
78 kDa glucose-regulated protein	HSPA5	Probably plays a role in facilitating the assembly of multimeric protein complexes inside the endoplasmic reticulum. Involved in the correct folding of proteins and degradation of misfolded proteins
Actin, cytoplasmic 1	ACTB	Involved in various types of cell motility
Adenyl cyclase-associated protein 1	CAP1	Directly regulates filament dynamics and has been implicated in a number of complex developmental and morphological processes, including mRNA localization and the establishment of cell polarity
Alcohol dehydrogenase	ADH	Catalyzes the conversion of primary unbranched alcohols to their corresponding aldehydes
Alfa-crystallin B chain	CRYAB	May contribute to the transparency and refractive index of the lens. Has chaperone-like activity, preventing aggregation of various proteins under a wide range of stress conditions
Alfa-tropomyosin and tropomyosin-4	TPM1 and TPM4	Binds to actin filaments. Plays a central role, in association with the troponin complex, in the calcium dependent regulation of vertebrate striated muscle contraction
Annexin A1 and A2	ANXA1 and ANXA2	Calcium/phospholipid-binding protein which promotes membrane fusion and is involved in exocytosis. This protein regulates phospholipase A2 activity. It seems to bind from two to four calcium ions with high affinity
Annexin A3	ANXA3	Inhibitor of phospholipase A2, also possesses anti-coagulant properties. Also cleaves the cyclic bond of inositol 1,2-cyclic phosphate to form inositol 1-phosphate

(continued)

Table 2.2 (continued)

Protein	Symbol	Function (UniProtKB/Swiss-Prot)
Annexin A5 and A8	ANXA5 and ANXA8	Anticoagulant protein that acts as an indirect inhibitor of the thromboplastin-specific complex, which is involved in the blood coagulation cascade
Apolipoprotein A-I	APOA1	Participates in the reverse transport of cholesterol from tissues to the liver
ATP synthase beta chain, mitochondrial precursor	ATP5F1	Mitochondrial membrane ATP synthase produces ATP from ADP in the presence of a proton gradient across the membrane which is generated by electron transport complexes of the respiratory chain
Calcyclin-binding protein	CACYBP	May be involved in calcium-dependent ubiquitination and subsequent proteasomal degradation of target proteins
Calcyphosine	CAPS	Calcium-binding protein. May play a role in cellular signaling events
Calgizzarin	S100A11	Facilitates the differentiation and the cornification of keratinocytes
Calmodulin	CaM	CaM mediates many crucial processes such as inflammation, metabolism, apoptosis, smooth muscle contraction, Intracellular movement, short-term and long-term memory, and the immune response. CaM can also make use of the calcium stores in the endoplasmic reticulum, and the sarcoplasmic reticulum
Carbonic anhydrase I	CA1	Reversible hydration of carbon dioxide. Can hydrates cyanamide to urea
Carbonic anhydrase II	CA2	Essential for bone resorption and osteoclast differentiation. Reversible hydration of carbon dioxide. Can hydrate cyanamide to urea
Cellular retinoic acid-binding protein II	CRABP2	Regulates the access of retinoic acid to the nuclear retinoic acid receptors
Chloride intracellular channel protein 1 and 3	CLIC1 and CLIC3	Channel activity depends on the pH. Membrane insertion seems to be redox-regulated and may occur only under oxydizing conditions. Involved in regulation of the cell cycle
Cofilin, non-muscule isoform	CFL1	Regulates actin cytoskeleton dynamics. Important for normal progress through mitosis and normal cytokinesis. Plays a role in the regulation of cell morphology and cytoskeletal organization

Table 2.2 (continued)

Protein	Symbol	Function (UniProtKB/Swiss-Prot)
Cytokeratin 1	KRT1	May regulate the activity of kinases such as PKC and SRC via binding to integrin beta-1 (ITB1) and the receptor of activated protein kinase C (RACK1/GNB2L1)
Cytokeratin 19	KRT19	Involved in the organization of myofibers
Cytosol aminopeptidase	LAP3	Presumably involved in the processing and regular turnover of intracellular proteins. Catalyzes the removal of unsubstituted N-terminal amino acids from various peptides
Destrin	DSTN	Actin-depolymerizing protein. Severs actin filaments (F-actin) and binds to actin monomers (G-actin)
Elongation factor 2	EEF2	Catalyzes the coordinated movement of the two tRNA molecules, the mRNA and conformational changes in the ribosome
Endoplasmic reticulum protein ERp29 precursor	ERP29	Plays an important role in the processing of secretory proteins within the endoplasmic reticulum (ER), possibly by participating in the folding of proteins in the ER
Endoplasmin precursor	HSP90B1	Molecular chaperone that functions in the processing and transport of secreted proteins. Functions in endoplasmic reticulum associated degradation (ERAD). Has ATPase activity
Esterase D	ESD	Serine hydrolase involved in the detoxification of formaldehyde
Ezrin	EZR	Involved in connections of major cytoskeletal structures to the plasma membrane
Fascin	FSCN	Plays a role in the organization of actin filament bundles and the formation of microspikes, membrane ruffles, and stress fibers
Fatty acid-binding protein, epidermal	FABP5	High specificity for fatty acids. Highest affinity for C18 chain length. Decreasing the chain length or introducing double bonds reduces the affinity
Ferritin light chain	FTL	Stores iron in a soluble, non-toxic, readily available form. Important for iron homeostasis

(continued)

Table 2.2 (continued)

Protein	Symbol	Function (UniProtKB/Swiss-Prot)
Flavin reductase	BLVRB	Catalyzes the NADPH-dependent reduction of a variety of flavins, such as riboflavin, FAD or FMN, biliverdins, methemoglobin and PQQ (pyrroloquinoline quinone)
Fructose-bisphosphate aldolase A	ALDOA	Plays a key role in glycolysis and gluconeogenesis. May also function as scaffolding protein
Galactokinase	GALK1	Major enzyme for galactose metabolism
Galectin-1	LGALS1	May regulate apoptosis, cell proliferation and cell differentiation. Binds beta-galactoside and a wide array of complex carbohydrates
Galectin-3-binding protein	LGALS3BP	Promotes intergrin-mediated cell adhesion. May stimulate host defense against viruses and tumor cells
Glutathione S-transferase P	GSTP1	Conjugation of reduced glutathione to a wide number of exogenous and endogenous hydrophobic electrophiles
Glutathione transferase omega 1	GSTO1	Exhibits glutathione-dependent thiol transferase and dehydroascorbate reductase activities
Glyceraldehyde 3-phosphate dehydrogenase liver	GAPDH	Play a role in glycolysis. Participates in nuclear events including transcription, RNA transport, DNA replication and apoptosis
GTP-binding nuclear protein RAN	RAN	GTP-binding protein involved in nucleocytoplasmic transport. Required for the import of protein into the nucleus and also for RNA export. Involved in chromatin condensation and control of cell cycle
Guanine nucleotide-binding protein beta subunit-like protein 12,3	GNB2L1	Involved in the recruitment, assembly and/or regulation of a variety of signaling molecules. Interacts with a wide variety of proteins and plays a role in many cellular processes
Heat shock 27 kDa protein	HSPB1	Involved in stress resistance and actin organization
Hemoglobin alpha, beta, gamma-A and gamma-G chain	HBA1, HBB, HBG1 and HBG2	Involved in oxygen transport
Histidine triad nucleotide-binding protein 1	HINT1	Hydrolyzes purine nucleotide phosphoramidates with a single phosphate group
Isocitrate dehydrogenase [NADP] cytoplasmic	IDH1	Participates in the NADPH regeneration and in the isocitrate metabolic process

Table 2.2 (continued)

Protein	Symbol	Function (UniProtKB/Swiss-Prot)
Leukocyte elastase inhibitor	SERPINB1	Regulates the activity of the neutrophil proteases elastase, cathepsin G, proteinase-3, chymase, chymotrypsin, and kallikrein-3. Also functions as a potent intracellular inhibitor of granzyme H
L-lactate dehydrogenase A chain	LDHA	Involved in the glycolysis
Macrophage migration inhibitory factor	MIF	Pro-inflammatory cytokine. Involved in the innate immune response to bacterial pathogens
Macrophage-capping protein	CAPG	May play an important role in macrophage function. May play a role in regulating cytoplasmic and/or nuclear structures through potential interactions with actin
Malate dehydrogenase, cytoplasmic	MDH1	Participates in the malate and NADH metabolic process
Maspin	SERPINB5	Tumor suppressor that blocks the growth, invasion, and metastatic properties of mammary tumors
Myosin light chain alkali, non-muscle isoform	MYL6	Regulatory light chain of myosin
N(G),N(G)-dimethylarginine dimethylaminohydrolase 2	DDAH2	Hydrolyzes N(G),N(G)-dimethyl-L-arginine (ADMA) and N(G)-monomethyl-L-arginine (MMA) which act as inhibitors of NOS. Has therefore a role in the regulation of nitric oxide generation
Peptidyl-prolyl cis-trans isomerase A	PPIA	PPIases accelerate the folding of proteins. It catalyzes the cis-trans isomerization of proline imidic peptide bonds in oligopeptides
Peroxiredoxin-1,- 2,- 4 and -6	PRDX1, PRDX2, PRDX4 and PRDX6	Involved in redox regulation of the cell
Phosphatidylethanolamine-binding protein	PEBP	Binds ATP, opioids and phosphatidylethanolamine. Has lower affinity for phosphatidylinositol and phosphatidylcholine. Serine protease inhibitor which inhibits thrombin, neuropsin and chymotrypsin but not trypsin, tissue type plasminogen activator and elastase
Phosphoglycerate mutase 1 and phosphoglycerate kinase 1	PGAM1 and PGK1	Important role as a glycolytic enzyme

(continued)

Table 2.2 (continued)

Protein	Symbol	Function (UniProtKB/Swiss-Prot)
Placental calcium-binding protein	S100A4	Involved in the regulation of cell cycle progression, modulating intercellular adhesion, and invasive and metastatic properties of cancer cells
Platelet-activating factor acetylhydrolase IB beta subunit	PAFAH1B2	Inactivates PAF by removing the acetyl group at the sn-2 position. This is a catalytic subunit
Pre-B cell enhancing factor precursor	NAMPT	Catalyzes the condensation of nicotinamide with 5-phosphoribosyl-1-pyrophosphate to yield nicotinamide mononucleotide, an intermediate in the biosynthesis of NAD. It is the rate limiting component in the mammalian NAD biosynthesis pathway
Profilin I	PFN1	Binds to actin and affects the structure of the cytoskeleton
Proteasome subunit alpha type-1 and -2	PSMA1 and PSMA2	Multicatalytic proteinase complex with ability to cleave peptides with Arg, Phe, Tyr, Leu, and Glu adjacent to the leaving group at neutral or slightly basic pH
Protein disulfide isomerase A3 precursor	PDIA3	Catalyzes the rearrangement of -S-S- bonds in proteins
Pyruvate Kinase, M1 isozyme	PKM-M1	Glycolytic enzyme that catalyzes the transfer of a phosphoryl group from phosphoenolpyruvate (PEP) to ADP, generating ATP. Plays a general role in caspase independent cell death of tumor cells and survival
Rho GDP-dissociation inhibitor 1 and 2	ARHGDIA and ARHGDIB	Regulates the GDP/GTP exchange reaction of the Rho proteins by inhibiting the dissociation of GDP from them, and the subsequent binding of GTP to them
Serotransferrin	TF	Responsible for the transport of iron from sites of absorption and heme degradation to those of storage and utilization. Serum transferrin may also have a further role in stimulating cell proliferation
Stress-induced-phosphoprotein 1	STIP1	Mediates the association of the molecular chaperones HSC70 and HSP90
Superoxide dismutase [Mn] mitochondrial and superoxide dismutase [Cu-Zn]	SOD2 and SOD1	Destroys superoxide anion radicals

Table 2.2 (continued)

Protein	Symbol	Function (UniProtKB/Swiss-Prot)
Thioredoxin	TXN	May participate in various redox reactions. Modulates TNF-alpha signaling and NF-kappa-B activation. Has peroxidase activity and may contribute to the elimination of cellular hydrogen peroxide
Transforming growth factor-β-induced protein IG-H3	TGFβI	Binds to type I, II, and IV collagens. This adhesion protein may play an important role in cell-collagen interactions. In cartilage, may be involved in endochondral bone formation
Translationally controlled tumor protein	TPT1	Involved in calcium binding and microtubule stabilization
Triosephosphate isomerase	TPI1	Enzyme involved in the glycolysis and gluconeogenesis
UDP-glucose pyrophosphorylase 2	UGP2	Plays a central role as a glucosyl donor in cellular metabolic pathways
Vimentin	VIM	Involved in the stabilization of type I collagen mRNAs
α-enolase	ENO1	Multifunctional enzyme that, as well as its role in glycolysis, plays a part in various processes such as growth control, hypoxia tolerance and allergic responses

2.6 Conclusions and Future Perspectives

Amniotic membrane expresses several biochemical components, which are involved in numerous physiologic processes, such as transportation of molecules/metabolites across the cell membrane, modulation of immune system, extracellular matrix remodeling, tissue regeneration, regulation of angiogenesis and cell cycle. Considering these effects on cells and tissues, several clinical applications have been evaluated using the amniotic membrane such as ocular surface reconstruction and skin application. In addition, many other applications have been hypothesized and are currently being explored, namely in human cancer. Although some studies of transcriptomic and proteomic analysis have been performed, it is crucial to carry out more studies in order to: (1) clarify the proteome; (2) determine the levels of expression of each protein; (3) determine the secretome; and (4) characterize the genetic variability between amniotic membranes. Moreover, it is imperative to identify which molecules and cellular mechanisms are involved in physiological effects that are observed in several in vitro and in vivo studies.

Table 2.3 Proteins identified from membrane fraction of human amniotic membrane

Protein	Symbol	Function
Integrin alpha 6 precursor	ITGA6	Receptor for laminin in epithelial cells and it plays a critical structural role in the hemidesmosome
Vimentin	VIM	Involved in the stabilization of type I collagen mRNAs
Lamin A/C	LMNA	Plays an important role in nuclear assembly, chromatin organization, nuclear membrane and telomere dynamics
Heat shock cognate 71 kDa protein	HSPA8	Acts as a repressor of transcriptional activation
Ezrin	EZR	Involved in connections of major cytoskeletal structures to the plasma membrane
Heat shock 70 kDa protein 1	HSPA1	Stabilize preexistent proteins against aggregation and mediate the folding of newly translated polypeptides in the cytosol as well as within organelles
Voltage-dependent anion-selective channel protein 1 and 2	VDAC1 and VDAC2	Forms a channel through the mitochondrial outer membrane that allows diffusion of small hydrophilic molecules
Collagen alpha 1 (VI) and 2 (VI) chain precursor	COL6A1 and COL6A2	Collagen VI acts as a cell-binding protein
Laminin gamma-2 chain precursor	LAMC2	Binding to cells via a high affinity receptor, laminin is thought to mediate the attachment, migration and organization of cells into tissues during embryonic development by interacting with other extracellular matrix components
Guanine nucleotide-binding protein G(I)/G(S)/G(T) beta subunit 1 and 2	GNB1 and GNB2	Involved as a modulator or transducer in various transmembrane signaling systems
Calpactin I light chain	S100A10	Regulator of protein phosphorylation in that the ANXA2 monomer is the preferred target of tyrosine-specific kinase
Mimecan precursor	OGN	Induces bone formation in conjunction with TGF-β-1 or TGF-β-2
Cytokeratin 5, 8, 19 and 24	KRT5, KRT8, KRT19 and KRT24	Involved in the organization of cytoskeleton

References

1. Méhats C, Schmitz T, Marcellin L, Breuiller-Fouché M (2011) Biochemistry of fetal membranes rupture. Gynécol Obstet Fertil 39:365–369
2. Higa K, Shimmura S, Shimazaki J, Tsubota K (2005) Hyaluronic acid-CD44 interaction mediates the adhesion of lymphocytes by amniotic membrane stroma. Cornea 24:206–212
3. Hao Y, Ma DH-K, Hwang DG et al (2000) Identification of antiangiogenic and antiinflammatory proteins in human amniotic membrane. Cornea 19:348–352
4. Solomon A, Rosenblatt M, Monroy D et al (2001) Suppression of interleukin 1alpha and interleukin 1beta in human limbal epithelial cells cultured on the amniotic membrane stromal matrix. Br J Ophthalmol 85:444–449

5. Kim JS, Kim JC, Na BK et al (2000) Amniotic membrane patching promotes healing and inhibits proteinase activity on wound healing following acute corneal alkali burn. Exp Eye Res 70:329–337
6. King A, Paltoo A, Kelly R et al (2007) Expression of natural antimicrobials by human placenta and fetal membranes. Placenta 28:161–169
7. Akle C, Adinolfi M, Welsh K et al (1981) Immunogenicity of human amniotic epithelial cells after transplantation into volunteers. Lancet 2:1003–1005
8. Cornwell KG, Landsman A, James KS (2009) Extracellular matrix biomaterials for soft tissue repair. Clin Podiatr Med Surg 26:507–523
9. Tseng SC, Li D-Q, Ma X (1999) Suppression of transforming growth factor-beta isoforms, TGF-beta receptor type II and myofibroblast differentiation in cultured human corneal and limbal fibroblasts by amniotic membrane matrix. J Cell Physiol 179:325–335
10. Lee S, Li D, Tan D et al (2000) Suppression of TGF-beta signaling in both normal conjunctival fibroblasts and pterygial body fibroblasts by amniotic membrane. Curr Eye Res 20:325–334
11. Niknejad H, Yazdanpanah G, Mirmasoumi M et al (2013) Inhibition of HSP90 could be possible mechanism for anti-cancer property of amniotic membrane. Med Hypotheses 81:862–865
12. Li W, He H, Kawakita T et al (2006) Amniotic membrane induces apoptosis of interferon-gamma activated macrophages in vitro. Exp Eye Res 82:282–292
13. Seo JH, Kim YH, Kim JS (2008) Properties of the amniotic membrane may be applicable in cancer therapy. Med Hypotheses 70:812–814
14. Ganatra M (2003) Amniotic membrane in surgery. J Pakistan Med Assoc 53:29–32
15. Mermet I, Pottier N, Sainthillier JM et al (2007) Use of amniotic membrane transplantation in the treatment of venous leg ulcers. Wound Repair Regen 15:459–464
16. Niknejad H, Peirovi H, Jorjani M et al (2008) Properties of the amniotic membrane for potential use in tissue engineering. Eur Cell Mater 15:88–99
17. Dua HS, Gomes JA, King AJ, Maharajan VS (2004) The amniotic membrane in ophthalmology. Surv Ophthalmol 49:51–77
18. Amer MI, Abd-El-Maeboud KH (2006) Amnion graft following hysteroscopic lysis of intrauterine adhesions. J Obstet Gynaecol Res 32:559–566
19. Jin CZ, Park SR, Choi BH et al (2007) Human amniotic membrane as a delivery matrix for articular cartilage repair. Tissue Eng 13:693–702
20. Toda A, Okabe M, Yoshida T, Nikaido T (2007) The potential of amniotic membrane/amnion-derived cells for regeneration of various tissues. J Pharmacol Sci 105:215–228
21. Chen Z, Tortella FC, Dave JR et al (2009) Human amnion-derived multipotent progenitor cell treatment alleviates traumatic brain injury-induced axonal degeneration. J Neurotrauma 26:1987–1997
22. Fernandes M, Sridhar MS, Sangwan VS, Rao GN (2005) Amniotic membrane transplantation for ocular surface reconstruction. Cornea 24:643–653
23. Yang L, Shirakata Y, Shudou M et al (2006) New skin-equivalent model from de-epithelialized amnion membrane. Cell Tissue Res 326:69–77
24. Wilshaw S-P, Kearney JN, Fisher J, Ingham E (2006) Production of an acellular amniotic membrane matrix for use in tissue engineering. Tissue Eng 12:2117–2129
25. Sadler TW (2012) Langman's: medical embryology, 12th edn. Wolters Kluwer Health, Philadelphia, p 400
26. Pollards SM, Aye N, Symonds E (1976) Scanning electron microscope appearances of normal human amnion and umbilical cord at term. Br J Obstet Gynaecol 83:470–477
27. Miki T, Lehmann T, Cai H et al (2005) Stem cell characteristics of amniotic epithelial cells. Stem Cells 23:1549–1559
28. Miki T, Marongiu F, Ellis EC et al (2007) Isolation of amniotic epithelial stem cells. Curr Protoc Cell Biol Chapter 1:Unit 1E 3
29. Streuli C (1996) Basement membrane as a differentiation and survival factor. In: Ekblom P, Timpl R (eds) The laminins. CRC Press, The Netherlands, p 321

30. Fukuda K, Chikama T, Nakamura M, Nishida T (1999) Differential distribution of subchains of the basement membrane components type IV collagen and laminin among the amniotic membrane, cornea and conjunctiva. Cornea 18:73–79
31. Fukushima Y, Ohnishi T, Arita N et al (1998) Integrin alpha3beta1-mediated interaction with laminin-5 stimulates adhesion, migration and invasion of malignant glioma cells. Int J Cancer 76:63–72
32. Akashi T, Miyagi T, Ando N et al (1999) Synthesis of basement membrane by gastrointestinal cancer cell lines. J Pathol 187:223–228
33. Bryant-Greenwood G (1998) The extracellular matrix of the human fetal membranes: structure and function. Placenta 19:1–11
34. Malak T, Ockleford C, Bell S et al (1993) Confocal immunofluorescence localization of collagen types I, III, IV, V and VI and their ultrastructural organization in term human fetal membranes. Placenta 14:385–406
35. Parry S, Strauss JF (1998) Premature rupture of the fetal membranes. N Engl J Med 338:663–670
36. Fetterolf DE, Snyder RJ (2012) Scientific and clinical support for the use of dehydrated amniotic membrane in wound management. Wounds 24:299–307
37. Szpak P (2011) Fish bone chemistry and ultrastructure: implications for taphonomy and stable isotope analysis. J Archaeol Sci 38:3358–3372
38. Lodish H, Berk A, Lawrence Zipursky S et al (2000) Molecular cell biology, 4th edn. W. H. Freeman, New York, p 1184
39. Alberts B, Johnson A, Lewis J et al (2002) Molecular biology of the cell, 4th edn. Garland Science, New York
40. Prévost TP (2006) Biomechanics of the human chorioamnion. Massachusetts Inst Technol 115
41. Meinert M, Eriksen GV, Petersen AC et al (2001) Proteoglycans and hyaluronan in human fetal membranes. Am J Obstet Gynecol 184:679–685
42. Schmidt MB, Mow VC, Chun LE, Eyre DR (1990) Effects of proteoglycan extraction on the tensile behavior of articular cartilage. J Orthop Res 8:353–363
43. Danielson KG, Baribault H, Holmes DF et al (1997) Targeted disruption of decorin leads to abnormal collagen fibril morphology and skin fragility. J Cell Biol 136:729–743
44. Iozzo RV, Murdoch AD (1996) Proteoglycans of the extracellular environment: clues from the gene and protein side offer novel perspectives in molecular diversity and function. Faseb J 10:598–614
45. Yamada KM (1991) Fibronectin and other cell interactive glycoproteins. In: Hay ED (ed) Cell Biol. Extracell. matrix, 2nd edn. Springer US, New York, pp 111–146
46. Feinberg RF, Kliman H, Lockwoodt CJ (1991) Is oncofetal fibronectin a trophoblast glue for human implantation? Am J Pathol 138:537–543
47. Aota S, Nagai T, Olden K et al (1991) Fibronectin and integrins in cell adhesion and migration. Biochem Soc Trans 19:830–835
48. Champliaud M-F, Lunstrum GP, Rousselle P et al (1996) Human amnion contains a novel laminin variant, laminin 7, which like laminin 6, covalently associates with laminin 5 to promote stable epithelial-stromal attachment. J Cell Biol 132:1189–1198
49. Hieber A, Corcino D, Motosue J et al (1997) Detection of elastin in the human fetal membranes: proposed molecular basis for elasticity. Placenta 18:301–312
50. Malak T, Bell S (1994) Distribution of fibrillin-containing microfibrils and elastin in human fetal membranes: a novel molecular basis for membrane elasticity. Am J Obstet Gynecol 171:195–205
51. Cunningham FG, Leveno KJ, Bloom SL et al (2001) Williams obstetrics, 23rd edn. McGraw-Hill, New York, p 1404
52. Bara M, Moretto P, Durlach J, Guiet-Bara A (2003) Physiological importance of the connective tissue in the human amnion: role of magnesium. Magnes Res 16:35–42
53. Beall M, van den Wijngaard J, van Gemert M, Ross M (2007) Amniotic fluid water dynamics. Placenta 28:816–823
54. Beall M, Wang S, Yang B et al (2007) Placental and membrane aquaporin water channels: correlation with amniotic fluid volume and composition. Placenta 28:421–428

55. Zhu X, Jiang S, Zhu X et al (2009) Expression of aquaporin 1 and aquaporin 3 in fetal membranes and placenta in human term pregnancies with oligohydramnios. Placenta 30:670–676
56. Jiang S, Zhu X, Ding S et al (2012) Expression and localization of aquaporins 8 and 9 in term placenta with oligohydramnios. Reprod Sci 19:1276–1284
57. Prat C, Blanchon L, Borel V et al (2012) Ontogeny of aquaporins in human fetal membranes. Biol Reprod 86:1–8
58. Ishibashi K, Kondo S, Hara S, Morishita Y (2011) The evolutionary aspects of aquaporin family. Am J Physiol Regul Integr Comp Physiol 300:R566–R576
59. Silini A, Parolini O, Huppertz B, Lang I (2013) Soluble factors of amnion-derived cells in treatment of inflammatory and fibrotic pathologies. Curr Stem Cell Res Ther 8:6–14
60. He H, Li W, Tseng DY et al (2009) Biochemical characterization and function of complexes formed by hyaluronan and the heavy chains of inter-alpha-inhibitor (HC-HA) purified from extracts of human amniotic membrane. J Biol Chem 284:20136–20146
61. Fortunato S, Menon R, Lombardi S (1998) Presence of four tissue inhibitors of matrix metalloproteinases (TIMP-1, -2, -3 and -4) in human fetal membranes. Am J Reprod Immunol 40:395–400
62. Fortunato SJ, Menon R, Lombardi SJ (1998) The effect of transforming growth factor and interleukin-10 on interleukin-8 release by human amniochorion may regulate histologic chorioamnionitis. Am J Obstet Gynecol 179:794–799
63. Li H, Niederkorn JY, Neelam S et al (2005) Immunosuppressive factors secreted by human amniotic epithelial cells. Invest Ophthalmol Vis Sci 46:900–907
64. Apte R, Sinha D, Mayhew E et al (1998) Cutting edge: role of macrophage migration inhibitory factor in inhibiting NK cell activity and preserving immune privilege. J Immunol 160:5693–5696
65. Okazaki T, Casey M, Okita J et al (1981) Initiation of human parturition. XII: biosynthesis and metabolism of prostaglandins in human fetal membranes and uterine decidua. Am J Obstet Gynecol 139:373–381
66. Bryant-Greenwood G, Rees M, Turnbull A (1987) Immunohistochemical localization of relaxin, prolactin and prostaglandin synthase in human amnion, chorion and decidua. J Endocrinol 114:491–496
67. Toth P, Li X, Lei Z, Rao C (1996) Expression of human chorionic gonadotropin (hCG)/luteinizing hormone receptors and regulation of the cyclooxygenase-1 gene by exogenous hCG in human fetal membranes. J Clin Endocrinol Metab 81:1283–1288
68. Yañez R, Oviedo A, Aldea M et al (2010) Prostaglandin E2 plays a key role in the immunosuppressive properties of adipose and bone marrow tissue-derived mesenchymal stromal cells. Exp Cell Res 316:3109–3123
69. Parolini O, Alviano F, Bagnara GP et al (2008) Concise review – isolation and characterization of cells from human term placenta: outcome of the first international workshop on placenta derived stem cells. Stem Cells 26:300–311
70. Pappa KI, Anagnou NP (2009) Novel sources of fetal stem cells: where do they fit on the developmental continuum? Regen Med 4:423–433
71. Cai J, Li W, Su H et al (2010) Generation of human induced pluripotent stem cells from umbilical cord matrix and amniotic membrane mesenchymal cells. J Biol Chem 285:11227–11234
72. Parolini O, Soncini M (2006) Human placenta: a source of progenitor/stem cells? J Reprod Med Endocrinol 3:117–126
73. Ilancheran S, Michalska A, Peh G et al (2007) Stem cells derived from human fetal membranes display multilineage differentiation potential. Biol Reprod 77:577–588
74. Marongiu F, Gramignoli R, Sun Q et al (2010) Isolation of amniotic mesenchymal stem cells. Curr Protoc Cell Biol 12:1E.5.1–1E.5.11
75. Barbati A, Mameli MG, Sidoni A, Di Renzo GC (2012) Amniotic membrane: separation of amniotic mesoderm from amniotic epithelium and isolation of their respective mesenchymal stromal and epithelial cells. Curr Protoc Stem Cell Biol 20:1E.8.1–1E.8.15

76. Tsai M-S, Hwang S-M, Chen K-D et al (2007) Functional network analysis of the transcriptomes of mesenchymal stem cells derived from amniotic fluid, amniotic membrane, cord blood and bone marrow. Stem Cells 25:2511–2523
77. Hopkinson A, McIntosh RS, Layfield R et al (2005) Optimised two-dimensional electrophoresis procedures for the protein characterisation of structural tissues. Proteomics 5:1967–1979
78. Hopkinson A, McIntosh RS, Shanmuganathan V et al (2006) Proteomic analysis of amniotic membrane prepared for human transplantation: characterization of proteins and clinical implications. J Proteome Res 5:2226–2235
79. Adams JC, Lawler J (2004) The thrombospondins. Int J Biochem Cell Biol 36:961–968
80. Clezardin P, Lawler J, Amiral J et al (1997) Identification of cell adhesive active sites in the N-terminal domain of thrombospondin-1. Biochem J 321:819–827
81. Zaslavsky A, Baek K-H, Lynch RC et al (2010) Platelet-derived thrombospondin-1 is a critical negative regulator and potential biomarker of angiogenesis. Blood 115:4605–4613
82. Funderburgh JL, Corpuz LM, Roth MR et al (1997) Mimecan, the 25-kDa corneal keratan sulfate proteoglycan, is a product of the gene producing osteoglycin. J Biol Chem 272:28089–28095
83. Tasheva ES, Koester A, Paulsen AQ et al (2002) Mimecan/osteoglycin-deficient mice have collagen fibril abnormalities. Mol Vis 8:407–415
84. Kampmann A, Fernandez B, Deindl E et al (2009) The proteoglycan osteoglycin/mimecan is correlated with arteriogenesis. Mol Cell Biochem 322:15–23
85. Kim M-O, Yun S-J, Kim I-S et al (2003) Transforming growth factor-beta-inducible gene-h3 (beta(ig)-h3) promotes cell adhesion of human astrocytoma cells in vitro: implication of alpha-6beta4 integrin. Neurosci Lett 336:93–96
86. Corsini NS, Martin-Villalba A (2010) Integrin alpha 6: anchors away for glioma stem cells. Cell Stem Cell 6:403–404
87. Baradaran-Rafii A, Aghayan H-R, Arjmand B, Javadi M-A (2007) Amniotic membrane transplantation. Iran J Ophthalmic Res 2:58–75
88. Baharvand H, Heidari M, Ebrahimi M et al (2007) Proteomic analysis of epithelium-denuded human amniotic membrane as a limbal stem cell nich. Mol Vis 13:1711–1721
89. Park S-J, Yoon W-G, Song J-S et al (2006) Proteome analysis of human amnion and amniotic fluid by two-dimensional electrophoresis and matrix-assisted laser desorption/ionization time-of-flight mass spectrometry. Proteomics 6:349–363

Chapter 3
Biophysical Properties of Amniotic Membrane

Ana Margarida Abrantes, João Casalta-Lopes, and Maria Filomena Botelho

Abstract Placenta constitutes the barrier between the fetus and the mother that allow the transport of nutrients as well as waste products between the fetus and the mother. The transport, can occur by different ways: simple diffusion, facilitated diffusion and receptor-mediated endocytosis, or by paracellular flow. The way nutrients are transported is influenced by several characteristics namely permeability, nutrient concentration gradients, placental blood flow and metabolism. Physical properties such as elasticity, stiffness and other rheological properties of amniotic membrane and extracellular matrix which are also influenced by the variation of the placenta composition and gestational age.

Keywords Placenta • Amniotic membrane • Biophysics • Physical properties • Mechanical tests

3.1 Introduction

During pregnancy, it is necessary to supply the nutritional needs for the harmonious growth of the fetus, and placenta, the barrier between the fetus and the mother, must simultaneously allow the transport of these nutrients as well as the waste products, between the fetus and the mother [1]. Thereby, for the transport through the placenta occurs, all the molecules must be able to cross the syncytiotrophoblast barrier, which is the first one to be permeated. The success of the placenta as a barrier depends on

---

A.M. Abrantes • M.F. Botelho (✉)
Biophysics Unit, Institute for Biomedical Imaging and Life Sciences (IBILI), Center of Investigation in Environment, Genetics and Oncobiology (CIMAGO), Faculty of Medicine, University of Coimbra, Coimbra, Portugal
e-mail: mfbotelho@fmed.uc.pt

J. Casalta-Lopes
Biophysics Unit, Institute for Biomedical Imaging and Life Sciences (IBILI), Radiotherapy Department, Centro Hospitalar e Universitário de Coimbra (CHUC), Faculty of Medicine, University of Coimbra, Coimbra, Portugal

A.C. Mamede, M.F. Botelho (eds.), *Amniotic Membrane: Origin, Characterization and Medical Applications*,
DOI 10.1007/978-94-017-9975-1_3

the establishment of this physiological boundary [1]. According to placenta structure, each layer is rich in different components, such as collagen, proteoglycans, fibronectin and laminin. Concerning gestational age, with increasing time there is an increase of the surface area of the uterine cavity especially important in the last weeks of pregnancy, which in turn determines the enhancement of elastic tension of the membranes [2]. Therefore, the amniotic membrane at the end of pregnancy is stretched in a way that its surface area increases approximately to twice considering their initial surface area [2]. However, the stress of an individual fetus' membrane is not homogeneous over all its surface area [3]. Hence, during pregnancy there are biomechanical changes occurring that influences and conditionates the pregnancy evolution.

3.2 Transport Through Amniotic Membrane

Mass transport through amniotic membrane can occur by different manners, and as the amniotic membrane controls the transcellular movement of water and solutes, it helps to maintain the normal growth and the fetus homeostasis [1, 4]. The syncytiotrophoblast that separates the blood from the mother from the fetus' blood, is composed by two polarized membranes, a microvillous membrane directed towards the maternal blood, and a basal membrane, which faces the fetal capillary [5]. Placenta may act as a nutrient sensor and coordinates nutrient transport functions with maternal nutrient availability [5]. In general, the transfer of a molecule through a barrier can be limited by two reasons, the diffusion process itself (diffusion-limited transport) or by the mass or the rate at which the molecule is supplied and/or removed by blood flow (blood flow-limited transport) [5].

Amniotic fluid volume increases up to 1 L by 36 weeks of gestation and then declines. Severe deficiency of amniotic fluid, named by oligohydramnios, can be caused by premature failure of the membrane or by decreased urine production by the fetal kidneys. Polyhydramnios defined as the excess of amniotic fluid, or simply hydramnios, can present with amniotic fluid amounts of more than 10 L. It can occur by excessive amniotic fluid production or by failure of its normal circulation. In these situations the massive amniotic fluid volume and pressure can cause significant membrane stress, increasing the risk of membrane failure [6].

Transport can occur by simple diffusion of lipophilic molecules such as oxygen or of hydrophilic molecules through transmembrane channels, which is the common mechanism for membrane water flow. Other types of transport can be observed such as the facilitated diffusion as occurs with D-glucose, the active transport for certain electrolytes, and receptor-mediated endocytosis, a mechanism for transfer large molecules, such as IgG. In addition to transcellular flow across the cell membrane, water and solutes may also cross biological membranes between cells, nominated paracellular flow [7–9].

Considering solute movement by simple diffusion, it is possible to quantify diffusion flow (represented by J_s), which accounts for the quantity of solutes that moves across a membrane surface per unit of area and per unit of time. Its units are mol cm^{-2} s^{-1}, using CGS units system.

As known the mean velocity of the particles moving across the membrane $\overline{v}$ and the solute concentration (C_s), the amount of solute that crosses 1 cm^2 of the membrane, normal to its axis, can be given by

$$J_s = C_s \cdot \overline{v} \tag{3.1}$$

The stationary situation can be considered, in which the solute concentrations in both sides of the membrane are constant over time. In these circumstances, J_s is also constant over time and can be assumed that $\overline{v}$ is also constant.

The motor force involved in the process of diffusion (diffusion force) can be interpreted as a result of a gradient of chemical potential, μ, a form of potential energy. For diluted solutions

$$\mu = \mu_0 + RT \ \ln \ C_s \tag{3.2}$$

where μ_0 is a constant that depends on temperature and pressure.

The relationship between diffusion force and chemical potential can be given by

$$F_D = -\frac{d\mu}{dx} \tag{3.3}$$

where F_D represents the diffusion force per mol of solute and $\frac{d\mu}{dx}$ and represents the gradient of chemical potential.

Equation 3.3 can, therefore, be rewritten as

$$F_D = -\frac{d\left(\mu_0 + RT \ \ln \ C_s\right)}{dx} = -\frac{RT}{C_s} \cdot \frac{dC_s}{dx} \tag{3.4}$$

Where T represents the absolute temperature (in Kelvin degrees) and R is the universal gas constant. Instead of considering the diffusion force per mol of solute, can be considered the force that is applied per particle of solute (represented by f). The relationship between f and F_D is

$$F_D = f \cdot A \tag{3.5}$$

where A is the Avogadro number. Considering the conditions imposed the velocity of a solute particle is directly proportional to the force f. The proportionality constant is called molecular mobility and it is represented by u'. Consequently is possible to write

$$\overline{v} = u' \cdot f = u' \frac{F_D}{A} \tag{3.6}$$

Using the equations above, Eq. 3.1 can be rewritten:

$$J_s = C_s \cdot \bar{v} = C_s \cdot u' \frac{F_D}{A} = -\frac{R \cdot T \cdot u'}{A} \cdot \frac{dC_s}{dx} \tag{3.7}$$

Since R, T, u' and A are constants for a certain condition, the above equation can be rewritten as

$$J_s = -D \cdot \frac{dC_s}{dx} \tag{3.8}$$

Equation 3.8 is known as the Fick's First Law of diffusion, which relates diffusion flow of a solute with its concentration gradient [7, 10]. D is the diffusion coefficient and is dependent on the solute concentration, solvent, temperature and pressure.

Simple diffusion across membranes can occur through the membrane itself or through transmembrane channels. In the first case, it is needed to consider the partition coefficient k that relates concentrations at the membrane interface to those of the compartments that the membrane separate, as in

$$k = \frac{C_s^{membrane}}{C_s^{compartment}} \tag{3.9}$$

Also, the phase by which the solute is diffusing is the membrane itself, so the diffusion constant as to account for that and it will be represented by D_m. Considering these conditions, the First Fick's Law of diffusion can be rewritted as

$$J_s = \frac{D_m k}{\Delta x} \cdot \Delta C_s = P_s \cdot \Delta C_s \tag{3.10}$$

where Δx is the membrane thickness, ΔC_S the difference of solute concentration between the two sides of the membrane and P_S the membrane's permeability to the solute.

Considering simple diffusion across transmembrane channels, membrane can be considered as a porous separation between compartments, where only a portion of it allows the solute to cross. Consequently, Eq. 3.8 has to account for the portion of the membrane that is not permeable to the solute. If it is considered

$$\phi = \frac{S_{pores}}{S_{membrane}} \tag{3.11}$$

where S_{pores} represents the total sectional area occupied by channels (pores) and $S_{membrane}$ represents the whole membrane surface, the First Fick's Law can again be rewritten as

$$J_s = \frac{D \cdot \phi}{\Delta x} \cdot \Delta C_s = \omega' \cdot \Delta C_s \tag{3.12}$$

With ω' being the membranes permeability to the solute.

Depending on the solute characteristics, should be considered a different type of model to describe simple diffusion across the membrane. If a solute can be diffused by both way described above, should also be considered a mixed model, where the pores and the membrane have different diffusion constants. In a mixed model where the portion of membrane that has no pores allows the solute to cross, can be determined the diffusion flow by the following equation

$$J_s = \frac{D \cdot \phi}{\Delta x} \cdot \Delta C_s + (1-\phi)\frac{D_m k}{\Delta x} \cdot \Delta C_s = \left[D \cdot \phi + D_m k(1-\phi)\right] \cdot \frac{\Delta C_s}{\Delta x} \quad (3.13)$$

Water is also transported across this membrane, and its transport is done primarily by channel proteins called aquaporins. This type of membrane proteins increases cell membrane permeability to water by 5- to 50-fold compared to the membranes in which water moves essentially through the lipid membrane [1].

Water flux early in placental development has not been extensively studied. However, there is available information related to the last period of pregnancy. In late gestation, the growth of the fetus is exponential, while the placenta grows slowly if there is growing. As a result, in the late period of pregnancy, there is a need to increase relative water flow through the placenta, probably because of the increased surface area and increased water permeability which increases the efficiency of water flux through the placenta. This fact may be associated with the increased fetal growth rate seen in late time of gestation. Normally, passage of water and solutes from the maternal to the fetal circulation in the placenta is limited by the membrane's permeability to water, as well as by the surface area available for exchange. Therefore and also taking into account some data about human trophoblast, membrane's water permeability appears to increase with gestation. Still, human placental membrane surface area is another matter which is not studied yet [7].

J_w can be defined as water flow across the membrane, given in $mol\ cm^{-2}\ s^{-1}$ (CGS units system). Considering hydrostatic pressure differences between the two sides of a membrane, can be established water movement due to a pressure gradient. The equation is similar to the First Law of Fick

$$J_w = -B \cdot \frac{dP}{dx} \quad (3.14)$$

where B is constant. Equation 3.14 can be rewritted, considering a stationary system, with the difference of hydrostatic pressure ΔP, and membrane thickness, Δx:

$$J_w = \frac{B}{\Delta x} \cdot \Delta P = L_p \cdot \Delta P \quad (3.15)$$

with L_p representing the membrane's filtration coefficient.

Water movement is also due to osmosis, where differences in the solute concentration between both sides of the membrane create a difference of osmotic pressure that leads to water motion from the side with the lowest solute concentration to the side with the highest one.

Osmotic pressure is a colligative property of solutions, which can be determined by

$$\pi_s = RT \cdot C_s \tag{3.16}$$

if it is a diluted solution.

Water flow, J_w, is directly proportional to the difference of osmotic pressure of the solutions. Since there is proportionality between water flow and osmotic pressure difference, the proportionality constant can be defined as L_p. However, oppositely to hydrostatic pressure, in osmosis the water moves from the side where the osmotic pressure is lower pressure to the side where it is higher. Therefore

$$J_w = -L_p \cdot \Delta\pi_s \tag{3.17}$$

As mentioned before, some solute molecules can be transported along with water. Although the Eq. 3.17 need to be corrected by including a reflexion coefficient, σ, since only the solute molecules that reflect on the membrane and do not cross it, will produce osmotic pressure, and therefore, are able to induce water movements:

$$J_w = -L_p \cdot \sigma\Delta\pi_s \tag{3.18}$$

Merging Eqs. 3.16 and 3.18, it is possible to obtain another equation that accounts for both hydrostatic and osmotic pressure contributions for water flow:

$$J_w = L_p \cdot (\Delta P - \sigma\Delta\pi_s) \tag{3.19}$$

Water flow can also be given as the volume of water that crosses a membrane per unit of surface and unit of time ($cm^3\,cm^{-2}\,s^{-1}$ or $cm\,s^{-1}$ in CGS units system), which is represented by J_v. J_w and J_v can be related using the average molar volume of water $(\bar{V}_w)$:

$$J_v = \bar{V}_w \cdot J_w \tag{3.20}$$

Knowing J_v and the average concentration of a solute inside the membrane, $\bar{C}_s$, the amount of solute that crosses the membrane dragged by water movements can be determined. To do so, it also need to be considered that a fraction σ of solute molecules is reflected:

$$J_s = (1-\sigma) \cdot \bar{C}_s \cdot J_v \tag{3.21}$$

Joining now the equations shown above, the final equation of transport of solute molecules, dragged by water through a membrane is

$$J_s = (1-\sigma) \cdot \bar{C}_s \cdot \bar{V}_w \cdot L_p \cdot (\Delta P - \sigma\Delta\pi_s) \tag{3.22}$$

Concerning nutrients transport, the transplacental transport characteristics and mechanisms have been widely studied. However the majority of studies are essentially in the third trimester, because the difficulties of the quantitative experimentation during the first and second trimester [11]. Although these difficulties, in vitro models have been developed in order to have deeper informations about feto-maternal placental transport. Concerning literature review there are three types of in vitro models. Perfusion models using placental cotyledones have been used to study placental transfer of nutrients, drugs, chemicals or carcinogens, the placental tissue metabolism, the role of placental transporters, and the perfusion of endogenous substances. Besides all the disadvantages concerning this model which includes the type of placenta used plus the technical considerations, it is important because it maintains the whole organ complexity being also easily available [11–13]. The second model describes tissue preparations in which placental explants and membrane vesicles are used. This model is used to study syncytiotrophoblast uptake of different substances, transport of regulating factors, enzyme function and to identify receptors and binding sites. The advantages associated with this model are related to the maintenance the intact micro-architecture, the possibility to study cell-cell interactions, the transporters function, the transfer of amino acid, drugs and toxicants across membrane, besides that can be performed at any stage of gestation [11, 14]. The third and last model type is based on cell cultures which uses primary placental cells and cell lines. This model allows the study of parasites and virus infections, the apoptosis signaling pathways, the xenobiotics uptake and release, the transport across cell membrane, the intracellular metabolism and the control of the endocrine function. Concerning all the disadvantages related with cell culture procedures, this model can be used at any stage of gestation, is possible to replicate rapidly in culture and the cells are easily cloned [11, 14–16].

The way how nutrients are transported is influenced by several biophysics characteristics namely membrane permeability, nutrient concentration gradients, placental blood flow and metabolism [17, 18]. Glucose is transported by facilitated diffusion according to the maternal-fetal concentration gradients [11]. Other nutrients such as calcium, are transported by primary active transport with energy consumption. Nutrients as amino acids, phosphorous, and lactate are transported across the membrane by secondary active transport with energy consumption. The energy consumed by these transfers is provided by ions gradients such as sodium, chloride and protons [17].

3.3 Mechanical Properties of Amniotic Membrane

The strength demonstrated by an intact fetal membrane is determined by its amniotic component. Regarding this, the elastic deformation is one characteristic that amniotic membrane presents and it is related with the presence of elastin fibers that are detected in the fetal amnion [19]. Concerning the different components of amniotic membrane which have different rheological properties related with stress-strain relationships, there are inconsistencies on which portion ruptures first [3]. Physical properties such as elasticity, stiffness and other biomechanical characteristics of amniotic membrane or extracellular matrix depend on the variation of the placentary compounds [20]. These physical properties, such as mechanical strength, make it an attractive scaffold to be used in tissue engineering, for example, as surgical graft [21].

In rheology, it is possible to split the materials into several categories, depending on their response over time to different sheer forces: non-deformable, Hooke's materials (linear elasticity), plastic, viscous and viscoelastic materials.

In order to understand physical properties of elastic deformations, first should be defined relative deformation, also known as strain. If we consider a solid bar with an initial length l_0, we can deform the bar by applying stretching forces (tension, T) that increases length up to l (with $l > l_0$). Strain is defined by

$$\varepsilon = \frac{l - l_0}{l_0} \tag{3.23}$$

For non-deformable materials, strain is null over time ($\varepsilon = 0$), regardless of the tension (T) applied. The Hooke's materials, on the other hand, have properties of purely elastic deformations, where there is a linear relationship between T and ε; when the strain force ceases, ε returns to zero, which means that the material returns to its initial shape.

In this type of deformations, proportionally can be correlated tensile stress and strain using Hooke's law, where the proportionality constant is called Young's modulus (E).

$$T = \frac{F}{S} = E \cdot \varepsilon = E \cdot \frac{l - l_0}{l_0} \tag{3.24}$$

Instead of applying stretching forces, can also be compressed the same bar in a similar way. In this case, $l < l_0$, but if the deformation is elastic

$$T = \frac{F}{S} = E \cdot \frac{l_0 - l}{l_0} \tag{3.25}$$

where the factor E is the elasticity modulus by compression, which is similar to Young's modulus.

We can also consider the deformation of a volume and relate it to the difference of pressure that is applied. In that case

$$\Delta P = \in \cdot \frac{V - V_0}{V_0} \tag{3.26}$$

with ϵ being called compressibility modulus.

Looking at membranes as thin surfaces, with negligible thickness; this way should be only considered two dimensions: length (*l*) and width (*w*). If stretched the membrane along its length, the elastic deformation can be expressed by

$$T = \frac{F}{w} = E' \cdot \frac{l - l_0}{l_0} \tag{3.27}$$

where E' is called elastance of the membrane.

About Young's modulus, modulus of compressibility and elastance, the higher value, the greater the tensile force which must be applied to cause the same distortion; this means that it is more difficult to deform a body that has high values of these moduli.

Not all strains may be considered purely elastic, since not all materials return to their original shape after cessation of the deformation force.

Plastic materials are those that do not deform until a certain threshold of T is reached. Beyond that point, ε increases as long as the tension is maintained, that means that the deformation is permanent. When the tension force is stopped, the material does not return to its original shape, maintaining the maximum stain reached.

With respect to liquid materials, viscous materials can be considered in which it is necessary to apply a non-zero shear stress to obtain deformation. The deformation capacity will depend on the viscosity of the liquid and can be subdivided them as Newtonian and non-Newtonian liquids. The first follow Newton's equation

$$T = \eta \cdot \frac{dv}{dx} \tag{3.28}$$

Where η represents the liquid viscosity and $\frac{dv}{dx}$ the gradient of velocity for the fluid. Non-Newtonian fluids do not follow the relationship showed in Eq. 3.28, as their viscosity can vary with $\frac{dv}{dx}$.

Finally, with viscoelastic materials not all energy spent to create deformation is used to restore the material's initial shape; some of it is dissipated. Unlike elastic materials, that follow Hooke's law, strain is both tension- and time-dependent. To describe this type of materials, its creep can be studied, which accounts for variations in strain over time, for a constant tension.

For viscoelastic bodies, tension and strain are related over time, depending both on Young's modulus and viscosity:

$$T(t) = E \cdot \varepsilon + \eta \cdot \frac{d\varepsilon}{dt} \tag{3.29}$$

The creep function can also be obtained from Eq. 3.29:

$$\varepsilon(t) = \frac{T_0}{E} \cdot \left(1 - e^{-\frac{E}{\eta} \times t}\right) \tag{3.30}$$

For a constant tension T_0, viscoelastic materials have a maximum strain of T_0/E. When tension ceases, strain diminishes over time in an exponential manner. These materials can either return to their original shape (Kelvin-Voigt materials) or can maintain a residual deformation, ε_r (Maxwell materials).

Soft biological tissues demonstrate a mechanical response that is inherently time-dependent, sometimes described as viscoelastic. The immediate consequence of a time-dependent response is that the values of the mechanical properties measured may depend on the rate at which the material was loaded during the test, such that the tissue appears stronger and stiffer when tested at faster rates. It is thus of critical importance to report details about the testing rate or testing time frame for any mechanical study of a viscoelastic material and to consider the loading rate when results from different studies are compared [6].

According to Lavery and Miller, amniotic membranes have viscoelastic properties of creep, with an increase of deformation over time with a constant load; stress relaxation, in which a time decrease in load is required to maintain a constant deformation; and a non-recoverable deformation, in which the amniotic membrane remains thinned after load removal, not returning to its original configuration [3]. This viscoelastic characterization of amniotic membrane is described as time-dependent [19].

Experimental results confirm that the tensile strength of the human chorioamnion exhibits a triphasic curve. In the mid second trimester (18–20 weeks' gestation), an increase in tensile strength is noticed followed by a plateau from 20 to 39 weeks and then drops steeply, possibly in anticipation of delivery [22].

The effect of this mechanical loading is not well understood, although intra-amniotic pressures have been measured both when the amniotic fluid volume is deficient as when it is excessive. Amniotic fluid begins to form concomitantly with the development of the chorion of the amniotic membrane, and until 16–18 weeks is a maternal serum transudate. The fetal kidneys begin to produce urine at 14–16 weeks and fetal urine becomes a significant contributor to amniotic fluid volume. The circulation of this fluid is also facilitated by fetal deglutition. Besides this, there is a net water movement across the amnion, perhaps associated with the large surface area afforded by the microvilli on the amniotic cells facing the fetus. Nonphysiologic mechanical conditions of the pregnant uterus also include excessive amniotic fluid pressure or volume, resulting in straining of the chorion of the amniotic membrane. Aside to the development of the whole body of the fetus, the amniotic fluid is the major entity distending this membrane inside the uterus [6].

Concerning the biomechanical properties, there is a straight relation with amniotic membrane structure. The membrane thickness has been previously reported as closely correlated with mechanical stiffness if all other properties (density, organization, etc.) are held constant. Proximal, distal and denuded amniotic membrane has a mean thickness of 115.6 ± 20.7, 63.6 ± 13.0 and 9.8 ± 4.3 μm, respectively. The thickness of proximal amniotic membrane is significantly greater than that of distal and denuded amniotic membrane, and the distal amniotic membrane is significantly different from denuded distal amniotic membrane [23].

Besides viscoelastic properties, another important factor in the mechanical response of the chorion of the amniotic membrane is its behavior under multicycle loading. Concerning previous data about the variability of the soft tissues, measured mechanical properties are not necessarily identical for two samples of the same material where one piece is loaded monotonically to failure and the other piece is cyclically loaded. Although most laboratory failure testing consists of single tests, some authors have used cyclic tests with increasing loading amplitude. Given the potential for cyclic loading of the membranes during labor in vivo, future studies using cyclic testing protocols are likely justified [6].

The first studies developed concerning amniotic membrane physical properties were developed by Duncan in 1886 [19]. To study amniotic membrane physical characteristics there are different mechanical tests that can be performed to evaluate physical properties of the membranes, namely uniaxial tensile testing, biaxial inflation (burst) testing, biaxial puncture testing and planar biaxial testing. All of these tests are easily performed with the exception of planar biaxial testing [24].

The results of most biomechanical testing protocols are extensive quantities; that is, the measured results are a combination of the material's intrinsic properties and the geometry of the test. Intrinsic material quantities can be computed from extrinsic quantities with knowledge of the test setup and geometry of the sample tested, as will be discussed for the different loading conditions below. Only intrinsic material properties can be fairly compared between different types of tests [6].

The mechanical characterization tests are not meant to simulate in vivo failure of the membrane but to allow for controlled laboratory characterization of prefailure deformation behavior and failure behavior of the materials. The most commonly reported intrinsic mechanical properties are the elastic modulus (Young's modulus, *E*), which quantifies the stiffness of the tissue, and the tensile strength (failure strength, *F*). The Young's modulus can be experimentally determined as the force per unit cross-section of the material divided by the fractional increase in length resulting from the stretching of a standard piece of the material [19]. The mechanism for establishing these parameters (*E*, *F*) from tests done using different geometries will be considered briefly in this chapter. Finally, mechanical assays are useful in identifying potential interventions for weakened membranes. The introduction of needle puncture holes into the membranes reduces their mechanical integrity (rupture tension, work to rupture), however the sealing of these defects with fibrin glue, resulted in a partial restoration of the mechanical integrity [6].

Concerning Young's Modulus, this parameter for the amniotic membrane may have different values at different gestational ages. When preterm and term amniotic membranes are compared, Young's Modulus of the preterm was significantly higher, revealing that preterm amniotic membranes are stiffer than those at term [19]. However, this difference between the two amniotic membranes of different ages has nothing to do with the number or the type of cells that constitute each part, once no relationship was found between elasticity and the constitution of amniotic membrane in terms of type or number of cells [19, 20].

Regarding experimental procedures, the most common test is the tensile testing, where the ends of a membrane strip are placed in vice grips and pulled apart. This test was initially performed by Artal [25, 26] and Uldbjerg [27, 28] and it evaluates the stress in one longitudinal component. In this evaluation, as the sample is stretched, the transverse dimension decreases resulting in a transverse strain but no stress is developed in the transverse direction [3, 19].

In a straightforward uniaxial tension test, a rectangular strip of material is pulled monotonically to failure. The experimental variables are the displacement δ (extension of the sample) and the load (force) (F). The key sample geometrical variables for a tensile test are the sample gage section (initial width, w_0 and initial thickness, t_0) and the initial gage length of the sample (length between the grips, l_0). The extensive experimental variables, F and δ, are easily converted to intensive variables via the geometry factors: engineering stress $= \frac{F}{w_0 \cdot t_0}$, and engineering strain, $\varepsilon = \frac{\delta}{l_0}$. The failure strength is simply the stress at maximum load, F_{max} [6, 29].

The tensile strength of the chorioamnion varied with gestational age, and concerning experimental data there is the indication that the chorioamnion gains strength in the early to mid-second trimester, about 17–20 weeks' gestation. After this period, tensile strength appears as a plateau until 38–39 weeks' gestation. When the gestation reaches 39 weeks a dramatic and statistically significant decrease in chorioamnion membrane tensile strength is noticed [22].

Most soft biological materials have strengths in the kPa or MPa range. In the absence of suitable thickness measurements, the tensile "rupture tension" has been reported as $= F_{max}/w_0$, although this is not an intrinsic mechanical property. The elastic modulus, E, is the slope of the stress–strain (σ-ε) curve for engineering materials in which stress and strain are initially linearly related; since in soft tissues stress and strain typically are not linearly related, the pseudolinear or tangent modulus is frequently reported [6].

The second test is the burst testing, in which a piece of membrane is clamped in a ring and either air or fluid pressure is applied perpendicular to the plane of the membrane. This test is the one that most closely mimics physiology, approximating the in vivo mechanical deformation and failure of the chorioamnion [3]. The tissue under test is inflated from one side with air or fluid pressure until it bursts. The bursting pressure (P) is then related to the failure strength of the membrane concerning the radius of the "bubble" prior to bursting, r, and the tissue thickness, t, using a hemispherical bubble assumption [3, 6]:

$$\sigma_F = \frac{P_{max} \cdot r}{2t_0} \qquad (3.31)$$

The bubble radius r and height can be measured directly or assumed from the radius of the specimen holder ring according to what has been shown that these two parameters are quite similar for membrane bursting tests. Alternatively, some authors have calculated an effective radius from the specimen holder radius and the bubble height. As was seen above for the tension test, some researchers have reported that the bursting tension can be given by

$$T = \frac{P_{max} \cdot r}{2} \qquad (3.32)$$

Through this equation we can see that T is directly related to failure strength via the tissue thickness but is not itself an intrinsic material property. In the case of bursting tests performed using the same bubble radius, the peak pressures can be compared directly as an extrinsic quantity [3, 6].

The limitation in this test is the lack of membrane thickness information from the bursting studies. To overcome this restriction further burst studies with measurements of membrane thickness are needed [6].

The puncture test, is the third method described in bibliography, in which a piece of membrane is also clamped in a ring, but a spherical metal probe is used to displace the central portion of the membrane perpendicular to the plane of its surface [3].

In a biaxial puncture test, the membrane is held such that a circular region of free tissue is exposed, and a blunt probe is brought down on the center of the membrane. Load is applied until the probe punctures through the membrane. The analytical mechanic analysis for this test geometry is more complicated than for the tensile or burst tests, but it has been shown that the failure strength at probe puncture can be calculated by the followed equation:

$$\sigma_F = \sqrt{\frac{F_{max} \cdot E}{6\pi \cdot R \cdot t_0}} \tag{3.33}$$

where R is the radius of the sphere-tipped probe.

Inversion of this equation to focus on the extensive quantity F_{max} (puncture force) shows that the puncture force is related to both the intrinsic properties of modulus of elasticity and membrane strength as well as the tissue thickness. Sharp decreases in the puncture force, as occurs in relation with gestational age (GA), can then be associated with changes in any or all of these three parameters (E, F, t_0). Although the intrinsic properties of the chorion are approximately an order of magnitude lower than those of the amnion, and as the amnion is a smaller portion of the total membrane thickness, the puncture forces differ by about a factor of two, much less than the differences in the intrinsic properties of strength and elastic stiffness. Because the circumstance of clinical membrane rupture in vivo is related to the whole membrane structure, the thickness-weighted puncture test may more accurately reflect the bilayer membrane mechanical behavior in vivo compared to tensile tests that aim to evaluate the intrinsic strength and elasticity properties. Another important characteristic is that the puncture test is also simple to perform. Some studies have demonstrated that a handheld puncture testing device, which can be brought into a clinical setting for routine mechanical characterization of cellulose acetate membranes, gives puncture force values comparable to those reported by previous investigators using the same probe geometry but obtained with a mechanical testing machine. Estimates of the membrane strength have been made based on puncture tests, but again there was a lack of thickness and elastic modulus information for the samples analyze and the good agreement with tensile data must be taken as a first-order estimate only. There is a need for additional puncture studies in which a more detailed characterization of the membrane samples thickness and stiffness is performed independently of the puncture measurements [6].

In each test referred above, the force applied to the membrane samples and the displacement of the membranes (in the direction of the force) are recorded simultaneously. From the curve plot of the force vs. displacement curve there is the possibility of calculate several parameters such as the rupture strength, stiffness (related to elasticity), work to rupture, displacement at rupture, and displacement at maximum force. The viscoelastic properties of the membranes, such as potential for stress relaxation and non-elastic deformation, can also be determined using these test methodologies.

In this context, an increase in tensile strength around the 17–20 weeks could be a consequence of the fusion of the chorioamnion to the maternal decidual tissue, leading to an increased stimulation of collagen production by fibroblasts. It seems that if the fusion process occurs, a sharp decrease in fetal fibronectin is detected in mother vaginal secretions. The tensile strength of the chorioamnion appears to stabilize from 20 to 38 weeks of gestation. During this gestational period from 20 to 38 weeks many patients, show that the chorioamnion has a tensile strength greater than what is required to maintain its integrity. We may wonder if this biological redundancy exists in the gestational membrane to prevent the preterm premature rupture of membranes (PPROM). When we analyze a single sample of a membrane of approximately 100 g by multiple times, each time that we evaluate, the membrane exhibits its own profile, bursting above and below its mean. Beyond the 39 weeks of gestation, all membrane segments exhibit a reduction in the mass (in grams) needed to burst. These observations corroborate the data obtained by other investigators who have found that the healthy chorioamnion membrane, at term, exhibits an increase in the collagenolytic enzyme, the metalloproteinase-9, a process that correlates with the decrease in membrane tensile strength. If we think in mass needed to burst, it appears that, until term, an intact healthy chorioamnion membrane expresses a tensile strength that bursts only above 250 g. The distinct quantitative results reflect the variety of test units used and the different hypothesis on test. Artal et al. reported that elasticity of the membranes near the placenta of patients with membrane premature rupture was lower than in patients whose membranes ruptured during the labor [25]. However, it was also shown that in preterm membranes with less than 37 weeks of gestation, stress tolerance was more than double that the term membranes that ruptured spontaneously (premature rupture of membranes) or were artificially ruptured. In our series, clinical chorioamnionitis did not identify membrane segments with a consistently low tensile strength. This observation seems to support the concept that premature rupture of the membranes can result from a focal area of infection or inflammation that leads to collagen breakdown and, consequently, the amniotic epithelium disruption. Only if the membrane exhibits a generalized inflammatory process there is a predictable decrease in overall tensile strength. This study can provide quantitative information about the strength of the chorioamnion throughout gestation. Using this information as a reference, it will be possible to study the effects of several pathologic conditions on the membrane strength and, perhaps to better understand the processes of spontaneous rupture of membranes [22].

The effect of labor type can also be considered. For that, the comparison of labored and unlabored deliveries, as is the case of cesarean delivery, can be performed by tensile and puncture testes. In any of these two situations, no effect is observed on the chorion layer, and only a minor effect is observed on the chorioamnion and amnion as is shown by the decrease of strength. Similarly, gestational age has a relatively big effect on membrane rupture resistance. In fact, preterm membrane samples show substantially greater mechanical integrity than term ones as it is shown by puncture testing, being the puncture force clearly decreased in the amnion layer with advancing of gestational age. One potential underlying mechanism of the clear mechanical change can be related to the collagen content, although there are conflicting reports regarding whether the amnion collagen content decreases with the gestational age. Consistent with these results, the bursting tension obtained through a biaxial inflation test was 20 % lower in membranes with 40–42 weeks when compared to 38–40 weeks or 34–38 weeks [6].

As a clinical approach, an interesting question concerns the membranes that rupture spontaneously prior to the onset of labor compared with those that only rupture after the onset of labor. In these conditions, there was also no difference in tensile rupture tension for both a thinning of the membrane near the rupture site in prematurely ruptured membranes has been noticed [6].

As a conclusion we can say that the amniotic membrane can be studied applying biophysical approaches which can give inestimable information's that allows us to understand its biological behavior.

References

1. Damiano A (2011) Review: water channel proteins in the human placenta and fetal membranes. Placenta 32:S207–S211
2. Hieber A, Corcino D, Motosue J et al (1997) Detection of elastin in the human fetal membranes: proposed molecular basis for elasticity. Placenta 18:301–312
3. Moore R, Mansour J, Redline R et al (2006) The physiology of fetal membrane rupture: insight gained from the determination of physical properties. Placenta 27:1037–1051
4. Levkovitz R, Zaretsky U, Gordon Z et al (2013) In vitro simulation of placental transport – part I: biological model of the placental barrier. Placenta 34:699–707
5. Jansson T, Powell T (2006) Human placental transport in altered fetal growth: does the placenta function as a nutrient sensor? – A review. Placenta 27:91–97
6. Calvin SE, Oyen ML (2007) Microstructure and mechanics of the chorioamnion membrane with an emphasis on fracture properties. Ann N Y Acad Sci 1101:166–185
7. Beall M, van den Wijngaard J, van Gemert M, Ross M (2007) Amniotic fluid water dynamics. Placenta 28:816–823
8. Beall M, van den Wijngaard J, van Gemert M et al (2007) Regulation of amniotic fluid volume. Placenta 28:824–832
9. Zeuthen T (1995) Molecular mechanisms for passive and active transport of water. Int Rev Cytol 160:99–161

10. Menjoge ARA, Navatha RS, Asad A et al (2010) Transport and biodistribution of dendrimers across human fetal membranes: implications for intravaginal administration of dendrimers. Biomaterials 31:5007–5021
11. Levkovitz R, Zaretsky U, Jaffa A et al (2013) In vitro simulation of placental transport – Part II: glucose transfer across the placental barrier model. Placenta 34:708–715
12. Hutson J, Garcia-Bournissen F, Davis A, Koren G (2011) The human placental perfusion model: a systematic review and development of a model to predict in vivo transfer of therapeutic drugs. Clin Pharmacol Ther 90:67–76
13. Mathiesen L, Mose T, Mørck TJ et al (2010) Quality assessment of a placental perfusion protocol. Reprod Toxicol 30:138–146
14. Sastry BR (1999) Techniques to study human placental transport. Adv Drug Deliv Rev 38:17–39
15. Heaton SJ, Eady JJ, Parker ML et al (2008) The use of BeWo cells as an in vitro model for placental iron transport. Am J Physiol Cell Physiol 295:C1445–C1453
16. Poulsen MS, Rytting E, Mose T, Knudsen LE (2009) Modeling placental transport: correlation of in vitro BeWo cell permeability and ex vivo human placental perfusion. Toxicol Vitro 23:1380–1386
17. Lager S, Powell TL (2012) Regulation of nutrient transport across the placenta. J Pregnancy 179827:179827
18. Jones H, Powell T, Jansson T (2007) Regulation of placental nutrient transport: a review. Placenta 28:763–774
19. Benson-Martin J, Zammaretti P, Bilic G et al (2006) The Young's modulus of fetal preterm and term amniotic membranes. Eur J Obstet Gynecol Reprod Biol 128:103–107
20. Niknejad H, Peirovi H, Jorjani M et al (2008) Properties of the amniotic membrane for potential use in tissue engineering. Eur Cell Mater 15:88–99
21. Riau AK, Beuerman RW, Lim LS, Mehta JS (2010) Preservation, sterilization and de-epithelialization of human amniotic membrane for use in ocular surface reconstruction. Biomaterials 31:216–225
22. Pressman EK, Cavanaugh JL, Woods JR (2014) Physical properties of the chorioamnion throughout gestation. Am J Obstet Gynecol 187:672–675
23. Chen B, Jones RR, Mi S et al (2012) The mechanical properties of amniotic membrane influence its effect as a biomaterial for ocular surface repair. Soft Matter 8:8379
24. Chua WK, Oyen ML (2009) Do we know the strength of the chorioamnion? A critical review and analysis. Eur J Obstet Gynecol Reprod Biol 144:128–133
25. Artal R, Sokol R, Neuman M et al (1976) The mechanical properties of prematurely and non-prematurely ruptured membranes: methods and preliminary results. Am J Obstet Gynecol 125:655–659
26. Artal R, Burgeson R, Hobel C, Hollister D (1979) An in vitro model for the study of enzymatically mediated biomechanical changes in the chorioamniotic membranes. Am J Obstet Gynecol 133:656–659
27. Oxlund H, Helmig R, Halaburt J, Uldbjerg N (1990) Biomechanical analysis of human chorioamniotic membranes. Eur J Obstet Gynecol Reprod Biol 34:247–255
28. Helmig R, Oxlund H, Petersen LK, Uldbjerg N (1993) Different biomechanical properties of human fetal membranes obtained before and after delivery. Eur J Obstet Gynecol Reprod Biol 48:183–189
29. Sillero A, Selivanov VA, Cascante M (2006) Pentose phosphate and calvin cycles: similarities and three-dimensional views. Biochem Mol Biol Educ 34:275–277

Chapter 4
Stem Properties of Amniotic Membrane-Derived Cells

Maddalena Caruso, Anna Cargnoni, and Ornella Parolini

Abstract New avenues in the general area of research into stem cells have been opened in recent years through the discovery that cells with stem properties can be isolated from the human placenta, which represents an ethically sound, easily procured and plentiful source, and is therefore an attractive alternative to other stem cell sources. In particular, the human amniotic membrane, which already has a long history of use as a surgical material, is also proving to be a valuable reserve of cells for research and development of novel therapeutic approaches for regenerative/reparative medicine. In this chapter we will describe the two main amniotic membrane-derived cell populations, i.e. epithelial cells and mesenchymal stromal/stem cells, and some of their sub-populations, in terms of their stem properties which have been discovered to date. We will explore reasons why these amniotic membrane-derived cell populations cannot be considered "true" stem cells, at least in the classical sense of the term, even though they have some features of progenitor-like cells and certainly display some very interesting biological properties both for research purposes and for potential clinical applications. We will also bring into discussion the fact that much still remains to be further investigated and even discovered in this evolving research field, although despite this, the very promising results achieved so far are certainly very encouraging as a basis for future studies.

Keywords Amniotic epithelial cells • Amniotic mesenchymal stromal/stem cells • Stem cells • Regenerative medicine • Immunomodulation

M. Caruso • A. Cargnoni • O. Parolini (✉)
Centro di Ricerca E. Menni, Fondazione Poliambulanza – Istituto Ospedaliero, Brescia, Italy
e-mail: ornella.parolini@poliambulanza.it; ornella.parolini@tin.it

A.C. Mamede, M.F. Botelho (eds.), *Amniotic Membrane: Origin, Characterization and Medical Applications*,
DOI 10.1007/978-94-017-9975-1_4

4.1 Introduction

The fact that the human amniotic membrane could play beneficial roles beyond fetal development has been known for more than 100 years, through experimental and clinical evidences reported over time in different areas and applications. Indeed, by the early 1900s, Davis had already reported the successful use of human fetal membranes (amniotic and chorionic membranes) as a skin substitute [1], with this study spurring application of the amniotic membrane as a surgical biomaterial in a variety of settings ranging from the now well-established clinical treatment of pathological conditions including ocular disorders, skin wound healing and reconstructive surgery, to more prospective disease scenarios, as supported by current pre-clinical studies [2].

The early origin of fetal membranes, which begin to develop even before gastrulation, and the fact that the amniotic ectoderm derives from the epiblast, which, through the process of gastrulation, is also the source of all of the three germ layers (endoderm, mesoderm, and ectoderm) of the embryo, have more recently led scientists to postulate that cells in the amniotic membrane may have retained some of the plasticity of pre-gastrulation embryo cells and may therefore harbour some degree of stemness [3]. This aspect, together with the fact that the placenta is usually discarded as a waste product after birth, therefore rendering its use free of ethical controversy, as well as the consideration that it is plentiful and easily procured in a safe and non-invasive manner, has prompted researchers to investigate the presence of stem or stem-like cells within its tissues.

The general area of research into stem cells continues to capture scientists' attention at an increasing rate, thanks to a significant rise in interest in the therapeutic potential of these cells. As knowledge in the field continues to broaden, and as more and more cells with stem characteristics are isolated from different tissues, the concept of "stemness" itself has also become the object of fervent debate among scientists, with a consequent and continuous revision of the "stem cell" definition [4–8]. In this evolving scenario, the hurdle to define the criteria which a cell should meet in order to be termed a "stem cell" is made even more complex by the fact that, most likely, we ourselves still lack a comprehensive understanding of all of the properties of these cells. Moreover, the task of identifying "stem cells" is made no easier by the fact that there are several cell types which exhibit some, but not all of the characteristics which have been defined, but which nonetheless have biological properties that make them especially interesting in view of potential therapeutic application [5, 6, 9].

In this chapter, conscious that we are discussing an intricate and evolving field, but far from entering into the debate regarding the "stem cell" definition, we will attempt to highlight the characteristics of human amniotic membrane-derived cells on the basis of what is currently known from studies which have been undertaken to date, in order to assess whether these cells are endowed with the characteristics that are generally associated with "stemness". To this end, we will adhere to one of the most classical working definitions of "stemness", which refers to undifferentiated cells that are endowed with the ability of both self-renewal and also the capacity to differentiate into several cell different types, i.e. "stem cells are generally defined as single cells that are clonal precursors of both more stem cells of the same type, as well as a defined set of differentiated progeny" [10]. We will therefore describe the features of amniotic membrane-derived cells which have been reported after their stepwise characterization through experimental approaches that are generally employed when a working characterization of stem cells is made. Moreover, given that designation of a "stem cell" is usually made also considering both the ontogeny of the tissue from which it derives, according to which it is generally classified as being either embryonic, fetal or adult, as well as the potency of the cell, that is, its capacity to differentiate towards other cell types, according to which it is designated into a hierarchy that ranges from unipotency to pluripotency, we will herein refer to amniotic membrane-derived cells as being fetal cells, and we will discuss different aspects related to their potency.

4.2 Where Can We Find the Cells Which Will Be Discussed?

Firstly, to provide readers with a "map" on which to pinpoint the cells which we are discussing, we will take a moment to summarize some of the key structural elements of the human amniotic membrane, which instead will be described more extensively in other chapters of this book.

In the maternal womb, the amniotic membrane is the innermost portion of the placenta, which encloses the fetus in the fluid-filled amniotic sac. Upon isolation from the term placenta, the amniotic membrane appears as a nearly transparent avascular tissue ranging from 0.02 to 0.5 mm in thickness. Histological analysis of this membrane allows us to distinguish three principal layers: an epithelial monolayer, an interposing acellular basement membrane and a stromal layer.

The amniotic epithelium faces the amniotic cavity and is in direct contact with the amniotic fluid. The other side of the epithelium lies on a thick basement membrane composed of various collagen types, laminin, and fibronectin ([11], and references inside). Histologically recognized as a single layer of flat, cuboidal to columnar epithelial cells that have microvilli on their apical surface, the amniotic epithelium harbours cells with some stem/progenitor features that are referred to as human amniotic epithelial cells (hAECs) [12–14]. Indeed, in line with mounting evidence regarding the presence of hAECs within this layer, when different tissues from human term placenta were analysed through immunofluorescence staining techniques in the quest for cells positive for stem cell/pluripotency markers, stage specific embryonic antigen (SSEA) -4 positive cells were detected throughout virtually all of the amniotic epithelium, while some solitary cells positive for the pluripotent stem cell antigens tumor rejection antigen (TRA) 1-60 and TRA1-81 were also detected within this layer [15]. Other authors also reported the observation of a number of cells positive for the pluripotent stem cell – specific transcription factor octamer-binding protein (OCT) -4 clustered within a specific region of the amniotic epithelial layer [16]. Taken together, these findings suggest that the amniotic epithelium contains cells heterogeneous in terms of stem cell/pluripotency marker expression, as indeed supported by the characterization of naïve hAECs isolated from human amniotic membrane, which will be further discussed below.

On the other side of the basement membrane lies the amniotic stroma, which is composed of an acellular compact layer containing collagens and fibronectin, followed by a fibroblast/stromal layer, which contains dispersed fibroblast-like mesenchymal cells and rare macrophages embedded in a loose network of reticulum ([17], and references inside; [18], and references inside). The fibroblast/stromal layer is in turn connected to a spongy layer, which is in loose contact with the chorionic membrane, which sits under the amniotic membrane ([17], and references inside).

Although the immunostaining studies performed by Miki and colleagues [15] described above did not yield results which showed the presence of stem cell/pluripotency marker-positive cells in regions of the amniotic membrane other than the epithelial layer, as we will see below, mounting evidence supports the notion that the fibroblast/stromal layer of the amniotic membrane harbours cells with features of mesenchymal stromal/stem cells (MSCs), referred to herein as human amniotic mesenchymal stromal/stem cells (hAMSCs) [14].

4.3 Amniotic Membrane-Derived Cells: Morphology and Proliferative Ability In Vitro

Typically, a naïve heterogeneous population of hAECs is obtained after single or multiple trypsin digestions of either pieces of the whole human amniotic membrane which have been peeled away from the chorion [19], or of tissue fragments which remain after isolation of hAMSCs, which can be performed either by manual scraping out of the amniotic mesoderm without affecting the epithelial layer [20] or application of multiple enzyme digestion-based protocols [21]. Various factors related to the use of trypsin, including concentration, duration of incubation and the number of digestions with this enzyme, as well as the freshness of placenta, all influence the yield and viability of the isolated hAECs [20, 22], with reported yields from a single placenta ranging from 8–50 million cells [23] to 200–300 million cells [19]. Special attention is currently also being turned toward the development of isolation protocols which will allow the introduction of hAECs into clinical settings [24, 25].

Upon isolation and plating in a standard culture medium [e.g. DMEM supplemented with 10 % FBS (fetal bovine serum) and antibiotics], hAECs adhere to the culture dish without any feeder layer or any specific pre-treatment of the culture substrate. In the presence of epidermal growth factor (EGF), these cells proliferate robustly and form a confluent monolayer of cells with a cobblestone-shape/cuboid morphology that is typical of epithelial cells [12]. Despite some differences in the proliferative ability of hAECs which have been reported in vitro [12, 23, 26], generally, these cells can be maintained for two to six passages before proliferation ceases [14]. Meanwhile, karyotypic analyses performed at different passages and under various culture conditions have shown no expansion-related chromosomal alterations in these cells [12, 24, 25, 27].

Interestingly, differences in in vitro culture conditions have themselves been shown to impact on hAEC phenotype/morphology. In particular, besides causing proliferation to cease, the absence of EGF has also been associated with changes whereby cells become multinucleated giant cells reminiscent of trophoblastic differentiation by embryonic stem cells (ESCs) [12]. Moreover, when hAECs are kept in high-density culture for several days, small cell clusters or "spheroids" begin to appear over the cobblestone monolayer of epithelial cells, and, as we will describe in further detail below, the cells within these spheroids show different characteristics in terms of their stem cell/pluripotency marker expression when compared to the epithelial cells in the underlying monolayer [12, 28]. Furthermore, some authors have reported that maintenance in culture can cause alterations in hAEC morphology, which has been observed to change from a typical epithelial to a fibroblast-like morphology [26, 29]. Whether these changes are due to the presence of contaminating hAMSCs in the hAEC culture, or whether this may be due to an epithelial to mesenchymal transition (EMT) is still under debate [30], although changes in the expression of both surface antigens and genes usually associated with EMT are suggestive of a bona fide EMT to explain these observed changes [25].

hAMSCs are isolated from the amniotic membrane either after removal of hAECs [31], after scraping of the amniotic stroma from the amniotic membrane without affecting the epithelial layer [20], or directly from pieces of the whole amniotic membrane [21]. Generally, hAMSC isolation protocols entail a series of enzymatic digestions with dispase and collagenase, with or without the addition of DNase [14, 21, 31, 32]. As an example of hAMSC yield, Marongiu and colleagues have reported the isolation of 20–100 million cells from one term placenta, although they also noted that cell yield may be affected by unpredictable differences between different placenta donors, as well as differences in variables such as the time between delivery and the commencement of cell isolation [31].

When cultured in conventional media, such as DMEM supplemented with 20 % FBS, hAMSCs, like MSCs from other sources [33], readily adhere to plastic culture vessels and exhibit fibroblast-like morphology [14, 34]. hAMSCs can generally be kept in culture until passage 5 [21, 26, 29], even though higher passage numbers have been reported [31], particularly when individual colonies were selected from bulk cultures [21, 35]. Tamagawa and colleagues have shown that the majority of hAMSCs in a series of populations analysed exhibited a normal phenotype between passages 3 and 7 [36, 37].

4.4 In Vitro Molecular Signature of Amniotic Membrane-Derived Cells: Stem Cell/Pluripotency Marker Expression

Evaluation of a cell population's "stemness" includes its characterization in terms of the expression of stem cell/pluripotency markers.

Our own efforts have revealed the expression of OCT-4 in amniotic membrane-derived cells, as detected by RT-PCR [38]. Subsequently, other groups reported that hAECs express surface markers and transcription factors which are generally referred to when characterizing human ESCs and pluripotency [12, 39], given that ESCs are known to be embryonic pluripotent cells, i.e. cells that can differentiate into cell types of all three of the embryonic germ-layers: hAECs include cells which are heterogeneous in terms of the expression of these markers. Specifically, hAECs have been shown to express SSEA-3, SSEA-4, TRA1-60 and TRA 1-81, OCT-4, SRY-related HMG-box gene-2 (SOX 2), NANOG, reduced expression protein-1 (REX-1), fibroblast growth factor-4 (FGF-4), germ cell tumour marker-2 (GCTM-2), criptic family-1 (CFC-1), developmental pluripotency-associated protein-3 (DPPA-3), prominin-1 (PROM-1), and paired box gene-6 (PAX-6), as detected by RT-PCR and/or by immunostaining [12, 13, 40]. Some hAECs also express c-kit, the surface receptor for stem cell factor [12, 40]. Meanwhile, neither telomerase expression nor telomerase activity were detectable in these cells [12].

In terms of stem cell/pluripotency marker expression in hAMSCs, conflicting results have been reported to date. Kim and colleagues have shown that hAMSCs express SSEA-3, SSEA-4, OCT-4 and REX-1 genes over passages in culture [35], while Nogami and co-workers, using an immunocytochemical technique, observed that hAMSCs were positive for nuclear SOX-2, but that only a limited number of these cells were positive for OCT-3/4, Kruppel-like factor-4 (KLF-4) and SSEA-4 expression [41]. Meanwhile, Fatimah and colleagues observed that gene expression for OCT-3/4, SOX-2, NANOG-3, REX-1, FGF-4 and frizzled family receptor-9 (FZD-9) markedly decreased in these cells after serial passages [42]. This decrease was also observed by Bilic and colleagues, who specifically reported that after one passage in culture, hAMSCs showed a marked reduction in the number of cells positive for the expression of SSEA-4 with respect to the number observed at passage 0, while expression of SSEA-3 did not alter significantly [26]. It is to be underlined that while the authors observed expression of SSEA-3 and SSEA-4 through flow cytometry in every amniotic membrane tested, they were able to confirm the presence of the SSEA-4 epitope by immunocytochemistry only in two out of five amnion cases. Likewise, while the authors detected OCT-3/4 transcripts in every hAMSC culture tested, they did not detect any OCT-3/4 protein in hAMSCs by immunocytochemistry. Such discrepant findings may be attributable to the techniques used in these studies. Indeed, with regard to assessing the presence of OCT-4 as a pluripotency marker, Ryan et al. have underlined the challenges in detecting the correct OCT-4 isoform (OCT-4A) which is associated with pluripotency, without false positive results being obtained due instead to the expression of pseudogene OCT-4B [43]. These authors therefore underlined the potential need for reassessment of pluripotency markers in various cell types, with attention to gene and protein isoforms and pseudogenes.

4.5 In Vitro Molecular Signature of Amniotic Membrane-Derived Cells: Expression of Lineage-Associated and Other Markers

A comprehensive phenotypic characterization of amniotic membrane-derived cells cannot exclude an analysis of the expression of markers which define specific cell types. In this regard, it is routinely ascertained that nearly 100 % of freshly isolated hAECs react with antibodies to pan-cytokeratins, confirming their epithelial nature [12]. Meanwhile, according to the minimal criteria for defining hAMSCs which were established during the first international workshop on placenta-derived stem cells [14], hAMSCs at passages 2–4 should be positive (≥95 %) for cluster of differentiation/designation (CD) 90, CD73 and CD105, whilst they should be negative (≤2 %) for CD45, CD34, CD14 and HLA-DR.

However, hAECs have also been shown to express a range of mesenchymal and hematopoietic markers, including CD10, CD13, CD29, CD44, CD49e, CD73, CD90 (Thy-1), CD105, CD166 and stromal cell surface marker (STRO) -1, with low to moderate expression of the major histocompatibility complex (MHC) class I molecules human leukocyte antigen (HLA)-A, HLA-B, and HLA-C. Some variability has been reported among different research groups in terms of the levels of expression of these markers, which also appears dependent on the passage number and culture conditions employed [12, 14, 25, 44]. Meanwhile, hAECs show negligible levels of the MHC class II molecules HLA-DQ and HLA-DR, and are negative for CD14, CD34, CD45, CD49d [13, 14, 32, 44].

In terms of further phenotypic characterization of hAMSCs, it has been reported that in addition to CD73, CD90 and CD105, these cells are also positive for CD13, CD27low, CD29, CD44, CD49e, CD54, CD117 (weak), CD166, and STRO-1, while they are negative for CD3, CD14, CD34 and CD45 (reviewed in [14, 34, 44]. These cells are also reported to be HLA-A-B-C-positive (low levels) and HLA-DR-negative (reviewed in [14]). The low/absent level of expression of HLA class I molecules and the absence of HLA class II molecules suggests the potential utility of amniotic membrane-derived cells in transplantation settings, which may even allow them to be applied across a histocompatibility barrier.

Interestingly, both freshly isolated hAECs and hAMSCs express a repertoire of different lineage-associated markers, suggesting that these cell populations may include progenitors of various cell lineages. For example, undifferentiated hAECs express neural [nestin, glutamic acid decarboxylase (GAD), myelin basic protein (MBP), neurofilament medium chain (NFM), neuron-specific enolase (NSE), 2′, 3′-cyclic nucleotide 3′-phosphodiesterase (CNPase), proteolipid protein/DM-20 (PLP), microtubule-associated protein-2 (MAP-2), MAP-2 kinase, glial fibrillary acidic protein (GFAP), neurofilament proteins], hepatic [albumin, alpha-fetoprotein (FP), alpha-1 antitrypsin (AT), cytokeratin 18 (CK18), glutamine synthetase (GS), carbamoyl phosphate synthetase-1 (CPS), phosphoenolpyruvate carboxykinase (PEPCK), cytochrome 2D6 (CYP2D6), CYP3A4, CYP2C9, transthyretin (TTR), tyrosine aminotransferase (TAT), hepatic nuclear factor 3-alpha (HNF3α), CCAAT/enhancer binding protein-alpha (CEBP-alpha)], pulmonary [NK2 homeobox-1 (Nkx 2.1), mucin, occludin, aquaporin-5, caveolin-1], cardiomyogenic [gata binding protein-4 (GATA-4), Nk2 homeobox-5 (NKx 2.5), myosin light chain-2A (MLC-2A), MLC-2V, myosin regulatory light chain-7 (MYL-7), atrial natriuretic peptide (ANP), calcium channel voltage-dependent L type alpha 1C subunit (CACNA1C), potassium voltage-gated channel Shal-related family member 3 (KCND-3)] and pancreatic [pancreatic and duodenal homeobox-1 (PDX-1)] lineage-associated genes (all reviewed in [44]).

Meanwhile, undifferentiated hAMSCs have been shown to express genes associated with hepatocytes (albumin, α-FP, CK18, α1-AT and HNF4α), the pancreatic lineage-associate marker PDX-1, the cardiomyogenic lineage-associated markers

GATA-4, MLC-2A, MLC-2V, cTnI and cardiac troponin I and T (cTnT), the cardiac-specific ion channel genes alpha-1c and KCND3 (kv4.3), and the neural lineage-associated genes nestin, musashi-1, neuronal class III β-tubulin (TUJ-1), neurofilament medium (NF-M), MAP-2 and GFAP (all reviewed in [44]). Interestingly, although hAMSCs derive from the avascular stromal layer of the amniotic membrane, they express many endothelial and angiogenic genes, such as platelet/endothelial cell adhesion molecule (PECAM-1/CD31), von Willebrand Factor (vWF), vascular endothelial growth factor (VEGF), vascular endothelial growth factor receptor-2 (VEGFR-2), basic FGF and angiopoietin-1 [42].

The detection methods which were used to investigate expression of the lineage-associated markers described above ranged from RT-PCR to various protein detection techniques, with some variability between results reported by different groups.

4.6 hAEC and hAMSC Subpopulations

Considering the heterogeneity of naïve amniotic membrane-derived cell populations, it is not surprising that several research groups have been able to highlight the presence of cell subpopulations with special features within these broader isolates.

For example, as already mentioned above, small cell clusters or "spheroids" have been observed over the cobblestone monolayer of epithelial cells after weeks of hAEC culture under high-density conditions [12]. Characterization of the cells in these culture conditions in terms of stem cell/pluripotency marker expression revealed that expression of genes including SSEA-3, SSEA-4, TRA-1-60, TRA-1-81, OCT-4 and NANOG was mainly restricted to cells of the spheroid-like structures rather than to cells of the underlying adherent layer [12]. These findings suggest that the cells growing as spheroids over the monolayer of epithelial cells retain their stem cell nature to a higher degree when in culture [12].

Our group has reported the isolation of two subpopulations of cells derived from the mesenchymal layer of the amniotic membrane that differ in their expression of HLA-DR, CD45, CD14, and CD86 [45]. In particular, the HLA-DR-positive subpopulation presented with a monocyte-macrophage-like immunophenotype, coexpressing the markers CD45, CD14, CD11b, and CD86. Specific comparison of the immunomodulatory characteristics of these two subpopulations [i.e. HLA-DR-positive cells (>90 % HLA-DR-positive cells) and HLA-DR-negative cells (<5 % HLA-DR-positive cells)] with those of the unfractionated hAMSCs, with particular attention to their effects on proliferation of T cells which had been stimulated with allogeneic target cells (MLR) or via T-cell receptor engagement, showed suppressive versus stimulatory capabilities of the subpopulations. Specifically, while HLA-DR-negative cells, like the unfractionated hAMSCs, exerted an anti-proliferative effect, the HLA-DR-positive population induced marked proliferation of anti-CD3-primed allogeneic T cells in vitro [45].

Some evidence suggestive of the presence of cells with properties similar to the so called "side population" has been reported for cells isolated from both the epithelial and mesenchymal layers of the amniotic membrane. “Side population” refers to a rare subpopulation of cells functionally defined on the basis of their ability to efflux Hoechst 33342 dye, and which are enriched in cells with stem cell-like properties. It has been shown that hAECs express the ATP-binding cassette transporter G2 (ABCG2/BCRP), a multidrug resistance transport protein involved in Hoechst effluxing [28], however the significance of this finding remains to be assessed. ABCG2/BCRP is also expressed by hAMSCs, with no significant change in the expression of this marker after serial passaging [42]. Meanwhile, the presence of a side population has also been reported within cells isolated from the human amniotic mesenchymal layer by Kobayashi and colleagues [46]. Most of these cells expressed the stem cell markers OCT-4, SOX-2, nestin and REX-1, were highly positive for CD13, CD29, CD44, CD46, CD49b, CD49c, CD49e, CD59, CD140a, and CD166; weakly positive for CD49d and CD51, and were negative for CD34, CD45, CD90, CD105, CD271, CD49a, CD56, CD106, CD117, CD133 and fms-like tyrosine kinase-1 (FLK-1) [46]. Under appropriate culture conditions, these cells were also shown to undergo differentiation towards multiple cell lineages, such as neuroectodermal, osteogenic, chondrogenic, and adipogenic. Recently, it has also been reported that hAMSC-derived side population cells can be induced to differentiate toward the vascular endothelial lineage through culture under hypoxic conditions (1 % instead of 20 % O_2) [47]. These authors observed that the expression of endothelial-associated genes such as kinase domain region (KDR), FLT-1, vascular endothelial (VE)-cadherin and vWF peaked after 2 weeks under hypoxia, with a concomitant enhancement in the protein expression of KDR and VE-cadherin.

4.7 Toward Assessing Stemness of Amniotic Membrane-Derived Cells: In Vitro and In Vivo Assays

The potential “stemness” of a cell population is generally assessed through the application of some generally accepted experimental tests in vitro and in vivo, which are aimed at evaluating the cells’ capacity for self-renewal, formation of clonal colonies and teratomas, and in vitro differentiation.

4.7.1 Self-Renewal and Clonal Colony Formation Capacity

According to the classical concept of “stemness”, a hallmark of stem cells is the ability to clonally self-renew. This defining functional characteristic of stem cells refers to the process by which a stem cell perpetuates itself over many generations, undergoing mitotic divisions which give rise to daughter cells which maintain an undifferentiated phenotype, i.e. each division resulting in at least one of the two daughter cells being equipotent to the mother cell in terms of proliferation capacity and differentiation ability. Estimation of the self-renewal capacity of individual stem cells in vitro is generally performed through assessing in vitro clonogenicity, i.e. the ability of a single cell to form a clonal colony.

In terms of the self-renewal/clonogenic capacity of both hAECs and hAMSCs, conflicting results have been reported to date. Ilancheran and colleagues reported that hAECs are clonogenic, forming large, flattened, undifferentiated colonies containing several hundred cells within a few weeks of culture, even though these authors were unable to establish hAEC long-term self-renewal capabilities [13]. The ability to form clonal colonies was shown to be dependent on culture passages and conditions, i.e. while hAECs at passage 0 seeded at low density were clonogenic, hAECs which were expanded in culture in xenobiotic-free media until to P5 failed to form clonal colonies [25]. Meanwhile, Miki and colleagues reported that they were unable to grow cells from single-cell clones or show long-term self-renewal, and therefore argued that further work would be needed to determine whether the amniotic membrane is a heterogeneous mixture of progenitor cells with varied differentiation potential, or if a single stem-like cell can give rise to all germ layers [12].

In the case of hAMSCs, the definition of self-renewal/clonogenicity is applied to the even more complex scenario of MSCs in general. Indeed, the possibility of self-renewal by single cell-derived clonal populations from a mesenchymal cell source is still a highly debated issue [6, 48, 49].

However, some authors have reported that hAMSCs are indeed clonogenic, with a colony forming efficiency (colonies formed/number of cells plated) of 0.88 % at day 7 after seeding at culture passage 3 comparable with other cells including human umbilical cord perivascular cells and embryonic stem cells [21, 42], while others have failed to observe clonal expansion of these cells [26].

4.7.2 Teratoma Formation and In Vivo Pluripotency Assays

Teratoma formation, that is, the ability to give rise to tumour-like formations containing tissues belonging to all three germ layers, is a defining trait of pluripotent human ESCs and is considered in stem cell research as a valuable means of demonstrating pluripotency in vivo, even though it is also a major safety concern for the use of human ESCs in clinical applications. Following transplantation of hAECs into the testes [12, 13], the rear leg muscles, liver and/or the interscapular fat pad [12] of immunodeficient mice, neither teratomas nor other types of tumours were observed. These results, although suggesting that hAECs are not pluripotent like human ESCs, are in line with other studies which report no evidence of tumourigenicity following transplant of amniotic membrane-derived cells into patients [50–54]. However, at the very least, the absence of tumour formation following transplantation could actually be a safety advantage in clinical applications.

Meanwhile, a study to investigate tumour/and or teratoma formation by hAMSCs showed that transplantation of these cells into recipient hearts had not resulted in tumours or teratomas by 21 days after transplantation [55].

Another approach used to test pluripotency in vivo is through generation of chimeric animals by injecting a single stem cell into a blastocyst, followed by assessment of whether the injected cell then goes on to contribute to all germ layer cells in the chimeric animal. Circumventing ethical issues related to similar experiments using human blastocysts, Tamagawa and colleagues created a xenogeneic chimera using cells derived from the whole human amniotic membrane mixed with mouse ESCs in vitro [36]. This resulted in the formation of chimeric aggregates, with human cell contribution demonstrated in all three germ layers.

4.7.3 In Vitro Differentiation

Perhaps the most commonly used experimental approach for determining the potency of stem cells in vitro is through specific assessment of their capacity to undergo multilineage differentiation when cultured in media supplemented with specific growth factors, hormones and/or other additives which are known to stimulate differentiation. Evaluation of morphological changes and the expression of various lineage-specific genes, as well as assessment of acquired abilities to exert tissue-specific functions, all together allow evaluation of the level of in vitro differentiation achieved.

In the presence of specific stimuli during in vitro culture, both hAECs and hAMSCs have been shown to undergo multilineage differentiation, and even have the capacity to differentiate across germinal boundaries outside of their specific lineage, although, as for other stem/progenitor cells, conflicting results have been reported to date in terms of the level of maturation achieved in vitro, which is very likely due in part to the fact that it is impossible in vitro to precisely simulate the microenvironmental cues which lead to differentiation in vivo.

Since the demonstration by Sakuragawa and colleagues that hAECs not only express neuronal, glial and oligodendrocyte markers [56, 57], but are also able to synthesize catecholamines from L-tyrosine [58] and have the ability to convert 3, 4-dihydroxyphenylalanine (L-DOPA) into dopamine [59], other authors have provided further evidence of the ability of these cells to differentiate in vitro towards the neural lineage (e.g. [12, 13]). Meanwhile, Niknejad and colleagues reported that this capacity is affected by the presence of several factors, including serum, noggin, basic-FGF and retinoic acid [22].

The differentiation potential of hAECs towards hepatic cells has also been investigated. Sakuragawa and colleagues demonstrated that albumin and α-FP-producing hAECs are promising transgene carriers for allogeneic transplantation into liver [60], while Takashima and colleagues have shown that cultured hAECs express several hepatocyte-related genes and demonstrate some hepatocyte-specific functions including albumin production, glycogen storage and albumin secretion, even when cultured in the absence of hepatic differentiation stimuli [61]. Other groups have gone on to demonstrate that hAECs differentiate in vitro towards hepatocyte-like cells, although the inducing factors and criteria which were used to assess differentiation differed somewhat in these studies, as did the reported level of hepatic maturation achieved [12, 13, 62, 63].

When cultured under pancreatic differentiation conditions, hAECs present with increased expression of PDX-1, expression of PAX-6, insulin, NKx2.2, glucagon, alpha-amylase 2B (AMY2B), and features of exocrine acinar beta cells [12, 13]. When cultured in small airway growth medium, which induces differentiation of ESCs into type II pneumocytes, hAECs produce surfactant proteins SP-A, SP-B, pro–SP-C and SP-D, whilst also secreting SP-D and adopting an epithelial phenotype with lamellar body formation [40].

Moreover, it has been reported that hAECs also possess mesodermal differentiation potential, given that they are able to differentiate in vitro into cells with characteristics of adipocytes, osteocytes and chondrocytes (reviewed in [44]).

hAECs have also been reported to differentiate toward the myogenic lineage, as assessed by immunocytochemical and morphological analyses [13, 29], and towards cardiomyocyte-like cells, with expression of cardiac-specific genes and acquisition of features of relatively mature cardiomyocytes observed [12, 13].

As expected from a MSC-like cell population, it has been shown that hAMSCs differentiate in vitro towards osteogenic, adipogenic and chondrogenic cell lineages [21, 29, 32, 41, 64]. These morphological changes are paralleled by expression of lineage-specific markers, such as osteopontin and osteocalcin, collagen II, and lipoprotein lipase, for osteogenic, chondrogenic and adipogenic differentiation, respectively. However, studies to date have failed to identify cell clones differentiating toward all three lineages [21]. It has also been observed that the differentiation potential of hAMSCs declines after in vitro culture. In particular, whilst osteogenic differentiation occurred in these cells after one passage of culture, it could no longer be observed after subsequent passages. However, clonal hAMSC colonies have been shown to maintain their differentiation potential over several passages, at least up to passage 8 [21]. Interestingly, it has also been shown that hAMSCs display increased osteogenic differentiation potential when cultured on microcarriers [65], and, as also shown for chorionic membrane-derived MSCs [66], osteogenic differentiation is observed when the cells are cultured on slowly degradable polyurethane foams (our personal demonstration), suggesting their potential application for future in vivo bone regeneration and tissue engineering approaches.

Myogenic differentiation of hAMSCs has also been observed, both toward skeletal and cardiac lineages. Portmann-Lanz et al. [29] and Alviano et al. [64] showed induction of mRNA expression of myogenic transcription factors such as MyoD and Myogenin, as well as the expression of desmin, in hAMSCs which had been cultured in myogenic differentiation medium. Meanwhile, Zhao and colleagues showed that hAMSCs are able to undergo cardiac differentiation in response to growth factors such as bFGF or activin A, expressing the cardiomyocyte-related genes Nkx2.5, ANP, and alpha-MHC [67].

Furthermore, Alviano et al. also demonstrated angiogenic potential in hAMSCs, reporting that hAMSCs cultured in semisolid medium (Matrigel) spontaneously form capillary-like structures [64]. The formation of these structures was enhanced by the addition of VEGF to the culture medium, which also resulted in increased expression of VEGF receptor 1 and 2 (FLT-1 and KDR), and was also associated with vWF expression. Conversely, Konig et al. reported that whilst hAMSCs cultured in endothelial growth medium underwent an alteration from a fibroblast-like toward an endothelial-like morphology, and were able to take up acetylated low-density lipoprotein and form endothelial-like networks, these cells did not show any expression of mature endothelial markers such as vWF and VE-cadherin [68]. Nonetheless, under differentiation-inducing conditions, a significant downregulation in the expression of pro-angiogenic genes and proteins (tenascin C, angiopoietin receptor Tie-2, VEGF-A, CD146, FGF-2, IL-8, MMP-1 and urokinase type plasminogen activator receptor) was observed, with a concomitant up-regulation in the expression of anti-angiogenic factors (endostatin, serpinF1, the FGF-2 signaling antagonist sprout-1 and angioarrestin).

hAMSCs have also been shown to differentiate into cells with characteristics of hepatocytes [37, 69] and of pancreatic islet cells [70]. In particular, under hepatic differentiation conditions, hAMSCs expressed glucose-6-phosphatase and ornithine transcarbamylase and showed glycogen storage [37], while under pancreatic induction, they produced insulin, glucagon and somatostatin [70].

hAMSC differentiation towards the ectodermal lineage has also been investigated. Indeed, when cultured in specific neurogenic media, these cells have been observed to increase their expression of neuronal markers (such as nestin, Musashi1, NSE, NFM, MAP-2 and Neu-N) and glial markers such as GFAP, which are even expressed, albeit at lower levels, in freshly isolated not-induced hAMSCs [35, 70–72].

4.7.4 Reprogramming Amniotic Membrane-Derived Cells to Pluripotency

The pioneering study of Takahashi and Yamanaka, who first demonstrated the possibility of obtaining stem cells with properties similar to ESCs [referred to as iPS (induced pluripotent stem cells)] by reprogramming mouse fibroblasts through simultaneous retroviral introduction of four genes (OCT-4, SOX-2, KLF-4 and c-MYC) [73], opened a promising new avenue of research aimed at investigating the potential of iPS-based cellular therapies for regenerative medicine, while avoiding ethical objections and providing new hope toward the possibility of taking a patient's own cells and reprogramming these for personalized disease treatments which would not be subject to immunological rejection. Since this landmark study, an increasing number of scientists have reported the possibility of generating iPS from a variety of somatic cells ([74] and references inside), even though some problems remain, relating both to the methods used for reprogramming these cells and to the efficiency of iPS generation. Concerns also remain regarding the safety of in vivo application of these cells, which among other properties, have tumour-generating features, whilst debate also continues regarding whether iPS are indeed indistinguishable from ESCs in terms of all of their characteristics, and in particular, if these two types of cells are actually functionally equivalent (reviewed in [75]).

Although the feasibility of reprogramming human amniotic membrane-derived cells has only recently been investigated, the results so far obtained are very promising and warrant further studies, especially considering that, as a starting population for reprogramming, these cells offer significant advantages over other adult/postnatal somatic cells. In particular: (i) they can be obtained in larger quantities compared to cells of other sources; (ii) being less differentiated than adult somatic cells, these cells may be more amenable to reprogramming; (iii) their collection soon after birth could allow the undertaking of cellular therapies later in the life of the donor, which would be based on the use of autologous cell-derived iPS.

To date, it has been shown that both hAECs and hAMSCs can be rapidly and efficiently reprogrammed into iPS [76–79]. In particular, Zhao and colleagues reported that iPS obtained from hAECs were similar to human ESCs in morphology, proliferative ability, surface marker expression, gene expression and epigenetic status of pluripotent cell-specific genes [78]. Furthermore, when tested for their ability to spontaneously differentiate through embryoid body formation, these cells were able to differentiate into various cell types in vitro and after injection into testis of SCID mice, and formed teratoma-like masses containing tissues of all three embryonic germ layers. hAECs have been shown to reprogram faster and more efficiently than adult and neonatal somatic dermal fibroblasts, likely due to their lower global DNA methylation state when compared to dermal fibroblasts, and their endogenous expression of the four genes (OCT-4, SOX-2, KLF-4, and c-MYC) which are used in cellular reprogramming [79]. Meanwhile, Cai and colleagues have reported the successful generation of iPS from cells derived from the mesenchymal region of the amniotic membrane, showing that selected clones also formed embryonic bodies and teratomas containing derivatives of all three germ layers, and could also be readily differentiated into functional motor neurons [77]. Interestingly, Ge and colleagues have very recently demonstrated the feasibility of generating a robust population of iPS from hAMSCs, which were characterized by stem cell surface marker and pluripotency gene expression, and which, under appropriate in vitro culture and stimulation, exhibited spontaneous contractility, calcium transience across the membrane, high expression of cardiac genes and a mature cardiac phenotypes [80].

4.8 Stemness and In Vivo Studies

In summary, we have so far seen that in terms of the stemness properties which have been analysed, the amniotic membrane consists of heterogenous hAEC and hAMSC cell populations that present with expression of different stem cell/pluripotency markers and debated self-renewal/clonogenicity properties. Neither cell type has been shown to form teratomas or other types of tumours in vivo, whilst both display the ability to differentiate in vitro towards multiple cell lineages, even though it remains to be confirmed as to whether the in vitro multipotency of these cells emanates from a single clonal population [3]. With the vision to developing future therapeutic applications, it is also important to consider what we have so far discovered regarding the ability of these cells to undergo multilineage differentiation upon in vivo transplantation into animal models, given that stemness is also partly ascertained through in vivo transplantation experiments which help us to deepen our understanding of the stemness and the therapeutic potential of a cell.

To this end, Moodley et al. have demonstrated that naïve hAECs differentiate into cells with features of type II pneumocytes 2 weeks after parenteral injection into a mouse model of bleomycin-induced lung injury [40]. The transplanted hAECs were shown to reduce inflammation and fibrosis post-injury. Moreover, it has also been shown that after in utero transplantation into foetal sheep, hAECs differentiated into alveolar cells and were also able to mitigate ventilation-induced preterm lung injury [81]. Meanwhile, Marongiu and colleagues have shown that after naïve hAEC transplantation into the livers of retrorsine-treated SCID/beige mice, extracts from the livers of these animals were found by qRT-PCR to be positive for the expression of mature human liver genes, including the major CYP genes, other metabolic enzymes, plasma proteins, hepatocyte enriched transcription factors and genes encoding hepatic transported proteins, which together suggest that in vivo differentiation of hAECs towards hepatocyte-like cells had occurred [62].

In terms of hAMSCs, Tsuji et al. observed that transplantation of these cells into the hearts of nude rats with chronic myocardial infarction resulted in their transdifferentiation into cardiomyocyte-like cells, even though the authors also reported that 2 weeks after transplantation, the survival rate of hAMSC-derived cardiomyocytes was low [55]. Another in vivo study showed that after transplantation into the ischaemic hindlimbs of mice, hAMSCs integrated into the ischemic tissue and a low number of these cells also displayed the endothelial cell-specific marker biotinylated isolectin B4 (ILB4), whilst vascular-like structures were also formed, suggesting a vasculogenic potential of these cells in vivo [82]. Furthermore, hAMSC transplantation in the presence of BMP-2 into non-cartilage tissue (sub-fascial space of the abdominal muscle) of mice, or implantation with a collagen-scaffold into osteochondral defects generated on the femoral condyles in rats, have both been shown to result in morphological changes in these cells, with deposition of collagen type II in the cytoplasm and in the pericellular matrix, suggesting in vivo differentiation toward the chondrogenic cell lineage [83]. Finally, other authors have reported keratinocyte-like differentiation of hAMSCs after their injection into the epidermis around wound areas created by skin excision in diabetic animals [84].

However, in addition to the few lines of evidence presented above regarding the ability of amniotic membrane-derived cells to undergo specific differentiation in vivo, several solid experimental results obtained to date also support the fact that these cells have beneficial effects when transplanted into animal models of various diseases (e.g. [44, 85]). Whether these effects are independent of the ability of the cells to differentiate in vivo, as supported by the recently emerging concept that they may exert beneficial effects through paracrine actions mediated by soluble factors which they secrete [85], or if instead differentiation is in fact occurring but current methods are simply not yet able to reveal this, or whether other yet unknown mechanisms are involved, all remain at the forefront of the challenges which lie ahead for future research toward a better understanding of these cells.

4.9 Conclusions

Here we have provided a profile of some of the cell types that can be isolated from the human amniotic membrane, namely hAECs, hAMSCs and some of their subpopulations, in terms of their stem properties. Whilst these cell populations have been shown to share some characteristics, they also differ in many respects, and although they answer some of the criteria which would allow them to be described as "stem cells", other characteristics which they display would argue against the use of this term. Overall, the collective properties hereto described for hAECs and hAMSCs are suggestive of cells which are more commonly defined as "progenitor cells" rather than "stem cells". Nonetheless, we cannot exclude the possibility that in the heterogeneity of the initial cell populations obtained, "true" stem cells may be present which current methods simply do not yet allow us to isolate. In any case, even though there may be uncertainty as to the term which would be most appropriate to describe these cells: be it "stem cells" or "progenitor cells", the absence of a proper nomenclature cannot prevent us from taking special note of all of the interesting biological properties of these cells and of the beneficial effects which they exert, by mechanisms which remain to be defined, after transplantation in animal models of different diseases. Indeed, these aspects warrant further efforts toward either proving or refuting the "stemness" of these cells and gaining a better understand of their therapeutic potential.

Acknowledgments We are grateful for financial support over recent years from Fondazione Poliambulanza, Fondazione Cariplo grant n. 2011-0495, Fondazione Cariplo grant n. 2012-0842, and Fondazione della Comunità Bresciana Onlus. We sincerely thank Dr. Marco Evangelista for help in editing the chapter.

References

1. Davis J (1910) Skin transplantation. Johns Hopkins Hosp Rep 15:307–396
2. Caruso M, Silini A, Parolini O (2013) The human amniotic membrane: a tissue with multifaceted properties and different potential clinical applications. In: Cetrulo KJ, Cetrulo CL, Taghizadeh RR (eds) Perinatal stem cells, 2nd edn. Wiley, Hoboken, pp 177–195
3. Parolini O, Soncini M (2006) Human placenta: a source of progenitor/stem cells? J Reprod Med Endocrinol 3:117–126

4. Blau H, Brazelton T, Weimann J (2014) The evolving concept of a stem cell: entity or function? Cell 105:829–841
5. Seaberg RM, van der Kooy D (2014) Stem and progenitor cells: the premature desertion of rigorous definitions. Trends Neurosci 26:125–131
6. Bianco P, Robey PG, Simmons PJ (2008) Mesenchymal stem cells: revisiting history, concepts and assays. Cell Stem Cell 2:313–319
7. Lander AD (2009) The "stem cell" concept: is it holding us back? J Biol 8:70
8. Brunt KR, Weisel RD, Li R (2012) Stem cells and regenerative medicine – future perspectives. Can J Physiol Pharmacol 90:327–335
9. Tajbakhsh S (2009) Stem cell: what's in a name? Nat Rep Stem Cells doi:10.1038/stemcells.2009.90
10. Weissman IL, Anderson DJ, Gage F (2001) Stem and progenitor cells: origins, phenotypes, lineage commitments and transdifferentiations. Annu Rev Cell Dev Biol 17:387–403
11. Riau AK, Beuerman RW, Lim LS, Mehta JS (2010) Preservation, sterilization and de-epithelialization of human amniotic membrane for use in ocular surface reconstruction. Biomaterials 31:216–225
12. Miki T, Lehmann T, Cai H et al (2005) Stem cell characteristics of amniotic epithelial cells. Stem Cells 23:1549–1559
13. Ilancheran S, Michalska A, Peh G et al (2007) Stem cells derived from human fetal membranes display multilineage differentiation potential. Biol Reprod 77:577–588
14. Parolini O, Alviano F, Bagnara GP et al (2008) Concise review – isolation and characterization of cells from human term placenta: outcome of the first international workshop on placenta derived stem cells. Stem Cells 26:300–311
15. Miki T, Mitamura K, Ross MA et al (2007) Identification of stem cell marker-positive cells by immunofluorescence in term human amnion. J Reprod Immunol 75:91–96
16. Echigoya T, Takeuchi M, Hori H et al (2008) Human amniotic epithelium as an unlimited source of Oct4-expressing totipotent stem cell subset. Open Biotechnol J 2:102–110
17. Niknejad H, Peirovi H, Jorjani M et al (2008) Properties of the amniotic membrane for potential use in tissue engineering. Eur Cell Mater 15:88–99
18. Dua HS, Gomes JA, King AJ, Maharajan VS (2004) The amniotic membrane in ophthalmology. Surv Ophthalmol 49:51–77
19. Miki T, Marongiu F, Ellis EC et al (2007) Isolation of amniotic epithelial stem cells. Curr Protoc Cell Biol Chapter 1:Unit 1E 3
20. Barbati A, Mameli MG, Sidoni A, Di Renzo GC (2012) Amniotic membrane: separation of amniotic mesoderm from amniotic epithelium and isolation of their respective mesenchymal stromal and epithelial cells. Curr Protoc Stem Cell Biol 20:1E.8.1–1E.8.15
21. Soncini M, Vertua E, Gibelli L et al (2007) Isolation and characterization of mesenchymal cells from human fetal membranes. J Tissue Eng Regen Med 1:296–305
22. Niknejad H, Peirovi H, Ahmadiani A et al (2010) Differentiation factors that influence neuronal markers expression in vitro from human amniotic epithelial cells. Eur Cell Mater 19:22–29
23. Toda A, Okabe M, Yoshida T, Nikaido T (2007) The potential of amniotic membrane/amnion-derived cells for regeneration of various tissues. J Pharmacol Sci 105:215–228
24. Murphy S, Rosli S, Acharya R et al (2010) Amnion epithelial cell isolation and characterization for clinical use. Curr Protoc Stem Cell Biol 13:1E.6.1–1E.6.25
25. Pratama G, Vaghjiani V, Tee JY et al (2011) Changes in culture expanded human amniotic epithelial cells: implications for potential therapeutic applications. PLoS One 6:1–12
26. Bilic G, Zeisberger SM, Mallik AS et al (2008) Comparative characterization of cultured human term amnion epithelial and mesenchymal stromal cells for application in cell therapy. Cell Transplant 17:955–968
27. Evron A, Goldman S, Shalev E (2011) Human amniotic epithelial cells cultured in substitute serum medium maintain their stem cell characteristics for up to four passages. Int J Stem Cells 4:123–132

28. Miki T, Strom SC (2006) Amnion-derived pluripotent/multipotent stem cells. Stem Cell Rev 2:133–142
29. Portmann-Lanz CB, Schoeberlein A, Huber A et al (2006) Placental mesenchymal stem cells as potential autologous graft for pre- and perinatal neuroregeneration. Am J Obstet Gynecol 194:664–673
30. Stadler G, Hennerbichler S, Lindenmair A et al (2008) Phenotypic shift of human amniotic epithelial cells in culture is associated with reduced osteogenic differentiation in vitro. Cytotherapy 10:743–752
31. Marongiu F, Gramignoli R, Sun Q et al (2010) Isolation of amniotic mesenchymal stem cells. Curr Protoc Stem Cell Biol. 2010 Mar; Chapter 1:Unit 1E.5. doi: 10.1002/9780470151808.sc01e05s12
32. Díaz-Prado S, Muiños-López E, Hermida-Gómez T et al (2010) Multilineage differentiation potential of cells isolated from the human amniotic membrane. J Cell Biochem 111:846–857
33. Dominici M, Le Blanc K, Mueller I et al (2006) Minimal criteria for defining multipotent mesenchymal stromal cells: the international society for cellular therapy position statement. Cytotherapy 8:315–317
34. Díaz-Prado S, Muiños-López E, Hermida-Gómez T et al (2011) Human amniotic membrane as an alternative source of stem cells for regenerative medicine. Differentiation 81:162–171
35. Kim J, Kang HM, Kim H et al (2007) Ex vivo characteristics of human amniotic membrane-derived stem cells. Cloning Stem Cells 9:581–594
36. Tamagawa T, Ishiwata I, Saito S (2004) Establishment and characterization of a pluripotent stem cell line derived from human amniotic membranes and initiation of germ layers in vitro. Hum Cell 17:125–130
37. Tamagawa T, Oi S, Ishiwata I et al (2007) Differentiation of mesenchymal cells derived from human amniotic membranes into hepatocyte-like cells in vitro. Hum Cell 20:77–84
38. Bailo M, Soncini M, Vertua E et al (2004) Engraftment potential of human amnion and chorion cells derived from term placenta. Transplantation 78:1439–1448
39. Brvanlou AH, Gage FH, Jaenisch R et al (2003) Setting standards for human embryonic stem cells. Science 300:913–916, 80-
40. Moodley Y, Ilancheran S, Samuel C et al (2010) Human amnion epithelial cell transplantation abrogates lung fibrosis and augments repair. Am J Respir Crit Care Med 182:643–651
41. Nogami M, Tsuno H, Koike C et al (2012) Isolation and characterization of human amniotic mesenchymal stem cells and their chondrogenic differentiation. Transplantation 93(12):1221–8
42. Fatimah SS, Tan GC, Chua K et al (2013) Stemness and angiogenic gene expression changes of serial-passage human amnion mesenchymal cells. Microvasc Res 86:21–29
43. Ryan JM, Pettit AR, Guillot PV et al (2013) Unravelling the pluripotency paradox in fetal and placental mesenchymal stem cells: Oct-4 expression and the case of the emperor's new clothes. Stem Cell Rev Rep 9:408–421
44. Manuelpillai U, Moodley Y, Borlongan C, Parolini O (2011) Amniotic membrane and amniotic cells: potential therapeutic tools to combat tissue inflammation and fibrosis? Placenta 32:S320–S325
45. Magatti M, De Munari S, Vertua E et al (2008) Human amnion mesenchyme harbors cells with allogeneic T-cell suppression and stimulation capabilities. Stem Cells 26:182–192
46. Kobayashi M, Yakuwa T, Sasaki K et al (2008) Multilineage potential of side population cells from human amnion mesenchymal layer. Cell Transplant 17:291–301
47. Maruyama N, Kokubo K, Shinbo T et al (2013) Hypoxia enhances the induction of human amniotic mesenchymal side population cells into vascular endothelial lineage. Int J Mol Med 32:315–322
48. Sarugaser R, Hanoun L, Keating A et al (2009) Human mesenchymal stem cells self-renew and differentiate according to a deterministic hierarchy. PLoS ONE 4:1–4
49. Keating A (2014) Mesenchymal stromal cells: new directions. Cell Stem Cell 10:709–716
50. Akle C, Adinolfi M, Welsh K et al (1981) Immunogenicity of human amniotic epithelial cells after transplantation into volunteers. Lancet 2:1003–1005

51. Tylki-Szymańska A, Maciejko D, Kidawa M et al (1985) Amniotic tissue transplantation as a trial of treatment in some lysosomal storage diseases. J Inherit Metab Dis 8:101–104
52. Yeager AM, Singer HS, Buck JR et al (1985) A therapeutic trial of amniotic epithelial cell implantation in patients with lysosomal storage diseases. Am J Med Genet 22:347–355
53. Scaggiante B, Pineschi A, Sustersich M et al (1987) Successful therapy of Niemann-Pick disease by implantation of human amniotic membrane. Transplantation 44(1):59–61
54. Sakuragawa N, Yoshikawa H, Sasaki M (1992) Amniotic tissue transplantation: clinical and biochemical evaluations for some lysosomal storage diseases. Brain Dev 14:7–11
55. Tsuji H, Miyoshi S, Ikegami Y et al (2010) Xenografted human amniotic membrane-derived mesenchymal stem cells are immunologically tolerated and transdifferentiated into cardiomyocytes. Circ Res 106:1613–1623
56. Sakuragawa N, Thangavel R, Mizuguchi M et al (1996) Expression of markers for both neuronal and glial cells in human amniotic epithelial cells. Neurosci Lett 209:9–12
57. Ishii T, Ohsugi K, Nakamura S et al (1999) Gene expression of oligodendrocyte markers in human amniotic epithelial cells using neural cell-type-specific expression system. Neurosci Lett 268:131–134
58. Sakuragawa N, Misawa H, OhsugI K et al (1997) Evidence for active acetylcholine metabolism in human amniotic epithelial cells: applicable to intracerebral allografting for neurologic disease. Neurosci Lett 232:53–56
59. Elwan MA (1998) Synthesis of dopamine from l-3,4-dihydroxyphenylalanine by human amniotic epithelial cells. Eur J Pharmacol 354:R1–R2
60. Sakuragawa N, Enosawa S, Ishii T et al (2000) Human amniotic epithelial cells are promising transgene carriers for allogeneic cell transplantation into liver. J Hum Genet 45:171–176
61. Takashima S, Ise H, Zhao P et al (2004) Human amniotic epithelial cells possess hepatocyte-like characteristics and functions. Cell Struct Funct 29:73–84
62. Marongiu F, Gramignoli R, Dorko K et al (2011) Hepatic differentiation of amniotic epithelial cells. Hepatology 53:1719–1729
63. Tee JY, Vaghjiani V, Liu YH et al (2013) Immunogenicity and immunomodulatory properties of hepatocyte-like cells derived from human amniotic epithelial cells. Curr Stem Cell Res Ther 8:91–99
64. Alviano F, Fossati V, Marchionni C et al (2007) Term amniotic membrane is a high throughput source for multipotent mesenchymal stem cells with the ability to differentiate into endothelial cells in vitro. BMC Dev Biol 7:1–14
65. Chen M, Wang X, Ye Z et al (2011) A modular approach to the engineering of a centimeter-sized bone tissue construct with human amniotic mesenchymal stem cells-laden microcarriers. Biomaterials 32:7532–7542
66. Bertoldi S, Farè S, Denegri M et al (2010) Ability of polyurethane foams to support placenta-derived cell adhesion and osteogenic differentiation: preliminary results. J Mater Sci Mater Med 21:1005–1011
67. Zhao P, Ise H, Hongo M et al (2005) Human amniotic mesenchymal cells have some characteristics of cardiomyocytes. Transplantation 79:528–535
68. König J, Huppertz B, Desoye G et al (2012) Amnion-derived mesenchymal stromal cells show angiogenic properties but resist differentiation into mature endothelial cells. Stem Cells Dev 21:1309–1320
69. Paracchini V, Carbone A, Colombo F et al (2012) Amniotic mesenchymal stem cells: a new source for hepatocyte-like cells and induction of CFTR expression by coculture with cystic fibrosis airway epithelial cells. J Biomed Biotechnol 575471:1–15
70. Tamagawa T, Ishiwata I, Sato K, Nakamura Y (2009) Induced in vitro differentiation of pancreatic-like cells from human amnion-derived fibroblast-like cells. Hum Cell 22:55–63
71. Sakuragawa N, Kakinuma K, Kikuchi A et al (2004) Human amnion mesenchyme cells express phenotypes of neuroglial progenitor cells. J Neurosci Res 78:208–214
72. Tamagawa T, Ishiwata I, Ishikawa H, Nakamura Y (2008) Induced in vitro differentiation of neural-like cells from human amnion-derived fibroblast-like cells. Hum Cell 21:38–45

73. Takahashi K, Yamanaka S (2006) Induction of pluripotent stem cells from mouse embryonic and adult fibroblast cultures by defined factors. Cell 126:663–676
74. Yamanaka S (2009) A fresh look at iPS cells. Cell 137:13–17
75. Yamanaka S (2014) Induced pluripotent stem cells: past, present and future. Cell Stem Cell 10:678–684
76. Nagata S, Toyoda M, Yamaguchi S et al (2009) Efficient reprogramming of human and mouse primary extra-embryonic cells to pluripotent stem cells. Genes Cells 14:1395–1404
77. Cai J, Li W, Su H et al (2010) Generation of human induced pluripotent stem cells from umbilical cord matrix and amniotic membrane mesenchymal cells. J Biol Chem 285: 11227–11234
78. Zhao H, Li Y, Jin H et al (2010) Rapid and efficient reprogramming of human amnion-derived cells into pluripotency by three factors OCT4/SOX2/NANOG. Differentiation 80:123–129
79. Easley CA, Miki T, Castro CA et al (2012) Human amniotic epithelial cells are reprogrammed more efficiently by induced pluripotency than adult fibroblasts. Cell Reprogram 14:193–203
80. Ge X, Wang I-NE, Toma I et al (2012) Human amniotic mesenchymal stem cell-derived induced pluripotent stem cells may generate a universal source of cardiac cells. Stem Cells Dev 21:2798–27808
81. Hodges RJ, Jenkin G, Hooper SB et al (2012) Human amnion epithelial cells reduce ventilation-induced preterm lung injury in fetal sheep. Am J Obstet Gynecol 206:448.e8–448.e15
82. Kim S-W, Zhang H-Z, Kim CE et al (2012) Amniotic mesenchymal stem cells have robust angiogenic properties and are effective in treating hindlimb ischaemia. Cardiovasc Res 93:525–534
83. Wei JP, Nawata M, Wakitani S et al (2009) Human amniotic mesenchymal cells differentiate into chondrocytes. Cloning Stem Cells 11:19–26
84. Kim S-W, Zhang H-Z, Guo L et al (2012) Amniotic mesenchymal stem cells enhance wound healing in diabetic NOD/SCID mice through high angiogenic and engraftment capabilities. PLoS ONE 7:e41105
85. Parolini O, Caruso M (2011) Review – preclinical studies on placenta-derived cells and amniotic membrane: an update. Placenta 25:S186–S195

Chapter 5
Amniotic Membrane in Health and Disease: An Obstetrical Perspective

Ana Luísa Areia and Paulo Moura

Abstract Amnion, chorion and decidua form a complex unit, both anatomically and functionally. In normal pregnancy they constitute a border that simultaneously separates and relates the foetal and maternal environments. The balance between formation and degradation of membrane components is a physiological phenomenon that can be found throughout pregnancy. Moreover, during pregnancy there is an equilibrated balance between membrane degradation and formation; under some circumstances, an imbalance may ensue, resulting in membrane rupture. Immunological interactions between mother and foetus occur not only at the placental level, but also where the membranes (amnion and chorion) contact the deciduas. The amniotic membrane appears to be an immune privileged tissue and to contain some immunoregulatory factors. Therefore, a successful pregnancy seems to be the consequence of numerous interactions between the receptive uterus and the mature blastocyst, under immunohumoral control. Implantation and parturition are specifically characterized by mechanisms of local inflammatory activity. Sequential processes lead to a common pathway of parturition involving increased uterine contractility, cervical ripening and decidual/foetal membrane activation. In human amnion and decidua there is increased production of prostaglandins during parturition; in fact, labour may result from changes in prostaglandin availability within the uterus. Therefore, the process of giving birth may be abnormally initiated out of time by any interference, as membrane mechanical rupture or the action of infectious agents. Chorioamnionitis is the microbial invasion of the amniotic cavity through amnion, which induces a systemic inflammatory reaction. Foetal exposure to infection may lead to perinatal death, neonatal sepsis and other postnatal complications.

Keywords Preterm birth • Amniotic membrane • Chorioamnionitis • Fetal inflammatory response syndrome • Reproductive immunology

---

A.L. Areia (✉) • P. Moura
Faculty of Medicine, University of Coimbra, Coimbra, Portugal

Obstetric Unit A, Coimbra Hospital and Universitary Center, Coimbra, Portugal
e-mail: analuisareia@gmail.com

A.C. Mamede, M.F. Botelho (eds.), *Amniotic Membrane: Origin, Characterization and Medical Applications*,
DOI 10.1007/978-94-017-9975-1_5

5.1 Introduction

The amnion is a membranous structure of embryonic origin that early in pregnancy intimately adheres to the chorionic membrane derived from extra-placental trophoblast, to form what is commonly referred to as the "foetal membranes". Together, they are firmly juxtaposed and connected with the maternal decidua (term for the endometrium modified during pregnancy and shed at parturition).

Amnion, chorion and decidua thus form a complex unit, both anatomically and functionally. During pregnancy and labour, whether in normal or in pathological circumstances, the role of the amniotic membrane cannot be conceived or apprehended without considering this context.

In normal pregnancy they constitute a border, or interface, that simultaneously separates and relates the foetal and maternal environments. They not only maintain an anatomical barrier that preserves the intra-amniotic space from external interference, but also actively cooperate in the control of maternal inflammatory and immunological local reactions to the conceptus and, ultimately, in the triggering and the course of labour.

From an obstetrical perspective, the main pathological problems with which the amnion is related are those of rupture before term and/or local mechanisms of infection and inflammation. If and when these closely intertwined events occur, intrauterine infection and/or preterm labour may ensue.

Perinatal infection and prematurity are very important causes of severe perinatal morbidity and of foetal and neonatal mortality and, thus, the most relevant clinical and epidemiological implications of amnion lesion or dysfunction. For this reason we will focus our attention on these topics.

5.2 The Amniotic Membrane and Normal Pregnancy

5.2.1 Amnion, Chorion and Decidua: A Functional Unit

Foetal membranes development is a complex process that starts with the formation of amniotic and chorionic cavities. The rapid growth of the amniotic cavity leads to the disappearance of the exocoelomic cavity and juxtaposes the amnion to the chorion, which in turn is in close anatomical and functional contact with the decidua.

Thus, the commonly designed "foetal membranes", that delimitate an extensive portion of the border between the foetal and the maternal environments, are composed of three layers: amnion and chorion, of ovular origin, and the deciduas, of maternal origin [1].

The amnion is itself composed of five layers (Fig. 5.1) and doesn't contain either vessels or nervous terminations. The innermost epithelial layer secretes collagen type III and IV, as well as glycoproteins that constitute the basement amniotic membrane; under this basement membrane, the compact layer is composed by collagen

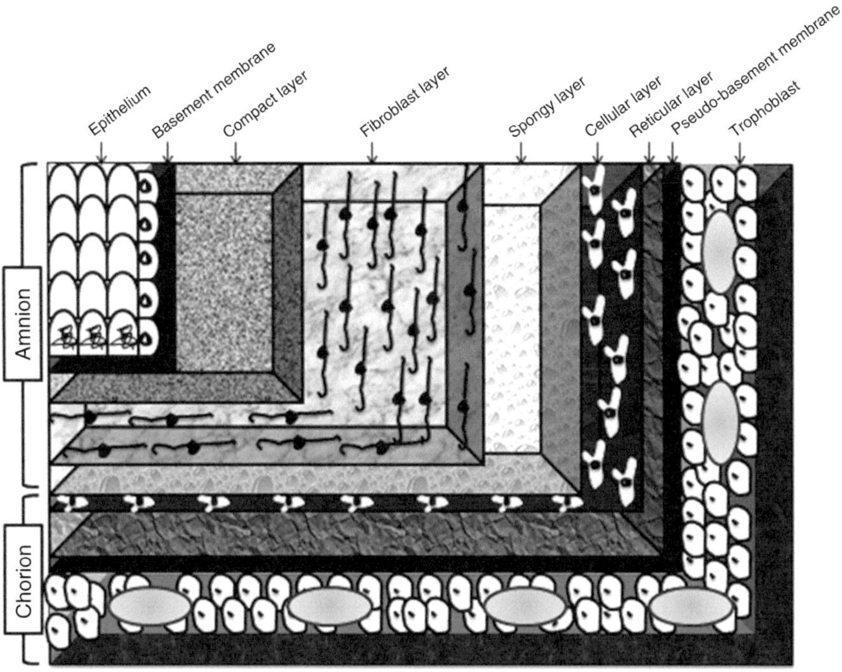

Fig. 5.1 Amnion and chorion layers

type I and III, synthesized by mesenchymal cells. The fibroblastic layer is the thinnest layer and contains macrophages; finally, the spongy layer, with few fibroblasts, permits the amnion to slide upon the underlying chorion [1].

Amnion mesenchymal cells synthetize interstitial collagen that makes up the compact layer of the amnion, source of the majority of tensile strength and are also active producers of interleukin (IL) -6, IL-8 and monocyte chemoattractant protein (MCP-1) [2]. As a site of prostaglandin production, the amniotic epithelium participates in the so called final common pathway of the initiation of labour [2].

Placental amnion (covering the foetal face of the placenta) could be involved in the modulation of chorionic vessel tone and blood flow [2]. The chorion is composed of three layers: one reticular layer, rich in collagen and proteoglicans, in contact with the deep layer of the amnion (innermost epithelial layer); one basal membrane, and one layer of trophoblast cells, that persist after villous atrophy [1].

Human decidua contains abundant immune cells during gestation, with more than 30 % of stromal cells in the first trimester expressing the leukocyte common antigen CD45 [3, 4].

Decidual population of natural killer (NK) cells, macrophages, decidual stromal cells (DSC) and T cells (CD4+) constitute 30–40 % of decidual cells [4], but B cells are absent [5]. The precocious elevation of lymphocyte number suggests that the influx and the proliferation of these cells are under hormonal influence. Special techniques like electron microscopy and immunohistochemistry underline the intimate contact between the trophoblast and these immune cells [5, 6]. The major cellular component of the decidua is DSC. These cells exert different immune activities that have emerged as relevant to the immunologic interaction between mother and fetus and may lead to either a normal pregnancy or abortion [3].

There are four major populations of decidual leukocytes present in early pregnancy: uterine NK, macrophages, dendritic cells (DC) and T cells (CD8+ and rare CD4+) [3].

In human hemochorial placenta, fetal trophoblast cells appear to be in extremely close contact with the maternal immune cells [3]. Thus, immunologic interactions between mother and foetus during pregnancy are thought to occur in the decidua [4, 6]. Moreover, it seems that there are two maternal-foetal interfaces: one made of an immunologically neutral population (in contact with the maternal immune system), and another, immunologically active population of trophoblast cells migrating to the decidua [7].

Spiral arteries' remodelling is one of the most important features of human placental development [5]. Anomalies of this phenomenon are linked to pregnancy complications (like preeclampsia) and complete remodelling involves trophoblast, but NK cells also intervene before trophoblast is able to do it [8].

Trophoblast cells, NK cells and DSC participate in angiogenesis' regulation at fetal-maternal interface level [9]. Moreover, uterine DC secrete stimulating and inhibiting angiogenesis molecules [10].

Recent evidence points out to the existence of a bidirectional trafficking across the maternal-foetal interface. Foetal cells have the potential to infiltrate maternal tissues and to differentiate into different types of cells (liver, muscle, skin and so on), transforming the mother to a chimera. Also, these foetal cells play a role in repairing maternal tissues that are damaged by a pathologic process [7].

The amniochorion is recognized as a leaky structure, with an extremely low transepithelial potential and high conductance. However, there are marked differences between the amniotic layer, which appears to be a more diffusional barrier, and the underlying chorionic layer. In addition, inflammatory mediators appear to weaken the amniotic membrane barrier trough disruption of tight junctions [11].

5.2.2 *Preservation of Anatomical Integrity and Rupture*

The balance between formation and degradation of membrane components is a physiological phenomenon that can be found throughout pregnancy. It seems that at least two mechanisms are responsible for membrane degradation: apoptosis in the cellular compartment and action of matrix metalloproteinases (MMP) in the extracellular matrix [12, 13]. Regulation of MMP depends upon factors that increase their expression (as cytokines) and factors that inhibit their activity (tissue inhibitor metalloproteinases – TIMPS) [1].

Apoptosis, or programmed cell death, is an active mechanism through which superfluous or non-functional cells are eliminated in order to maintain tissue normality. During pregnancy, apoptosis plays an important role in the induction of maternal tolerance and in trophoblast differentiation and turnover [5]. Destruction of activated B and T cells through apoptosis induces a specific tolerance [14].

Membranes are quite elastic and can expand to twice normal size during pregnancy [2]. The amnion is responsive both acutely and chronically to mechanical stretch, which alters amniotic gene expression. This in turn, potentially triggers both autocrine and paracrine responses, with the production of factors such as MMP and IL-8 that may modulate changes in membrane proprieties during term and preterm labour [2].

During pregnancy, there is an equilibrated balance between membrane degradation and formation which assures membrane firmness and permits adaptation to foetal growth and amniotic fluid expansion [1]. Under some circumstances, an imbalance between membrane formation and degradation may exist, leading to membrane weakening and resulting in membrane rupture. Chorion-decidual interface appears as a key element of preterm premature rupture of membranes (PPROM) pathophysiology. Oxidative stress, membrane distension and infection are examples that can contribute to weaken membranes [1]. Therefore, there are several factors that can promote this imbalance: infection, hormonal factors, membrane fusion' default, oxidative stress and mechanical factors [15].

(a) Infection

Infection has been considered over the last 20 years as one of the major mechanisms of PPROM [16, 17]. Intra-uterine contamination can occur through four routes:

1. Vaginal or ascending route, by the progression of microorganisms throughout the vaginal canal;
2. Haematogenic route, by means of the placenta;
3. Peritoneal route, with contamination through the Fallopian tubes in the context of an intra-abdominal infection;
4. Trans-uterine route, after an invasive procedure like amniocentesis [18].

(b) Hormonal factors

Progesterone has a major role in pregnancy maintenance and evidence demonstrates its secretion at the amnion, chorion and decidua in the human species [19, 20]. Relaxin has been studied as a potential candidate to explain preterm birth (PTB), without yet definitive answers [21–24]. Relaxin H2 is a systemic hormone produced by the corpus luteum, whose elevated levels are associated with PTB. On the contrary, decidual relaxin only acts locally and elevated levels are associated with PPROM [25]. PPROM is hence the result of a complex process that appears to start before the clinical episode of membrane rupture [1].

(c) Defects in membrane apposition

The close juxtaposition of the different membrane layers is essential for membrane development and stability [1]. Defects in strict membrane apposition have been associated with premature birth, abortion and intrauterine foetal death [26]. Nevertheless, the origin of these defects is currently unidentified, although it can sometimes be attributed to acute or mechanical stress or to invasive techniques of prenatal diagnosis [1].

(d) Oxidative stress

Oxidative stress reflects an imbalance between the systemic manifestation of oxidants-antioxidants [27]. The oxidants are represented by oxygen free radicals whose free electron gives them an important reactivity. Oxygen metabolites have a role in pathogen' and cancer cell' destruction, but can also have deleterious effects in normal cells and tissues when overly produced. Free radicals are permanently being generated in an organism (endogenous origin), but can also come from exogenous sources, like cigarette smoke or drugs (cocaine) [1].

(e) Mechanical factors

Hydramnios and multiple pregnancies are important risks factors for PPROM, through the increase in membrane tension [28, 29].

In vitro observations describe a five-step sequence in membrane rupture: membrane distension; amnion and choriodecidual separation; choriodecidual rupture; amnion non-elastic distension and amnion rupture [30].

5.2.3 Local Immune Regulation

Medawar propositions of 1953 explained the maternal immunological acceptance of the conceptus considering the foetal-maternal relationship as an allograft. The success of pregnancy could result from foetal antigenic neutrality or immaturity, absence of contact between foetal cells and the mother's immune system or maternal decrease of immunological activity throughout gestation [31]. They established a theoretical framework for scientific research and clinical reasoning for the next three decades.

None of them was to be confirmed in its original form but, nevertheless, they all contained some inherent truth and insight as the development of immunology has highlighted in more recent years: the foetus is not antigenically neutral, but the trophoblastic cells that contact maternal tissues have scant antigenic expression; there is control, although selective, of cell traffic at the placental circulation; there is no maternal immunological suppression, but the modulation of systemic and local immune responses facilitates the survival, progression, differentiation and growth control of the trophoblast.

The foetus naturally expresses an antigenic identity that is in part of paternal origin and, thus, different from its mother. Actually, in cases of assisted reproduction technology (ART) recurring to oocyte donation or surrogate motherhood, it may be totally allogenic.

In any case, an immune rejection response would be inevitable. However, the fact is that it is not the foetus but rather the trophoblast, at the placental and membrane interfaces, that contacts the woman's immune system – and these trophoblastic cells have a remarkably neutral antigenic identity, expressing only sufficient

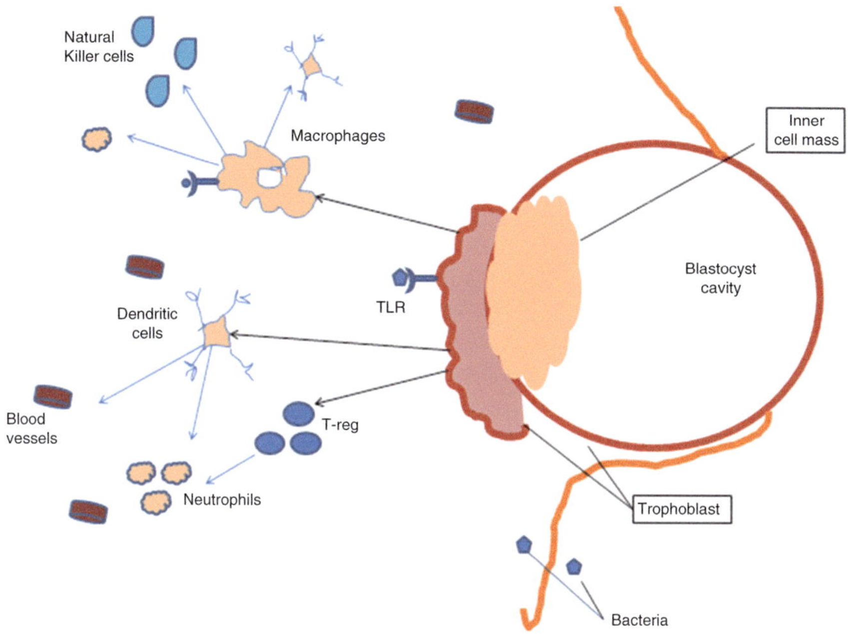

Fig. 5.2 Trophoblast-immune interaction

MHC antigens [like human leukocyte antigen (HLA) -G and HLA-E] and apoptosis-inducing ligands (such as FasL and TRAIL) to escape lysis by non-specific immune NK-cells or to suppress their activity [32].

In reality, the amniotic membrane seems to be an immune privileged tissue and to contain some immunoregulatory factors. Additionally, HLA-G molecule appears to be an important immunosuppressive factor during pregnancy. Expression of HLA-G in amniotic membrane may influence the host immune system in two ways: first it may play the role of a tolerogenic peptide and the host lymphocyte or DC may be inactivated by HLA-G's binding to inhibitory receptors; secondly, HLA-G may be recognized by some T cells and then serve as an activator of CD8+ T cells (because CD8 can bind to HLA-G), and these cells may have a suppressor function [33].

So, the theoretical paradigm has evolved from one of maternal-foetal tolerance to that of a decidual-trophoblastic-amniotic tolerance and active cooperation [7]: a successful pregnancy is the consequence of numerous complex interactions between the receptive uterus and the mature blastocyst, under immunohumoral control [3].

In fact, mother and foetus do not relate as host-allograft, but their complex relations are more similar to parasite and host, or tumour and host, consisting of not only support and nourishment, but also working together against common external dangers (Fig. 5.2).

Results of recent studies suggest that the trophoblast functions like the conductor of a symphony where the musicians are the cells of the maternal immune system.

Consequently, the success of pregnancy depends on how well the trophoblast communicates with each immune cell type and then how all of them work together [7].

Different immunological participants present in the decidua benefit from instructions sent by the trophoblast, which refines, educates, activates and sometimes neutralizes and even uses them to develop in harmony. On the other hand, the presence of an active local immune response is essential to trophoblastic development, as many cytokines act as growth factors and, ultimately, force differentiation and stop decidual invasion [5, 31].

Since the beginning of pregnancy, immunitary cells are present in the decidua, place of contact between the mother and the foetal-placental unit: these cells will suffer differentiation and specific education. The embryo expresses histocompatibility antigens (HLA) since the two-cell stadium but he won't be in direct contact with the mother, even in the foetal stage, as the trophoblast and the amnion separate them at the placental and membrane interfaces [5].

Endometrial decidualization affects all cellular populations: DSC, glandular cells and immunitary cells. DSC help the immunitary phenomenon in various ways, secreting immunosuppressive substances, producing cytokines, phagocytizing immune cells, presenting antigens and regulating macrophage activation [5].

The placenta, for a long time considered as a protective foetal barrier reinforced by the presence of sialic acid, mucopolysaccharides and the effects of hormones like human chorionic gonadotropin (hCG) and human placental lactogen (HPL), is in reality very porous [31].

In fact, there are five zones of interaction at the fetal-maternal level. During the first trimester, the contact between immunitary maternal cells and foetal cells is limited to the decidua. At the second trimester, maternal blood initiates perfusion of the intervillous space; syncytiotrophoblast microparticles can detach into the mother's circulation, enlarging immunitary contacts to the entire maternal organism. Moreover, villositary trophoblast is characterized by the absence of HLA antigens, class I and II [5].

Pregnancy is a Th2 phenomenon (Fig. 5.3); the shift away from type 1 cytokines production during pregnancy is beneficial for pregnancy, as these type of cytokines [e.g. interferon-alpha (INF-α) and tumor necrosis factor (TNF) -β] are harmful for pregnancy as they inhibit embryonic and fetal development, whereas type 2 cytokines stimulate trophoblast outgrowth and invasion [32].

Current research hypothesis propose that the potential of trophoblastic antigens to induce a natural and tolerogenic maternal response engages cytokines, chemokines, indoleamine 2,3-dioxygenase (IDO) and galectin-1 derived from the foetal-placental unit, which suggests a possible strategy to treat some forms of pregnancy pathologies via immune regulation [3].

5.2.4 *Pregnancy: A Controlled Inflammatory Process*

Implantation and placentation in the first trimester and early second trimester of pregnancy resemble an open wound and require a strong inflammatory response.

During the first stage, the embryo has to break through the epithelial lining of the uterus to implant; it must damage the underlying endometrial tissue to invade; and

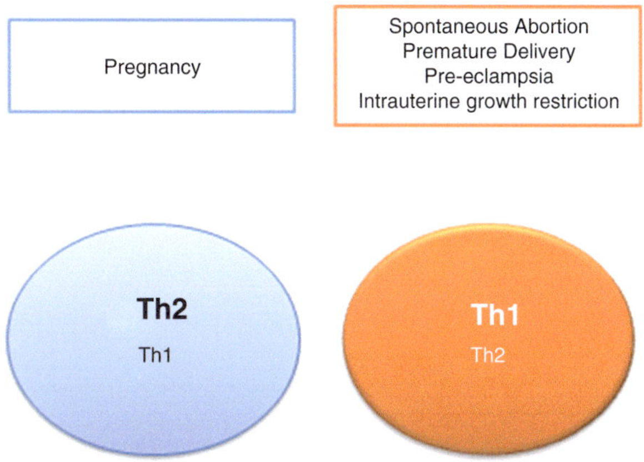

Fig. 5.3 Th1/Th2 balance in physiological pregnancy and in gestational diseases

it must replace the endothelium and vascular smooth muscle of the maternal blood vessels to secure an adequate blood supply. These activities create a veritable battleground of invading cells, dying cells and repairing cells. An inflammatory response is required and consequently triggered to secure the adequate repair of the uterine epithelium and the removal of cellular debris [7]. As so, during the early phases of pregnancy a successful implantation occurs in a pro-inflammatory microenvironment, and a Th1-type response is followed by a shift to Th2, to control endocrine and immune interactions [3].

Th1 responses may be suppressed during human pregnancy via downregulation of nuclear factor (NF)-KB and T-bet transcription. In addition, progesterone stimulates a Th2-type response, reduces inflammatory cytokines and represses allogeneic responses potentially deleterious, promoting pregnancy safety and foetal survival [3].

During pregnancy the cell mediated immune response of the maternal specific immune system is relatively suppressed, and this suppression seems to be compensated by an activation of the innate immune response. Although this innate response is essential in the response to extracellular bacterial infection, it is less efficient in clearing viruses and intracellular pathogens than the specific immune response; indeed, pregnant women are more sensitive to such infections [32].

Recent work redefined the Th1/Th2 theory of pregnancy immunomodulation of maternal responses, describing three biological periods that match up to mother's complaints [34, 35].

(i) Inflammatory period, first trimester: embryo implantation initiates endometrial and vascular uterine lesions. This situation imposes an inflammatory reaction to clean the wound, which may in part explain nausea and vomits felt by the mother [36];
(ii) Anti-inflammatory period, second trimester: throughout this period, characterized by growing and development of the conceptus, the environment is anti-inflammatory, Th2-type. During this trimester women usually feel better;
(iii) Inflammatory period, third trimester: in the natural and necessary ultimate end of pregnancy, reactivation of inflammatory activity is responsible for the initiation of labour, with the triggering of myometrial contractions and cervical dilatation – the placental membranes and the amnion in particular, are central intervenients in this essential biochemical process [34, 35].

Fig. 5.4 Th1/Th2 balance in physiological pregnancy

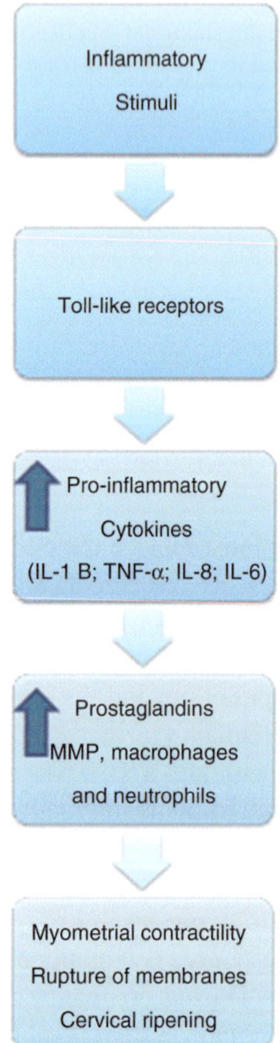

5.2.5 Labour: An Activation of Inflammatory Mechanisms

The pregnant uterus is replete with specialized immune cells primed to play roles in implantation, placentation and parturition. The major cell types comprise uterine NK cells, DC, T lymphocytes and macrophages [37].

Implantation and parturition are specifically characterized by mechanisms of local inflammatory activity [37–39]. In fact, pro-inflammatory cytokines, matrix degrading proteins, altered transcriptional factors, rapid hormonal changes and immune cell activity, are paramount for uterine activation and the onset of labour (Fig. 5.4) [40].

The safety of the gestational period, comprised of decidualization, placentation and foetal development, requires uterine quiescence guided by high levels of progesterone and the production of anti-inflammatory cytokines [36, 41].

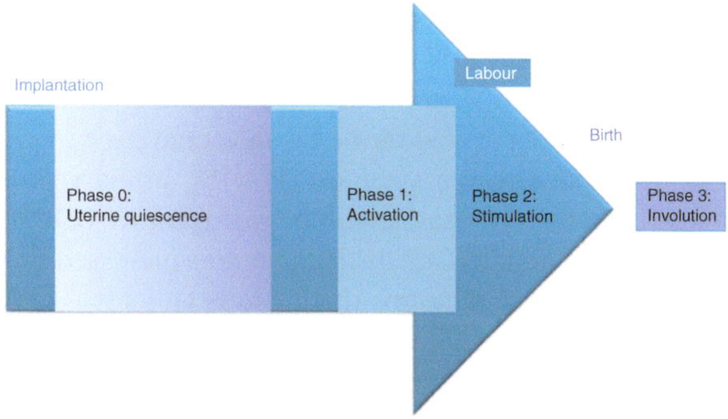

Fig. 5.5 Inflammation and parturition

The act of giving birth is the ultimate step in a pro-inflammatory signalling cascade that is orchestrated by an intrauterine milieu coupled to hormonal cues [37]. As so, parturition is characterized by an influx of immune cells into the myometrium to promote a recrudescence of an inflammatory process. This pro-inflammatory environment promotes the contraction of the uterus and the ultimate expulsion of the infant and of the placenta [42].

The physiological process of normal parturition at term can be divided in four phases (Fig. 5.5). During the long period of pregnancy the uterus remains quiescent, passively accepting distension in order to accommodate the increasing volume of gestational components and this corresponds to phase 0 (quiescence); phase 1 (activation) involves a level of uterine stretch that determines hypothalamic-pituitary-adrenal (HPA) activation. Phase 2 (stimulation) refers to stimulation of the activated uterus by various substances including corticotrophin-releasing hormone (CRH), locally produced prostaglandins (PG) and oxytocin. These sequential processes lead to a common pathway of parturition involving increased uterine contractility, cervical ripening and decidual and foetal membrane activation. Phase 3 (involution) corresponds to postpartum uterus's return to the pre-pregnant status [36, 43].

Temporal increase in inflammatory signals initiates labour. Inflammatory cytokines and chemokines such as TNF-α, IL-1 β and IL-8, increase in the decidual microenvironment including amniotic fluid and foetal membranes. This induces signals for innate immune cells to become activated [44, 45].

Upon initiation of a pro-inflammatory cascade including NF-κB activation, uterine immune cells produce inflammatory chemokines and cytokines. Increased uterine activation of transcription by NF-κB, leads directly to high levels of cyclooxygenase-2 (COX-2), prostaglandin E2 (PGE2), gap-junction protein connexion 43 and up-regulation of oxytocin receptors [46].

The role of PG, namely PGE2 and prostaglandin F2 alpha (PGF2α), is of central importance in the initiation of labour, as they promote the proliferation of gap-junctions at myometrial level (which permits rapid and generalized membrane depolarization and uterine global contractions) and modifications of the extracellular matrix in the cervix (that allows for passive dilatation) [47].

The amnion and the chorion are important local sources of arachidonic acid, and the intracellular activation of the enzymatic pathways of prostaglandin synthesis in the foetal membrane, with extension to the decidua, is an essential step in the process of labour. It is uncertain if the first stimulus goes from the decidua to the foetal membranes, or if it works the other way around.

The arachidonic acid formed is then converted to prostaglandin H (PGH) via the enzyme prostaglandin synthase (PGHS); primary prostaglandins (PG) are inactivated by hydroxyprostaglandin dehydrogenase (PGDH).

In human amnion and decidua there is increased production of PG during parturition. Even more, it seems that there is an increase of PG in amniotic fluid and in maternal plasma after labour has started [47].

Recent studies have clearly shown that, in both the amnion and chorion laeve, there is a marked increase in PGHS activity during labour and that this is due to increased expression of the PGHS-2 isomer in fetal tissues. But it seems that PGHS-2 expression in the decidua is absent, indicating that in this maternally derived tissue, PGHS-1 isoenzyme may be of more significance in relation to PG information than the PGHS-2 isoenzyme.

The chorion laeve possesses a very active PGDH and acts as a barrier between PG formed in the amnion and the chorion itself, and their transfer to the decidua and hence to the myometrium [47].

Challis et al. [36] suggested that in the chorion PGDH may be important in regulation of PG availability within the uterus. Indeed, subsequent studies demonstrated that there is a decrease in PGDH expression in fetal membranes in the lower uterine segment covering the cervix, suggesting that this may allow PG from this area of the membranes to access the cervix and result in cervical ripening. Changes may then occur in PGDH activity in the fundal portion of the uterus as labour progresses, allowing active PG to reach the myometrium and cause contractions [47].

Moreover, glucocorticoids may have a dual role increasing PG formation via stimulation of PGHS-2 expression and, at the same time, decreasing PG metabolism by inhibiting PGDH expression. Besides, exciting studies also revealed that progesterone produced locally within the chorion laeve is responsible for maintaining PGDH activity [47].

Or is it the foetus that ultimately controls the moment of birth in normal circumstances? It seems that Hypocrates thought that the baby ruptured the membranes and forced labour in search for the nourishment that the mother's womb could no longer adequately provide. Seemingly naive, or almost of magic nature, the explanation has found more modern formulations through the notion that foetus has direct control of amniotic fluid' composition (and thus indirectly of local PG synthesis), which could account for the essential chronological coordination between foetal maturity and birth.

5.3 From Inflammation to Preterm Birth

If, as stated above, normal parturition at term results from the activation of inflammatory mechanisms and prostaglandin synthesis by the amnio-chorio-decidual unit, this process may be abnormally initiated out of time by any interference

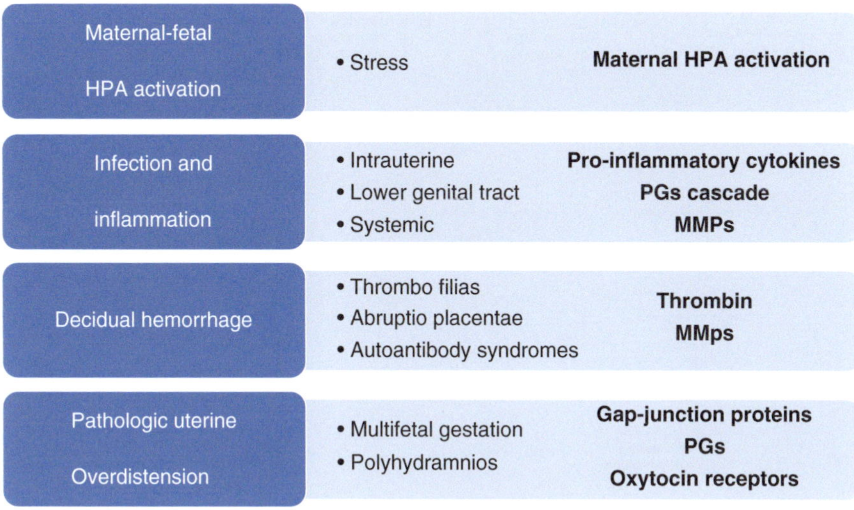

Fig. 5.6 Etiologies and pathways leading to spontaneous preterm birth

having this pro-inflammatory potential, as membrane mechanical rupture or infectious agents.

Due to the immediacy with which the onset of labour takes place and the necessary shift from anti to pro-inflammatory signal cascades, it is not surprising that unscheduled parturition is one of the most menacing complications of pregnancy, with the resultant adverse perinatal outcomes [37].

PTB accounts for 70 % of all perinatal mortality. Data from human studies provide data consistent with bacterial infection resulting in spontaneous PTB. Moreover, preterm deliveries (in the larger group of spontaneous PTB) and PPROM are often associated with intra-uterine inflammation or chorioamnionitis (CA) [48, 49].

PTB can result from a range of causes such as exposure to environmental triggers, maternal stress, foetal or maternal genetic abnormalities, or hormonal imbalance (Fig. 5.6). However, infection is one of the most heralded causes of PTB due to the drastic link between underlying infectious agents and their ability to promote inflammatory responses [37]. Nonetheless, in the last 10 years there is increasing evidence also for a genetic predisposition of the mother as a potential cause for preterm delivery, which of course could also be mediated through the characteristics of her innate inflammatory responses [50].

While the evidence for infection mediated PTB is substantial, the underlying mechanisms that induce early birth in response to pathogenic presence remain vague (Fig. 5.7). Investigation into the mechanisms that lead to PTB in response to pathogenic agents should take into account several factors [37]:

- Route of entry: determines where the agent will subsist and what pathways will be activated; the same pathogen delivered by alternative routes can lead to differential inflammatory responses;
- Different pathogens may elicit diverse inflammatory responses.

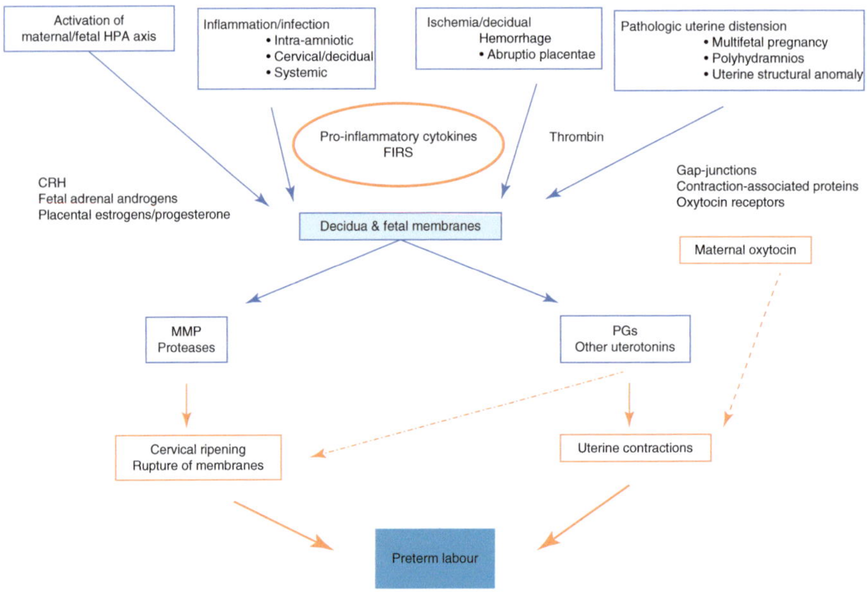

Fig. 5.7 Pathways to preterm birth

It is rarely the foreign organism that directly causes preterm birth; rather, it is the immune response of the host evoked by the pathogen that leads to aberrant pregnancy outcomes [37].

There are numerous data that support the hypothesis of an inflammation-triggered inadequate immunologic response with a consecutively increased risk of PT delivery. It seems that gene-environment interactions play a significant role in determining the risk of PTB. Polymorphisms of certain critical genes may be responsible for a harmful inflammatory response in those that possess them. Accordingly, polymorphisms that increase the magnitude or the duration of inflammatory response (TNF2 allele, IL-1 RA2) were associated with an increased risk of PTB [50].

Throughout the literature, bacteria show a stronger correlation with increased incidence of PTB compared to virus. This may be due to the differential sites of infection. Usually bacteria are found in mucosal membranes that surround the amniotic sac or those lining the intrauterine canal; on the other hand, viruses, needing the host cell machinery for replication, tend to infect trophoblast cells of the placenta, as these cells possess specific receptors needed for viral particle entry [51, 52].

A very plausible explanation for initiation of distinct immune pathways is probably the activation of Toll-like receptors (TLRs) [37].

TLRs are a diverse group of innate immune sentinel receptors evolutionarily conserved, with each TLRs (1–10) being specific for a different pathogen associated molecular pattern (PAMP). Importantly, TLRs are expressed on trophoblast and uterine immune cells. So, it is likely that differential uterine immune responses

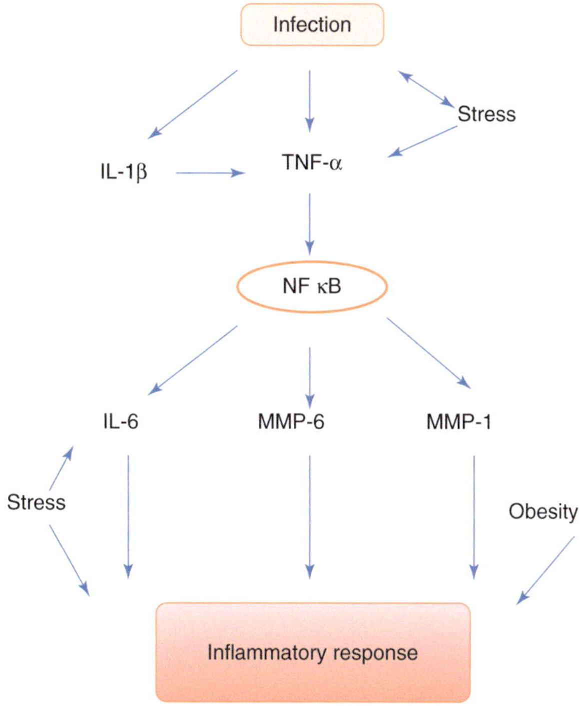

Fig. 5.8 Schema of the patho-physiological mechanisms leading to the induction of labour and dilatation of the cervix, in the presence of ascending infection

occur due to the diversity of pathogens that ensue activation of any one of these TLRs, ultimately leading to deleterious inflammation and PTB [37].

Accordingly, evidence demonstrates that the activated TLR pathways and the route of pathogenic entry (intrauterine ascension versus systemic infection) may determine the immunological cascade of aberrant cellular and cytokine activity that lead to PTB. The majority of these pathways lead to an increase in the NF-KB activity that allows the production of inflammatory cytokines and chemokines [46].

Current literature seems to suggest that systemic inflammatory responses might induce TNF-α trough NF-KB pathway to activate events leading to PTB (Fig. 5.8). In contrast, in the intra-uterine setting, the mode of action is mostly likely TNF-α independent [37].

5.4 Intra-uterine Infection

Ascending genital infection is the most frequent mechanism of intra-uterine infection and represents an important risk factor for preterm labour, PPROM and preterm delivery before 32 weeks of gestation. It occurs when pathogenic bacteria pass the cervical barrier causing decidual and chorioamniotic inflammation characterized by bacterial infection of the amniotic fluid [53].

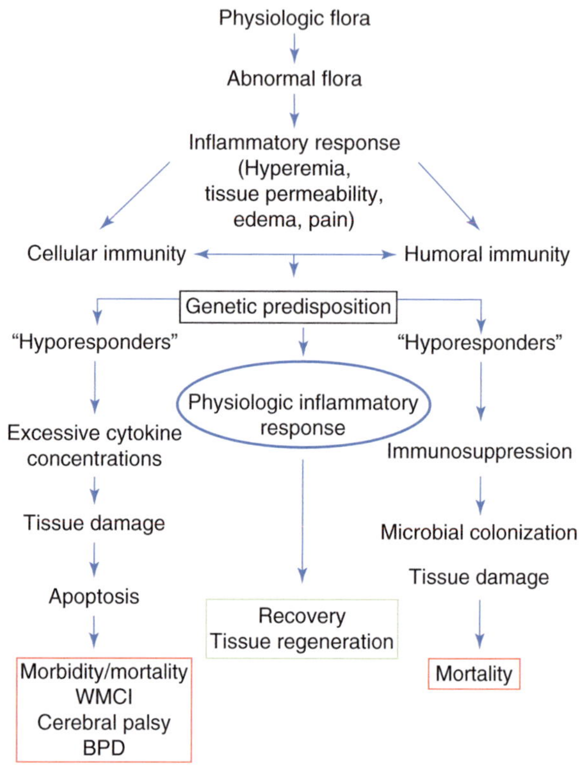

Fig. 5.9 Perinatal morbidity and mortality associated with inadequate inflammatory response to an infectious stimulus

Interestingly, several reports note that bacterial agents are rarely found at the placental level, in contrast to viral pathogens [18, 54]. Evidence suggests that viral entry into trophoblast cells induces trophoblast apoptosis and the resultant inflammatory events can lead to PTB (Fig. 5.9).

A four-stage process has been described by Romero and Mazor [55–57]. The first stage corresponds to modification in the vaginal and cervical flora, with proliferation of pathogenic microorganisms. The second stage corresponds to the contamination of the chorion-decidual space that can induce a local inflammatory reaction in the decidua. Infection of chorion-amniotic cavity through transmembrane passage of microorganisms corresponds to the third stage. Finally, it's possible that inhaling and/or swallowing the infected amniotic fluid infects the foetus.

Both maternal and foetal immune systems have important roles in CA, given the association established between polymorphisms in immunoregulatory genes and the risk of CA and PTB [50]. The presence of infectious agents in the chorioamnion engenders a maternal and fetal inflammatory response characterized by the release of a combination of pro-inflammatory and inhibitory cytokines and chemokines in the maternal and foetal compartments [58].

When infection is limited to the decidua or the amniochorion space (localized inflammation confined to chorion-decidua), the inflammatory process is detected within the membranes (histologic CA), is of maternal origin and is denoted amnionitis. The next stage is microbial invasion of the amniotic cavity through amnion (inflammation in amnion or chorionic plate without funisitis), being named CA [37]; the intra-amniotic inflammatory process appears to be of foetal rather than maternal origin. CA alone induces maternal systemic inflammatory reaction which clinically presents as amnion infection syndrome (AIS) causing pyrexia and elevated inflammatory markers in the mother alone; unfortunately, early detection is difficult [50].

In the final stage, funisitis, there is foetal invasion by microorganisms which elicits fetal inflammatory response characterized by infection or inflammation of the umbilical cord [37, 58, 59]. This is referred to foetal inflammatory response syndrome (FIRS) which can be proven histologically by the presence of funisitis or biochemically by the detection of elevated IL-6 serum levels during the perinatal period [50]. In the presence of FIRS, there is a dramatic increase of foetal and neonatal morbidity compared to CA alone [50].

5.4.1 Chorioamnionitis

Multiple studies have reported risk factors for CA, including longer duration of membrane rupture, prolonged labour, nulliparity, african american ethnicity, internal monitoring of labour, multiple vaginal exams, meconium-stained amniotic fluid, smoking, alcohol or drug abuse, immune-compromised states, epidural anaesthesia, colonization with group B streptococcus, bacterial vaginosis, sexually transmissible genital infections and vaginal colonization with ureaplasma [58].

CA is an acute inflammation of the membranes, typically due to ascending polymicrobial bacterial infection in the setting of membrane rupture [58]. It complicates as many as 40–70 % of PTB with PPROM or spontaneous labour [60] and 1–13 % of term births [61].

Infection can also occur with intact membranes and it appears to be very common for low virulence organisms including Ureaplasma spp. and Mycoplasma spp [48], found in the lower genital tract over 70 % of women. Rarely, as for Listeria monocytogenes, is haematogenous spread implicated in CA [58]. Foetal membranes aggressed by these bacteria produce cytokines; cytokines accelerate production and secretion of PGs and MMP. PGs cause myometrial contractions and MMP lyses chorioamniotic membranes, ultimately resulting in membrane rupture [49].

Amnionitis is the final stage of extraplacental chorioamniotic inflammation. Patients with amnionitis have a more advanced form of intrauterine infection/ inflammation, and thus a more intense fetal and intraamniotic inflammatory response than those with chorionitis alone. Amnionitis may reflect the presence of strong chemotactic stimuli located in the amniotic fluid rather than in the chorion/ decidua. Indeed, amnionitis is a better independent predictor of proven or suspected early-onset neonatal sepsis (OR 3.8; CI 95 % 1.1–13.3) than funisitis, considered as the final stage during ascending intrauterine infection [59].

Funisitis has the highest positive predictive value for intra-amniotic infection because it occurs late in the course of the infectious process and has been associated with an increased risk of neonatal infection-related complications and cerebral palsy [60].

Maternal inflammatory response may produce clinical CA and/or lead to prostaglandin release, ripening of the cervix, membrane injury and labour at term or premature birth at earlier gestational ages [58].

The presence of bacteria induces the release of pro-inflammatory cytokines (IL-1, TNF) by macrophages, amnion, decidua and myometrium. These cytokines, together with endotoxins released by Gram negative bacteria, induce an increase in the production of PG, endothelin and corticotropin – releasing hormone (CRH) in decidual, chorionic and amniotic cells that will further provoke uterine contractions [50, 55, 56].

Also, IL-1 and TNF- α can trigger the secretion of MMP from chorionic and cervical cells which induce the degradation of extracellular matrix of the lower uterine and cervix [50]. IL-1 in addition, mediates the release of IL-8 from decidual, chorionic, amniotic and cervical cells which leads to the activation and recruitment of elastase-producing granulocytes and further contributing to the modification of cervical extracellular matrix [62, 63]. In the end, through contractions and cervical dilatation, labour will ensue.

In order for these innate immune reactions to be possible, at least one pattern recognition receptor (PRR) must be present in the chorioamniotic membranes [49]. Moreover, TLR-4 can detect the lipopolysaccharide of Gram-negative bacteria and shows higher expression in the chorion and during the earlier stages of gestation [49].

Symptoms of local infection are uterine tenderness and purulent or foul amniotic fluid [58]. If and when CA induces a systemic inflammatory reaction in pregnant woman, clinical and laboratorial signs will become apparent, with fever, tachycardia and elevated inflammation protein markers in the maternal compartment [49].

The cervical mucus plug as well as the placenta and membranes provide a barrier to infection of the amniotic fluid and the foetus. Also, peroxide producing lactobacilli in the birth canal may induce changes in the flora that impairs the virulence of the pathogenic organism.

Notwithstanding, host defence mechanisms preventing intra-amniotic infection remain poorly elucidated [58].

5.4.2 Foetal Involvement

The proportion of preterm infants exposed to CA increases with decreasing gestational age, to up to 80 % below 28 weeks gestational age [64].

Foetal exposure to infection may lead to perinatal death, neonatal sepsis and other postnatal complications.

When intrauterine inflammation is present, the foetus may be exposed through direct contact with amniotic fluid or through the foetal-placental circulation. The consequent response to CA has been referred to as FIRS [65].

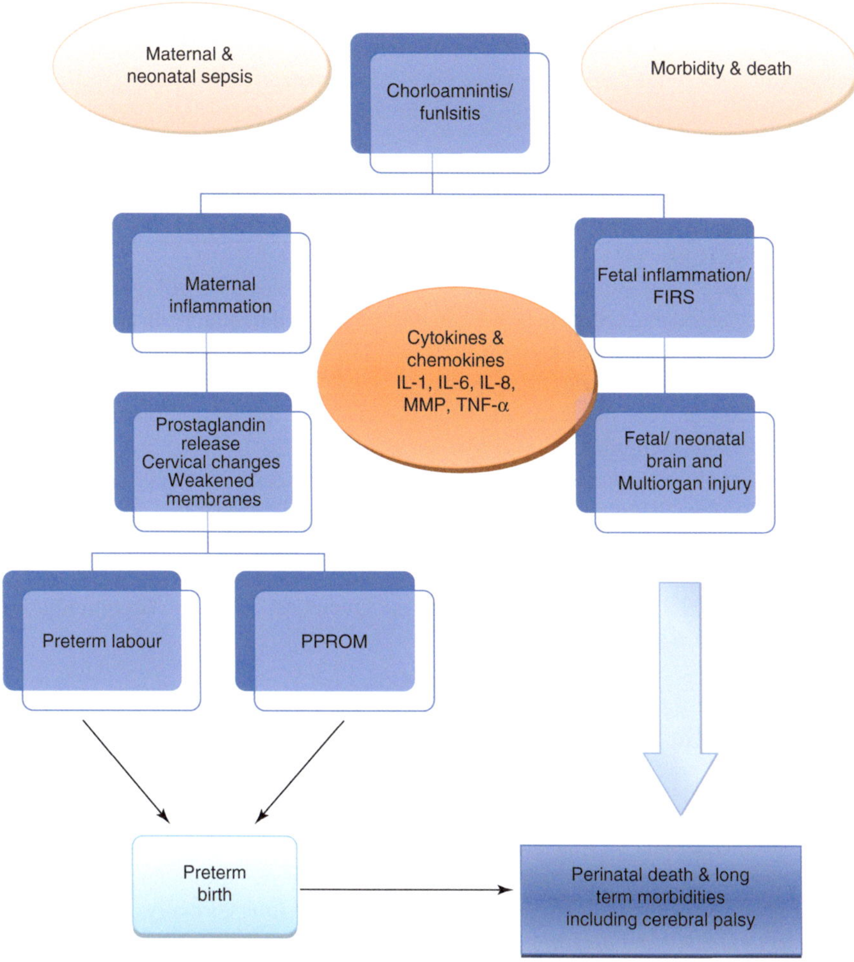

Fig. 5.10 Perinatal morbidity and mortality associated with inadequate inflammatory response

Numerous publications confirm the neurotoxic effect of bacterial endotoxins and proinflammatory cytokines on the fetal brain. Moreover, infection is associated with a reduced cardiovascular regulation, which contributes significantly to perinatal morbidity [50].

FIRS may induce cerebral white matter injury (CWMI), which may result in spastic cerebral palsy and short and long term neurological deficits [58]. In view of that, if FIRS occurs, the risk of developing CWMI or cerebral palsy is 11× greater [50].

So, cerebral lesions can have their origin before the rupture of membranes and be the consequence of the same process that initiates membrane lesions (Fig. 5.10) [1].

Although earlier studies focused mainly on neurological and respiratory outcomes, additional sequelae of CA related FIRS have more recently been described in several other areas of the foetal organism, turning it into a multi-organ disease of the foetus [49]. Moreover, evidence is increasing that the effects of CA/FIRS on health and disease may extend beyond the neonatal period [48].

In fact, direct and indirect exposure of the foetus to contaminated amniotic fluid may cause:

1. Respiratory problems: after intrauterine inflammation preterm lungs are more susceptible to secondary injury, leading to chronic lung disease (bronchopulmonary dysplasia);
2. Neurological problems: development delay and lifelong neurological disabilities, such as mental retardation, cerebral palsy, as well as school and behavioural difficulties. In addition, perinatal brain damage might increase the risk for the development of psychiatric disorders such as schizophrenia later in life. Diffuse or focal white matter disease has been shown to be the predominant feature of brain damage in premature newborns below 32 weeks;
3. Ocular problems: retinopathy of prematurity is a disease of the eye that is characterized by a disorganized growth of retinal blood vessels that may result in scarring and retinal detachment;
4. Endocrine problems: transient hypothyroxinemia of prematurity (temporarily low levels of T3 and T4) and foetal thymus involution (strongly associated with foetal infection and funisitis) [48].

5.5 Amniotic Fluid Volume Anomalies

Amniotic fluid (AF) is necessary for human growth and development. It protects the fetus from mechanical trauma and its bacteriostatic properties may help to maintain a sterile intrauterine environment. The space created by AF allows fetal movement and supports the normal development of both lungs and limbs. Finally, AF offers an access to fetal cells and metabolic products, useful for fetal diagnosis [66].

In the first trimester the amnion does not contact the chorion nor the placenta, and the amniotic cavity is surrounded by the exocoelomic cavity [67]. The exocoelomic fluid participates in the exchange of molecules between mother and fetus and at this stage, AF function is uncertain.

By the end of the first trimester of human gestation, the exocoelomic cavity has been obliterated and the amniotic cavity becomes the only significant deposit of extrafetal fluid. In fact, during the first half of pregnancy, amniotic fluid is an extension of foetal extracellular space. Later, as the foetal skin becomes keratinized, the production and renewal of amniotic fluid is mainly determined by foetal urine production and swallowing. AF volume increases between 10 and 30 weeks of gestation; after 30 weeks, this expansion slows down, and AF volume may remain unchanged until 36–38 weeks, after which the volume tends to decrease [66].

The volume of AF may be dramatically altered in pathologic states (Table 5.1), conveying to excessive AF (polyhydramnios) or reduced AF volume (oligohydramnios). Apart from ruptured membranes or extreme maternal fluid imbalance,

Table 5.1 Amniotic fluid volume anomalies

Conditions associated with polyhydramnios	Conditions associated with oligohydramnios
Fetal	Fetal
Chromosomal abnormalities	Chromosomal abnormalities
Congenital gastrointestinal anomalies	Congenital anomalies
Central nervous system problems	Growth restriction
Non-immune hydrops fetalis (infections, heart failure)	Demise
Placenta	Post-term pregnancy
Twin-twin transfusion	Ruptured membranes
Maternal	Placenta
Diabetes	Abruption
Idiopathic	Twin-twin transfusion
	Maternal
	Uteroplacental insufficiency
	Hypertension
	Preeclampsia
	Diabetes
	Drugs
	Prostaglandin synthase inhibitors
	Angiotensin-converting enzyme inhibitors
	Idiopathic

variations in amniotic fluid are mainly dependent on foetal pathology and their consequences are also deleterious to the foetus: like cord compressions resulting from fluid reduction or the initiation of premature labour through uterine distension and/or membrane rupture.

5.6 Conclusions

- Foetal membranes delimitate an extensive portion of the border between the foetal and the maternal environments. Amniotic epithelium participates in the common pathway of the initiation of labour and human decidua contains abundant immune cells whose activities are central to the immunological interaction between mother and foetus.
- The balance between formation and degradation of membrane components is a physiological phenomenon that can be found throughout pregnancy, that when altered, may lead to membrane weakening and result in membrane rupture. Infection (the major mechanism), oxidative stress and membrane distension can contribute to weaken membranes, leading to PPROM.
- The foetus naturally expresses an antigenic identity that is in part of paternal origin and, thus, different from its mother. In reality, the amniotic membrane seems

to be an immune privileged tissue and to contain some immunoregulatory factors. The theoretical paradigm has evolved from one of maternal-foetal tolerance to that of a decidual-trophoblastic-amniotic tolerance and active cooperation: a successful pregnancy is the consequence of numerous complex interactions between the receptive uterus and the mature blastocyst, under immunohumoral control.

- Implantation and placentation in the first trimester and early second trimester of pregnancy require a strong inflammatory response. As so, during the early phases of pregnancy, successful implantation occurs in a pro-inflammatory microenvironment, and a Th1-type response is followed by a shift to Th2, to control endocrine and immune interactions.
- As with Implantation, normal parturition is characterized by mechanisms of local pro-inflammatory activity. In human amnion and decidua there is increased production of prostaglandins immediately before and during labour. This pro-inflammatory environment promotes the contraction of the uterus and the ultimate expulsion of the infant and of the placenta.
- Ascending genital infection is the most frequent mechanism of intra-uterine infection and represents an important risk factor for preterm labour and PPROM.
- The presence of infectious agents in the chorioamnion engenders a maternal and fetal inflammatory response characterized by the release of a combination of pro-inflammatory and inhibitory cytokines and chemokines in the maternal and foetal compartments. This may lead to a sequential inflammatory process ensuing in amnionitis, CA and/or funisitis.
- Intrauterine infection can also occur with intact membranes and it appears to be very common for low virulence organisms.
- Amnionitis is the final stage of extraplacental chorioamniotic inflammation. Funisitis, considered as the final stage during ascending intrauterine infection, has been associated with an increased risk of neonatal infection-related complications and cerebral palsy.
- FIRS is an inflammatory response characterized by infection or inflammation of the umbilical cord when there is foetal invasion by microorganisms.
- FIRS may induce CWMI, which may result in spastic cerebral palsy and short and long term neurological deficits. Consequently, foetal exposure to infection may lead to perinatal death, neonatal sepsis and other postnatal complications. Moreover, evidence is increasing that the effects of CA/FIRS on health and disease may extend beyond the neonatal period.

References

1. Pasquier J, Doret M (2008) Fetal membranes: embryological development, structure and the physiopathology of the preterm premature rupture of membranes. J Gynecol Obstet Biol Reprod (Paris) 37:579–588
2. Cunningham FG, Leveno KJ, Bloom SL et al (2009) Implantation, embryogenesis and placental development. In: Cunningham FG, Leveno KJ, Bloom SL et al (eds) Williams Obstet, 23rd edn. McGraw-Hill Professional, Slock, London, p 1404

3. Chen S-J, Liu Y-L, Sytwu H-K (2012) Immunologic regulation in pregnancy: from mechanism to therapeutic strategy for immunomodulation. Clin Dev Immunol 258391:10
4. Bulmer J, Morrison L, Longfellow M et al (1991) Granulated lymphocytes in human endometrium: histochemical and immunohistochemical studies. Hum Reprod 6:791–798
5. Hanssens S, Salzet M, Vinatier D (2012) Immunological aspect of pregnancy. J Gynecol Obstet Biol Reprod (Paris) 41:595–611
6. Von Rango U (2008) Fetal tolerance in human pregnancy – a crucial balance between acceptance and limitation of trophoblast invasion. Immunol Lett 115:21–32
7. Mor G, Abrahams VM (2008) The immunology of pregnancy. Creasy and Resnik's Maternal-Fetal Medicine: Principles and Practice, 8th edition Elsevier Health Sciences, Philadelphia, 87–99
8. Craven C, Morgan T, Ward K (1998) Decidual spiral artery remodelling begins before cellular interaction with cytotrophoblasts. Placenta 19:241–252
9. Kammerer U, Kruse A, Barrientos G et al (2008) Role of dendritic cells in the regulation of maternal immune responses to the fetus during mammalian gestation. Immunol Invest 37:499–533
10. Plaks V, Birnberg T, Berkutzki T et al (2008) Uterine DCs are crucial for decidua formation during embryo implantation in mice. J Clin Invest 118:3954–3965
11. Ross MG (2011) Inflammatory mediators weaken the amniotic membrane barrier through disruption of tight junctions. J Physiol 589:5
12. Gaudin P, Troame C, Monier F et al (1998) Protéolyse matricielle, protéolyse focalisée et inflammation. Ann Biol Clin (Paris) 56:661–669
13. Menon R, Fortunato SJ (2004) The role of matrix degrading enzymes and apoptosis in repture of membranes. J Soc Gynecol Investig 11:427–437
14. Vinatier D, Dufour P, Subtil D (1996) Apoptosis: a programmed cell death involved in ovarian and uterine physiology. Eur J Obstet Gynecol Reprod Biol 67:85–102
15. Bryant-greenwood GD, Millar LK (2000) Human fetal membranes: their preterm premature rupture. Biol Reprod 63:1575–1579
16. Mercer BM, Goldenberg RL, Das AF et al (2003) What we have learned regarding antibiotic therapy for the reduction of infant morbidity after preterm premature rupture of the membranes. Semin Perinatol 27:217–230
17. Newton ER (2005) Preterm labor, preterm premature rupture of membranes and chorioamnionitis. Clin Perinatol 32:571–600
18. Goldenberg RL, Hauth JC, Andrews WW (2000) Intrauterine infection and preterm delivery. N Engl J Med 342:1500–1507
19. Chibbar R, Wong S, Miller F, Mitchell B (1995) Estrogen stimulates oxytocin gene expression in human chorio-decidua. J Clin Endocrinol Metab 80:567–572
20. Oh S-Y, Kim CJ, Park I et al (2005) Progesterone receptor isoform (A/B) ratio of human fetal membranes increases during term parturition. Am J Obstet Gynecol 193:1156–1160
21. Weiss G, Goldsmith LT (2005) Mechanisms of relaxin-mediated premature birth. Ann N Y Acad Sci 1041:345–350
22. Petersen L, Skajaa K, Uldbjerg N (1992) Serum relaxin as a potential marker for preterm labour. Br J Obstet Gynaecol 99:292–295
23. Vogel I, Goepfert AR, Møller HJ et al (2006) Early mid-trimester serum relaxin, soluble CD163 and cervical length in women at high risk for preterm delivery. Am J Obstet Gynecol 195:208–214
24. Lowndes K, Amano A, Yamamoto S, Bryant-Greenwood G (2006) The human relaxin receptor (LGR7): expression in the fetal membranes and placenta. Placenta 27:610–618
25. Rocha FG, Slavin TP, Li D et al (2013) Genetic associations of relaxin: preterm birth and premature rupture of fetal membranes. Am J Obstet Gynecol 209:258.e1–258.e8
26. Ulm B, Ulm M, Bernaschek G (1999) Unfused amnion and chorion after 14 weeks of gestation: associated fetal structural and chromosomal abnormalities. Ultrasound Obstet Gynecol 13:392–395
27. Woods JR, Plessinger MA, Miller RK (2001) Vitamins C and E: missing links in preventing preterm premature rupture of membranes? Am J Obstet Gynecol 185:5–10

28. Moore R, Mansour J, Redline R et al (2006) The physiology of fetal membrane rupture: insight gained from the determination of physical properties. Placenta 27:1037–1051
29. Iams J, Stilson R, Johnson F et al (1990) Symptoms that precede preterm labor and preterm premature rupture of the membranes. Am J Obstet Gynecol 162:486–490
30. Arikat S, Novince RW, Mercer BM et al (2006) Separation of amnion from choriodecidua is an integral event to the rupture of normal term fetal membranes and constitutes a significant component of the work required. Am J Obstet Gynecol 194:211–217
31. Medawar P (1953) Some immunological and endocrinological problems raised by the evolution of viviparity in vertebrate. Symp Soc Exp Biol 7:320–338
32. van Nieuwenhoven AV, Heineman M, Faas M (2003) The immunology of successful pregnancy. Hum Reprod Update 9:347–357
33. Kubo M, Sonoda Y, Muramatsu R, Usui M (2001) Immunogenicity of human amniotic membrane in experimental xenotransplantation. Invest Ophthalmol Vis Sci 42:1539–1546
34. Mor G (2011) Clinical aspect of reproductive immunology. Am J Reprod Immunol 66:451
35. Saito S, Nakashima A, Shima T, Ito M (2010) Th1/Th2/Th17 and regulatory T-cell paradigm in pregnancy. Am J Reprod Immunol 63:601–610
36. Challis JR, Lockwood CJ, Myatt L et al (2009) Inflammation and pregnancy. Reprod Sci 16:206–215
37. Thaxton JE, Nevers TA, Sharma S (2010) TLR-mediated preterm birth in response to pathogenic agents. Infect Dis Obstet Gynecol 378472:1–8
38. Van Mourik MS, Macklon NS, Heijnen CJ (2009) Embryonic implantation: cytokines, adhesion molecules and immune cells in establishing an implantation environment. J Leukoc Biol 85:4–19
39. Norman JE, Bollapragada S, Yuan M, Nelson SM (2007) Inflammatory pathways in the mechanism of parturition. BMC Pregnancy Childbirth 7:1–4
40. Mendelson CR (2009) Minireview: fetal-maternal hormonal signaling in pregnancy and labor. Mol Endocrinol 23:947–954
41. Kelly RW, King AE, Critchley HO (2001) Cytokine control in human endometrium. Reproduction 121:3–19
42. Creasy RK, Resnik R, Iams JD et al (2008) Creasy and Resnik's maternal-fetal medicine: principles and practice, 7th edn. Elsevier Health Sciences, Philadelphia
43. Behrman RE, Butler AS (2007) Biological pathways leading to preterm birth. In: Behrman RE, Butler AS (eds) Preterm birth causes, consequences prev. The National Academies Press, Washington, p 790
44. Protonotariou E, Chrelias C, Kassanos D et al (2010) Immune response parameters during labor and early neonatal life. In Vivo (Brooklyn) 24:117–124
45. Romero R, Brody DT, Oyarzun E et al (1989) Infection and labor. III. Interleukin-1: a signal for the onset of parturition. Am J Obstet Gynecol 160:1117–1123
46. Lindström TM, Bennett PR (2005) The role of nuclear factor kappa B in human labour. Reproduction 130:569–581
47. Gibb W (1998) The role of prostaglandins in human parturition. Ann Med 30:235–241
48. Gantert M, Been J, Gavilanes A et al (2010) Chorioamnionitis: a multiorgan disease of the fetus? J Perinatol 30:S21–S30
49. Choi SJ, Jung S-H, Eom M et al (2012) Immunohistochemical distribution of toll-like receptor 4 in preterm human fetal membrane. J Obstet Gynaecol Res 38:108–112
50. Holst D, Garnier Y (2008) Preterm birth and inflammation: the role of genetic polymorphisms. Eur J Obstet Gynecol Reprod Biol 141:3–9
51. Parry S, Holder J, Strauss JF III (1997) Mechanisms of trophohlast-virus interaction. J Reprod Immunol 37:25–34
52. Arechavaleta-Velasco F, Koi H, Strauss JF, Parry S (2002) Viral infection of the trophoblast: time to take a serious look at its role in abnormal implantation and placentation? J Reprod Immunol 55:113–121
53. Gomez R, Ghezzi F, Romero R et al (1995) Premature labor and intra-amniotic infection. Clinical aspects and role of the cytokines in diagnosis and pathophysiology. Clin Perinatol 22:281–342

54. Romero R, Sirtori M, Oyarzun E et al (1989) Infection and labor. V. Prevalence, microbiology and clinical significance of intraamniotic infection in women with preterm labor and intact membranes. Am J Obstet Gynecol 161:817–824
55. Romero R, Mazor M, Brandt F et al (1992) Interleukin-1 alpha and interleukin-1 beta in preterm and term human parturition. Am J Reprod Immunol 27:117–123
56. Romero R, Mazor M, Sepulveda W et al (1992) Tumor necrosis factor in preterm and term labor. Am J Obstet Gynecol 166:1576–1587
57. Romero R, Mazor M (1988) Infection and preterm labor. Clin Obstet Gynecol 31:553–584
58. Tita AT, Andrews WW (2010) Diagnosis and management of clinical chorioamnionitis. Clin Perinatol 37:339–354
59. Park C-W, Moon KC, Park JS et al (2009) The involvement of human amnion in histologic chorioamnionitis is an indicator that a fetal and an intra-amniotic inflammatory response is more likely and severe: clinical implications. Placenta 30:56–61
60. Yoon BH, Romero R, Bin MJ et al (2001) Clinical significance of intra-amniotic inflammation in patients with preterm labor and intact membranes. Am J Obstet Gynecol 185:1130–1136
61. Seong H, Lee S, Kang J et al (2008) The frequency of microbial invasion of the amniotic cavity and histologic chorioamnionitis in women at term with intact membranes in the presence or absence of labor. Am J Obstet Gynecol 199:1–9
62. Lockwood C (1994) Recent advances in elucidating the pathogenesis of preterm delivery, the detection of patients at risk, and preventative therapies. Curr Opin Obstet Gynecol 6:7–18
63. Rechberger T, Woessner JJ (1993) Collagenase, its inhibitors, and decorin in the lower uterine segment in pregnant women. Am J Obstet Gynecol 168:1598–1603
64. Lahra M, Beeby P, Jeffery H (2009) Maternal versus fetal inflammation and respiratory distress syndrome: a 10-year hospital cohort study. Arch Dis Child Fetal Neonatal Ed 94:13–16
65. Gotsch F, Romero R, Kusanovic J et al (2007) The fetal inflammatory response syndrome. Clin Obstet Gynecol 50:652–683
66. Ross MG, Beall MH (2008) Amniotic fluid dynamics. Creasy and Resnik's Maternal-Fetal Medicine: Principles and Practice, 8th edition Elsevier Health Sciences, Philadelphia, 47–54
67. Calvo RM, Jauniaux E, Gulbis B et al (2002) Fetal tissues are exposed to biologically relevant free thyroxine concentrations during early phases of development. J Clin Endocrinol Metab 87:1768–1777

Part II
Medical Applications of Amniotic Membrane

Chapter 6
Amniotic Membrane in Ophthalmology

Esmeralda Costa and Joaquim Neto Murta

Abstract Human amniotic membrane has been widely used in ophthalmology for decades, particularly in ocular surface reconstruction. This article reviews its mechanism of action, surgical principles and clinical applications. The fate of transplanted amniotic membrane in the eye and its limitations will also be addressed.

Keywords Amniotic membrane • Ophthalmology • Clinical applications • Ocular surface • Ocular reconstruction

6.1 Historical Background

Amniotic membrane has been used in medicine for therapeutic purposes since the beginning of the twentieth century. The first reported application was in skin reconstruction in 1910, when Davis used fetal membranes (i.e. both chorion and amnion) to treat burned and ulcerated skin [1, 2]. Its use in ophthalmology dates back to 1940 when de Rötth used fresh fetal membranes in conjunctival reconstruction [1, 2]. The rationale for its use was based on the recognition that two granulation surfaces tend to adhere to each other unless there is an epithelial layer between them [1]. The thinness, smoothness and transparency of the amniotic membrane made it an ideal surrogate for diseased conjunctiva [1]. Even so, de Rötth obtained only limited success, which he attributed to the deleterious effect of the presence of chorion on graft survival [1].

E. Costa (✉)
Faculty of Medicine, University of Coimbra, Coimbra, Portugal
e-mail: esmeraldaguedes@gmail.com

J.N. Murta
Faculty of Medicine, University of Coimbra, Coimbra, Portugal

Centre of Integrated Responsibility in Ophthalmology,
Coimbra Hospital and Universitary Center, Coimbra, Portugal

A.C. Mamede, M.F. Botelho (eds.), *Amniotic Membrane: Origin, Characterization and Medical Applications*,
DOI 10.1007/978-94-017-9975-1_6

In 1946, Sorsby and Symons reported the success of human amniotic membrane grafts in the treatment of severe ocular burns [3]. These authors were the first to suggest that amniotic membrane may have biological properties, in addition to acting as a mechanical barrier. Despite the good outcomes with a low index of complications, there were no reports on the ophthalmological use of human amniotic membrane until 1995, when Kim and Tseng used it to treat limbal deficiency in an experimental model [4]. Since then, interest in human amniotic membrane has seen a resurgence in ophthalmology, particularly in ocular surface reconstruction [1, 2].

6.2 Mechanism of Action

The exact mechanism of action of human amniotic membrane is still unknown [1, 5]. It is probably best regarded as a physical structure with biological properties. Amniotic membrane has anti-angiogenic, anti-inflammatory, antibiotic and anti-viral properties; it promotes epithelial migration, inhibits fibrosis and is non-immunogenic [1, 2, 6–8]. Some of these features, particularly the ability to reduce inflammation and fibrosis and promote epithelial migration, have motivated its use in ophthalmology. In vivo, human amniotic membrane is known to facilitate ocular surface healing with minimal inflammation, thus reducing scarring, adhesion of tissues and vascularization [1]. These properties probably have multiple mechanisms, some of which have been demonstrated, namely: human amniotic membrane can act as a basement membrane [1, 8, 9]; it can trap and induce apoptosis of immuno-inflammatory cells [1, 10, 11]; it can downregulate the transforming growth factor beta (TGF β) pathway and miofibroblastic differentiation of cultivated fibroblasts [1, 2]. These processes rely on the presence of several cytokines and growth factors in human amniotic membranes epithelial cells and stroma [1, 5, 12].

Clinically, human amniotic membrane may be used either as a scaffold or as a bandage for ocular surface reconstruction [5]. If placed within the lesion, stromal side down, deepithelialized human amniotic membrane functions as a basement membrane for corneal, conjunctival and limbal epithelial cell growth by facilitating epithelial migration [2, 5, 8, 9] and differentiation [2], thus preventing apoptosis [2, 8, 9] and reinforcing cell adhesions [1, 2, 5]. When placed over the ocular surface, human amniotic membrane acts as a biologic bandage that protects the healing epithelium from the mechanical action of the eyelids and, at the same time, its anti-inflammatory properties provide a barrier to inflammatory cells and proteins present in the tear film [2, 6, 13].

6.3 Clinical Applications

Human amniotic membrane has been used clinically in corneal and/or conjunctival reconstruction to promote epithelial healing and modulate inflammation and scarring. Its success in ocular surface reconstruction and its molecular similarities with

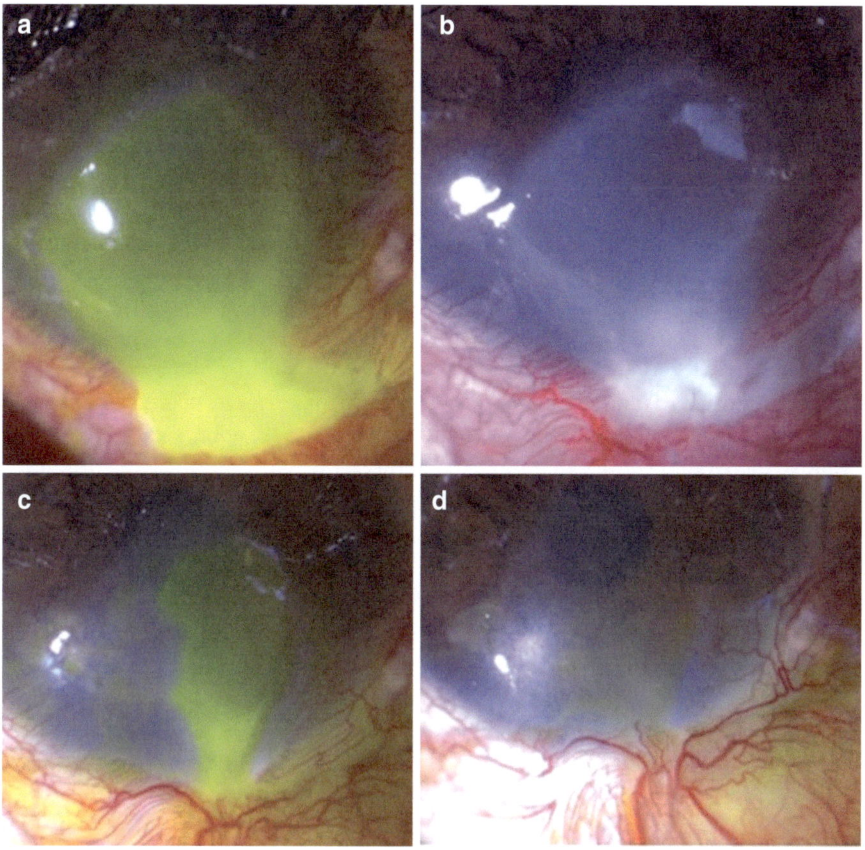

Fig. 6.1 Healing of a corneal and conjunctival ulcer after amniotic membrane transplantation (**a**–**d**)

the limbal niche prompted its use also as a substrate for limbal stem cell ex vivo expansion [8, 14].

Among the many numerous clinical conditions where human amniotic membrane can be used are ocular burns, corneal ulcers, pterygia, glaucoma surgery, and symblepharon. However, its therapeutic success has not been clearly demonstrated for most, if not all, of the current indications. In fact, we need to ascertain which pathologies can be considered true, clinically significant indications, by means of a meta-analysis of published reports and randomized clinical trials.

6.3.1 Corneal Applications

Corneal Ulcers and Perforations

Human amniotic membrane has been used in the treatment of persistent epithelial defects, corneal ulcers (including neurotrophic and shield ulcers), corneal melting, descemetoceles and perforations (Fig. 6.1) [1, 2, 5, 7, 12]. It is a surgical treatment option that is usually considered when medical therapy is insufficient [2]. Given its biological properties, human amniotic membrane helps restore corneal epithelium and stromal thickness, with reported success rates of 50–92.9 % [1, 2, 7, 12, 15–17]. The treatment of refractory corneal epithelial defects can prevent corneal perforation. Sealing corneal perforation with multilayered amniotic grafts (with or without tissue adhesive) may avoid the need for an emergency penetrating keratoplasty and, in some cases, it might even eliminate the need for a corneal transplant [13, 16]. However, not all corneal perforations can be treated with amniotic membrane grafts and only perforations up to 1.5–2 mm in diameter should be considered [2, 17]. Amniotic membrane transplantation has also been used as an adjunctive procedure in the treatment of infectious corneal ulcers and impending perforations, to promote wound healing and reduce inflammation [18]. In addition, amniotic membrane seems to have some degree of intrinsic antibiotic action [19]. Possible explanations include close adherence of the membrane to the wound surface that limits bacterial growth, the presence of antimicrobial peptides and protection from proteolytic enzymes released by bacteria and inflammatory cells [19]. Moreover, it has been demonstrated that human amniotic membrane does not interfere with the ocular penetration of topical antibiotics in corneas with epithelial defects, and antibiotic-soaked human amniotic membrane may function as an effective drug-delivery vehicle [19]. The amniotic membrane's lack of transparency is a potential drawback since it may hamper monitoring of the clinical evolution [2]. It is recommended that its application is preceded by a course of 2–7 days of intensive antimicrobial therapy [18, 19] and it is probably best if the membrane is soaked in antibiotic solution immediately prior to grafting. Its efficacy in fungal and acanthamoeba keratitis is doubtful since hyphae and cysts are very resistant to treatment and are known to persist in the stroma from early stages [2]. A major advantage of performing amniotic membrane grafting in cases of severe ulcerative keratitis is that it may help avoid an emergency keratoplasty and improves the prognosis of a posterior elective keratoplasty [20].

Painful Bullous Keratopathy

A healthy, properly functioning corneal endothelium is crucial to maintain corneal deturgescence and ensure good vision. If it is compromised, then excessive corneal hydration ensues and epithelial bullae form. The spontaneous rupture of these bullae can be very painful. The definite treatment is corneal transplant. Placing an amniotic membrane graft after epithelial debridement promotes the formation of a stable epithelium [21]. That is useful as a temporary measure to relieve symptoms until the transplant is done or as a permanent treatment in eyes with no visual potential. Epithelial healing is usually complete 2–3 weeks after surgery. Long-term success in pain relief ranges between 82.5 and 95.3 % [2, 22–28].

Limbal Stem Cell Deficiency (Acute or Chronic)

Human amniotic membrane has been used to modulate the inflammatory reaction, prevent tissue necrosis and minimize late sequelae (corneal and conjunctival scarring, vascularization, opacification and conjunctivalization of the cornea, recurrent and persistent epithelial defects) in the acute stage of inflammatory conditions that involve the ocular surface [2, 29]. These include chemical and thermal burns and toxic epidermal necrolysis/Stevens-Johnson syndrome [6].

Eventually, these diseases can lead to a state of chronic limbal stem cell deficiency that can present with variable degrees of severity. Mild cases are amenable to conservative measures. Severe cases usually require a limbal transplant to restore the limbal niche and a corneal transplant to restore corneal clarity. In this context, human amniotic membrane has been used as an adjunctive therapy to limbal and corneal grafts and as a substrate for ex vivo stem cell culture and subsequent transplantation [29].

Several case reports and clinical series have pointed out the beneficial effects of amniotic membrane transplantation (AMT) in the acute stage of ocular burns, especially when performed in the first 7–10 days [1, 2, 5–7, 12]. It has been suggested as a first-line therapeutic option, along with medical management [1]. Greater benefits were achieved in mild and moderate cases. AMT did not prevent late-onset complications in severe cases [12]. However, a recent meta-analysis on randomized trials comparing AMT with medical therapy alone failed to show definite evidence supporting its use [30]. This does not necessarily mean that AMT is devoid of interest. Heterogeneity in disease presentation, treatment, outcome measures and success definition probably affected the search for clear evidence regarding this treatment [30].

Encouraging results have been reported for amniotic membrane transplantation in the acute phase of Stevens-Johnson syndrome [31]. Despite the small number of treated eyes, AMT in conjunction with medical treatment seems to be effective in preserving an intact ocular surface with good visual outcomes in mid-to-long-term follow-up [31–42]. A case-control study of fifty-eight eyes concluded that AMT performed in the first 2 weeks was superior to medical management alone (success of 95.7 % versus 65.2 %, respectively), in moderate and severe cases [40]. The sutureless application of an amniotic membrane around a conformer or other medical device seems to be a good alternative, provided that the entire ocular surface is covered [37, 38, 42].

The treatment of established chronic limbal cell deficiency is challenging. Several therapeutic options are available and a treatment algorithm was proposed in accordance with disease severity and extent [29]. Amniotic membrane transplantation is useful as an adjunct to sequential sectoral epitheliectomy in partial limbal insufficiency and in limbal grafts since it promotes epithelial healing and acts as a bandage (Fig. 6.2) [5, 29]. The success of the treatment depends largely on the underlying pathology [5]. Chronic progressive diseases such as Stevens-Johnson syndrome and ocular cicatricial pemphygoid have a poor prognosis [5]. The success

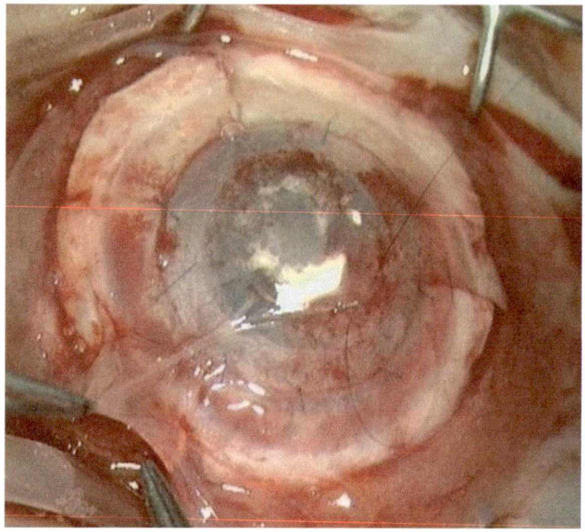

Fig. 6.2 Application of an amniotic membrane patch as an adjunct to keratolimbal allograft

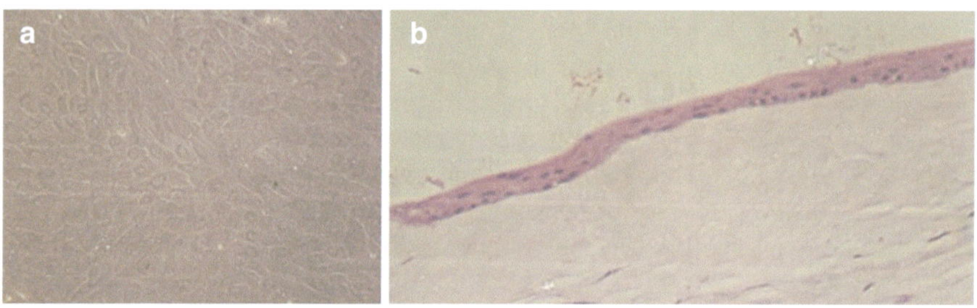

Fig. 6.3 Confluent sheet of ex vivo cultured limbal epithelial cells (**a** – inverted phase microscopy). In histology (**b**), amniotic membrane is seen underneath the multilayered epithelium

also differs according to the type of limbal transplant. Autologous grafts usually do better than allografts, but the availability of autologous tissue is conditioned by the extent of the disease and allogenic transplants are the only option for total bilateral deficient patients. Moreover, these procedures require large amounts of keratolimbal tissue and that can lead to iatrogenic limbal deficiency in living donors [12]. The advent of ex vivo limbal stem cell culture has enabled more patients to be treated with autologous tissue, since only a 1 mm^2 biopsy is required. In this case, amniotic membrane is used as a substrate for limbal epithelial precursors in ex vivo culture. When the culture has reached confluence, the cell sheet is grafted in the eye, together with the underlying amniotic membrane (Fig. 6.3).

Band Keratopathy

Band keratopathy is a chronic corneal degeneration characterized by the presence of calcium deposits throughout the epithelium, basement membrane, Bowman's layer and superficial stroma. It may be associated with systemic conditions such as sarcoidosis and chronic renal failure or with chronic ocular inflammations [43]. The whitish deposits cause vision loss and epithelial instability, leading to pain and photophobia. Treatment involves calcium removal but recurrence is common.

Amniotic membrane can be transplanted after surgical removal of calcific deposits, to promote epithelial healing. Superficial keratectomy can be aided by chemical chelation with ethylene-diamine-tetra acetic acid [44–47]. Alternatively, the deposits can be removed with excimer laser (phototherapeutic keratectomy) [43, 47]. The use of adjunctive amniotic membrane should be compared with keratectomy alone. Potential benefits seem to be limited to faster epithelialization, since it does not prevent recurrence of calcium deposition [6].

6.3.2 Conjunctival Applications

Pterygium

Human amniotic membrane is an alternative to conjunctival or limbal grafting associated with pterygium excision. It is usually considered when autografts are not feasible, such as in the presence of a filtering bleb, extensive areas of subconjunctival fibrosis and/or symblepharon, or when glaucoma surgery is likely. Also when large dissections are required, as in recurrent pterygia, there is not enough conjunctiva left to cover the defect and human amniotic membrane is an option [2, 13].

Reported recurrence rates after pterygium excision combined with amniotic grafts are higher than with conjunctival grafts [48, 49]. They range between 3.0 % 40.9 % for primary and 9.5 % 52.6 % for recurrent pterygia [50–56]. The superiority of conjunctival and limbal grafts in primary pterygia was demonstrated in a recent meta-analysis [48]. In recurrent pterygia, the study was inconclusive due to the reduced number of cases [48].

Symblepharon

Symblepharon is usually the result of an inflammatory process of the ocular surface. The aggression can be self-limited, such as an ocular burn, or chronically relapsing, such as Stevens-Johnson syndrome.

Amniotic membrane is useful both in symblepharon prevention and its treatment. A patch of amniotic membrane placed in the acute phase helps restrain the inflammatory reaction and may prevent the formation of adhesions. In general, self-limited conditions have a better prognosis [5, 57]. Once symblepharon is established, the treatment involves the lysis of the adhesions and placing a graft to reconstruct the conjuntiva. Otherwise, the adhesions would reform. Depending on the severity of the disease, symblepharon lysis can be associated with amniotic membrane alone or in combination with intraoperative mitomycin C and/or oral mucosal and conjunctival autografts [58–63]. Successful outcomes have been reported in 53.8–100 % of treated eyes [16, 62–64]. In chronic relapsing conditions, such as ocular cicatricial pemphigoid, initial effectiveness deteriorates slightly over time [65].

Conjunctival Reconstruction

Amniotic membrane has been used in ocular surface reconstruction after the excision of conjunctival tumors and other lesions [6, 66–70]. It is more cosmetically acceptable than mucosal grafts and its translucency allows post-operative tumor monitoring [2, 6, 12].

Amniotic membrane can be used as a substitute for tarsal conjunctiva after giant papillae resection and in the correction of cicatricial entropion, to help reepithelialization of the tarsal plate [2, 6, 7].

6.3.3 Glaucoma Surgery

Amniotic membrane's anti-scarring properties and its ability to act as a surrogate for conjunctiva have prompted its use in glaucoma surgery as an adjunct in trabeculectomy and in the treatment of bleb leakage [2, 6].

Attempts have been made to prevent fibrosis and failure of drainage procedures by placing amniotic membrane beneath the scleral flap [71, 72], between the flap and overlying conjunctiva [73–76] or simultaneously under and over the flap [77, 78].

Placing human amniotic membrane under the flap is associated with higher success rates, better intra-ocular pressure control and lower complication rates in primary trabeculectomies [72] and in refractory cases [71].

Amniotic membrane-shielded blebs showed favorable effects on wound healing and bleb survival, but there were no significant improvements compared with standard trabeculectomy [73–75]. In refractory cases, amniotic membrane over the flap also failed to improve the results obtained with trabeculectomy plus intra-operative mitomycin C [76].

Placing amniotic membrane under and over the scleral flap might be a useful option in refractory glaucoma, either isolated [78] or associated with intra-operative mitomycin C [77]. In these cases, glaucoma shunt surgery with adjunctive amniotic membrane may prove useful, as it seems to reduce the risk of bleb encapsulation [79]. Bleb leakage is a possible complication after glaucoma surgery and is usually addressed by conjunctival advancement. Amniotic membrane transplantation has been suggested as an alternative [80–84], but it seems prone to early releakage [82]. The single trial that compared the two surgical procedures demonstrated the superiority of conjunctival advancement [82–84]. A patch of hyperdry amniotic membrane using a tissue adhesive has been reported to successfully treat bleb leaks in two eyes [85].

6.3.4 *Strabismus, Orbit and Oculoplastic Surgery*

Amniotic membrane has been reported to be a useful adjunct in the management of restrictive strabismus by preventing the formation of adhesions [86–88], and in oculoplastic and orbital surgery, particularly in forniceal reconstruction, socket reconstruction for prosthetic fitting [5] and as a cover for extruded prosthesis [6, 89].

6.3.5 *Scleral Melt*

There are few reports on the use of amniotic membrane in the treatment of scleral ischaemia, melting and perforation. It has been applied as a patch to reduce scleral melting and inflammation and promote conjunctival healing [90] and as a multilayered graft to fill in scleral defects [91]. In addition, amniotic membrane can be combined with tenonplasty to promote revascularization of the ischemic areas [67, 92, 93], and with a tectonic corneal or scleral graft in cases of melting or perforation [90, 93, 94].

6.4 Surgical Technique

Depending on the desired therapeutic purpose, amniotic membrane can be placed as a graft (inlay) or as a patch (onlay or overlay) [5, 12, 13].

When human amniotic membrane is intended to act as a basement membrane for epithelial regeneration, it is placed as a graft within the limits of the diseased area, basement membrane side up. Epithelium will grow over it and the graft will become incorporated into the host tissue. The cellular debris and exudates within the defect must be removed, along with the epithelium from its borders [2], so that stromal surface of the human amniotic membrane adheres to the ulcer bed and the neighboring epithelium grows over the membrane [12, 15]. A lamellar pocket can be created around the corneal ulcer with a crescent blade to allow the insertion of the membrane and ensure the epithelium does not proliferate underneath the membrane [13]. A similar technique can be used in conjunctival reconstruction by overlapping the edges of the membrane with the recessed and peritomized conjunctiva [6].

The layered or fill in technique is a variation of the graft technique and is used to repair a deep corneal ulcer. It involves placing several layers of human amniotic membrane to fill in the entire depth of the defect. Either multiple individual layers or a folded human amniotic membrane can be used: the "blanket-fold" or "fluffed-up" technique [1]. In this case, the orientation of the bottom layers of human amniotic membrane is not important [1, 2, 13]. The most superficial one must be sutured with the basement membrane side up [2, 13].

Amniotic membrane can also be placed as a patch to cover the ocular surface. The membrane will protect the underlying healing epithelium but not become incorporated into the host tissue, and it will fall off or can be removed after few days [5, 13]. In this case, human amniotic membrane is placed stromal side up and leaving the basement membrane face down to stimulate epithelial growth. Depending on the disease extent, it can cover the entire ocular surface (cornea, bulbar, forniceal and palpebral conjunctiva) or just a part of it [6]. Graft and patch techniques are often used simultaneously [2].

The size of the membrane should be slightly larger than the tissue defect (around 1 mm). It should be spread over the surface, taking care to ensure there is no blood or fluid underneath, and the visual axis should be avoided, if possible [2].

Human amniotic membrane must be correctly oriented, so the ophthalmic surgeon must be able to recognize the two surfaces. Before cryopreservation, human amniotic membrane is flattened onto nitrocellulose paper, with the epithelium/basement membrane side up [2, 6]. Additionally, a suture can be placed, so that the knot marks the basement membrane surface [2]. Even so, the orientation can be lost during surgery. Intra-operatively, the stromal (mesenchymal) side is easily recognized since it sticks to a surgical sponge [2, 6, 13].

Amniotic membrane is usually kept in place with sutures. Non absorbable, 10-0 nylon sutures are generally used to anchor the human amniotic membrane onto the cornea [13]. This prevents premature detachment and allows a more controlled removal [1]. Interrupted or running sutures can be used as the surgeon prefers. We find it very useful to place one or two perpendicular crossed sutures from limbus to limbus to prevent the patch of human amniotic membrane from detaching (Fig. 6.4). Absorbable sutures are preferred for conjunctiva, such as 8-0 vicryl [2]. Trans-palpebral double-armed non-absorbable 4-0 silk sutures can be used to secure human amniotic membrane in the fornices, passing through the eyelid and exiting in the eyelid skin [1, 13]. This action is particularly important in diseases where fornix obliteration can develop.

More recently, sutureless alternatives of securing amniotic membrane have been introduced. human amniotic membrane can be kept in place with a tissue adhesive or mounted on a plastic structure. These techniques present several advantages

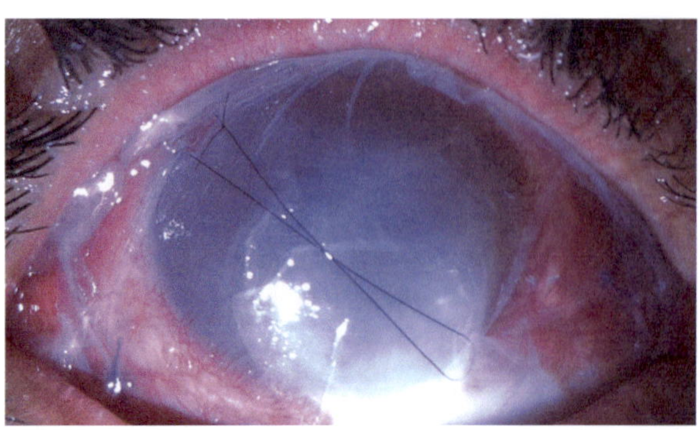

Fig. 6.4 Crossed sutures from limbus to limbus

inasmuch as they can be performed under topical anesthesia, surgery time is shorter and there are no suture-related complications.

Fibrin glue is a biological product composed of thrombin and fibrinogen. Once in contact, thrombin converts fibrinogen into fibrin and a fibrin clot forms within few seconds. Fibrin glue is effective in securing a graft of human amniotic membrane to the corneal or conjunctival surface [13].

Prokera® (Bio-tissue Inc, Miami, FL, USA) is a commercially available medical device that acts as a sutureless biological bandage. It is made of a cryopreserved amniotic membrane clipped into a termoplastic ring set and can be easily inserted with topical anesthesia [5, 13]. Suturing amniotic membrane around a conformer is another method of placing a patch of human amniotic membrane onto the ocular surface that is inexpensive, simple and easily accessible [5].

6.5 The Fate of Human Amniotic Membrane Grafted in the Eye

Histological and cytological findings, in vivo confocal microscopy and anterior segment optical coherence tomography (AS-OCT) studies have helped shed light on the fate of human amniotic membrane transplanted to the human eye. The clinical, subjective impressions have been objectively documented with in vivo and ex vivo imaging techniques [15, 21, 66].

As previously mentioned, an amniotic membrane patch usually degrades or falls off after 1–4 weeks [15]. It protects the underlying healing epithelium from the eyelid friction and retains some inflammatory cells that would compromise the regenerating process. If the ocular surface is heavily inflamed, the membrane disintegrates faster and should be replaced until the inflammation subsides [2, 13].

A graft of amniotic membrane, however, is incorporated into the host tissue (cornea and conjunctiva) [5]. An epithelial sheet originating in the host cells forms over the membrane and, in the cornea, is complete after 2 weeks, on average [6, 15, 21, 95–97]. Newly formed corneal epithelium is thinner, with lower cell density and larger cells [15]. Over the course of a month it gradually acquires a mature appearance: polarized, four to five layer structure, with columnar cells on the basal aspect, wing cells suprabasally and flat squamous cells on the surface [15, 21]. It exhibits well-defined desmosomes and hemidesmosomes [21, 96]. The formation of an epithelial layer is critical, since failure of proper epithelialization precludes graft integration [15].

When used as a substrate for conjunctival regeneration, amniotic membrane becomes covered with epithelium of normal conjunctival phenotype. Impression cytology shows uniformly smaller nongoblet epithelial cells with twice the cell density of normal control eyes. The goblet cell density was almost ten times higher [66].

In addition to acting as a basal membrane for epithelial cells, human amniotic membrane can be used as a surrogate for corneal stroma [21]. Multilayered grafts are incorporated into the corneal stroma and are invaded by presumed corneal stroma derived cells (CSDC) in areas of Bowman's zone rupture [15, 21]. These cells synthesize collagen, but its irregularly arranged fibers preclude the formation of a transparent medium [21]. Hence, amniotic stroma provides a scaffold for metabolically active CSDC that can help rebuild stromal tissue, thereby strengthening the damaged area [21].

The use of corneal confocal microscopy and AS-OCT allowed in vivo observation of transplanted amniotic tissue over time. This is particularly valuable for amniotic grafts that become integrated and undergo progressive modifications. By the end of the first month, multilayered grafts become more homogenous and lose amniotic epithelial cells. After around 3 months, CSDC can be seen in the amniotic stroma and their density increases as time passes. Amniotic membrane grafts contract over time anteroposteriorly [15] and centripetally [21] and its thickness stabilizes after 6 months. The graft tends to clear over time, but a residual opacity always remains and limits potential visual benefits [15].

6.6 Limitations

Amniotic membrane has been used in a growing number of ocular pathologies. However, evidence supporting its use remains limited [5, 7]. Despite the reported success in case reports and small case series, the heterogeneity in clinical presentation, severity, concomitant treatments and success criteria makes it difficult to draw conclusions and ascertain its true role [5].

The amniotic membrane's mechanism of action is not completely understood. However, we must keep in mind that it does not replace tissue, but promotes tissue repair, and that its success relies on the presence of healthy host cells capable of regeneration. Therefore, amniotic membrane as an isolated therapy won't succeed in cases of severe limbal insufficiency or extensive conjunctival scarring [2, 13]. Its efficacy is also compromised in the presence of palpebral abnormalities, dry eye and uncontrolled ocular surface inflammation [1, 2, 13]. These problems must be investigated and addressed prior to amniotic membrane transplantation [1].

The fact that amniotic membrane is a biological product raises a number of questions. Intra- and inter-donor variations make it a non-standardized product and that certainly affects its performance [5]. Variables include maternal health, age, race and diet, and also fetal sex, gestational age and health [5]. Despite serological testing and observation of a quarantine period, the risk of infectious disease transmission cannot be completely eliminated [5]. There is also a risk of microbial contamination during processing since the membrane cannot be sterilized [13]. Therefore, regular microbiologic testing is mandatory.

The processing and preservation method used and the storage time can also alter the amniotic membrane composition and further influence its biological properties and its clinical effect [5].

Finally, its lack of transparency often precludes the adequate visualization of the cornea and intra-ocular structures. That is not a concern in conjunctival reconstruction, since its translucency is similar to conjunctiva and provides a better visualization than alternative mucosal grafts.

6.7 Conclusions

Amniotic membrane is a useful adjunct in several ocular conditions. However, its efficacy should be compared with the standard treatment in randomized clinical trials, so that its role can be ascertained for each individual disease and its indications clearly defined. For that purpose, it is mandatory to establish criteria for disease severity, treatment protocol, outcomes and success, so that comparisons can be made within homogenous groups.

The biological heterogeneity of amniotic membrane introduces a bias that is very difficult, if not impossible, to eliminate. Each membrane is unique and its composition varies during storage time. The creation of a standardized product would give outcomes that are more predictable and eliminate this confusing factor. Building a synthetic membrane embedded with growth factors, cytokines, antibiotics and other molecules with specific functions may overcome this problem in the near future.

References

1. John T (2003) Human amniotic membrane transplantation: past, present, and future. Ophthalmol Clin N Am 16:43–65
2. Fernandes M, Sridhar MS, Sangwan VS, Rao GN (2005) Amniotic membrane transplantation for ocular surface reconstruction. Cornea 24:643–653
3. Sorsby A, Symons H (1946) Amniotic membrane grafts in caustic burns of the eye (burns of the second degree). Br J Ophthalmol 30:337–345
4. Kim J, Tseng S (1995) Transplantation of preserved human amniotic membrane for surface reconstruction in severely damaged rabbit corneas. Cornea 14:473–484
5. Rahman I, Said D, Maharajan V, Dua H (2009) Amniotic membrane in ophthalmology: indications and limitations. Eye 23:1954–1961
6. Dua HS, Gomes JA, King AJ, Maharajan VS (2004) The amniotic membrane in ophthalmology. Surv Ophthalmol 49:51–77
7. Gomes J, Romano A, Santos M, Dua H (2005) Amniotic membrane use in ophthalmology. Curr Opin Ophthalmol 16:233–240
8. Grueterich M, Espana EM, Tseng SC (2003) Ex vivo expansion of limbal epithelial stem cells: amniotic membrane serving as a stem cell niche. Surv Ophthalmol 48:631–646
9. Dua HS, Azuara-Blanco A (2000) Limbal stem cells of the corneal epithelium. Surv Ophthalmol 44:415–425
10. Liu T, Zhai H, Xu Y et al (2012) Amniotic membrane traps and induces apoptosis of inflammatory cells in ocular surface chemical burn. Mol Vis 18:2137–2146
11. Shimmura S, Shimazaki J, Ohashi Y, Tsubota K (2001) Antiinflammatory effects of amniotic membrane transplantation in ocular surface disorders. Cornea 20:408–413

12. Meller D, Pauklin M, Thomasen H et al (2011) Amniotic membrane transplantation in the human eye. Dtsch Arztebl Int 108:243–248
13. Kheirkhah A, Li W, Casas V, Tseng SC (2006) Sutureless amniotic membrane transplantation. Expert Rev Ophthalmol 1:49–62
14. Takács L, Tóth E, Berta A, Vereb G (2009) Stem cells of the adult cornea: from cytometric markers to therapeutic applications. Cytometry A 75:54–66
15. Nubile M, Dua HS, Lanzini M et al (2011) In vivo analysis of stromal integration of multilayer amniotic membrane transplantation in corneal ulcers. Am J Ophthalmol 151:809–822
16. Prabhasawat P, Tesavibul N, Komolsuradej W (2001) Single and multilayer amniotic membrane transplantation for persistent corneal epithelial defect with and without stromal thinning and perforation. Br J Ophthalmol 85:1455–1463
17. Rodriguez-Ares M, Tourino R, Lopez-Valladares M, Gude F (2004) Multilayer amniotic membrane transplantation in the treatment of corneal perforations. Cornea 23:577–583
18. Kim J, Kim J, Hahn T, Park W (2001) Amniotic membrane transplantation in infectious corneal ulcer. Cornea 20:720–726
19. Chen JH-C, Ma DH-K, Tsai RJ-F (2002) Amniotic membrane transplantation for pseudomonal keratitis with impending perforation. Chang Gung Med J 25:144–152
20. Hoffmann S, Szentmáry N, Seitz B (2013) Amniotic membrane transplantation for the treatment of infectious ulcerative keratitis before elective penetrating keratoplasty. Cornea 32(10):1321–1325
21. Said DG, Nubile M, Alomar T et al (2009) Histologic features of transplanted amniotic membrane: implications for corneal wound healing. Ophthalmology 116:1287–1295
22. Pires RT, Tseng SC, Prabhasawat P et al (1999) Amniotic membrane transplantation for symptomatic bullous keratopathy. Arch Ophthalmol 117:1291–1297
23. Mejía LF, Santamaría JP, Acosta C (2002) Symptomatic management of postoperative bullous keratopathy with nonpreserved human amniotic membrane. Cornea 21:342–345
24. Espana EM, Grueterich M, Sandoval H et al (2003) Amniotic membrane transplantation for bullous keratopathy in eyes with poor visual potential. J Cataract Refract Surg 29:279–284
25. Chansanti O, Horatanaruang O (2005) The results of amniotic membrane transplantation for symptomatic bullous keratopathy. J Med Assoc Thail 88:57–62
26. Georgiadis NS, Ziakas NG, Boboridis KG et al (2008) Cryopreserved amniotic membrane transplantation for the management of symptomatic bullous keratopathy. Clin Exp Ophthalmol 36:130–135
27. Altiparmak UE, Oflu Y, Yildiz EH et al (2009) Prospective comparison of two suturing techniques of amniotic membrane transplantation for symptomatic bullous keratopathy. Am J Ophthalmol 147:442–446
28. Chawla B, Sharma N, Tandon R et al (2010) Comparative evaluation of phototherapeutic keratectomy and amniotic membrane transplantation for management of symptomatic chronic bullous keratopathy. Cornea 29:976–979
29. Dua HS, Miri A, Said DG (2010) Contemporary limbal stem cell transplantation – a review. Clin Exp Ophthalmol 38:104–117
30. Clare G, Suleman H, Bunce C, Dua H (2012) Amniotic membrane transplantation for acute ocular burns. Cochrane Database Syst Rev 9:CD009379
31. Shay E, Kheirkhah A, Liang L et al (2009) Amniotic membrane transplantation as a new therapy for the acute ocular manifestations of Stevens-Johnson syndrome and toxic epidermal necrolysis. Surv Ophthalmol 54:686–696
32. Honavar SG, Bansal AK, Sangwan VS, Rao GN (2000) Amniotic membrane transplantation for ocular surface reconstruction in Stevens-Johnson syndrome. Ophthalmology 107:975–979
33. John T, Foulks GN, John ME et al (2002) Amniotic membrane in the surgical management of acute toxic epidermal necrolysis. Ophthalmology 109:351–360
34. Kobayashi A, Yoshita T, Sugiyama K et al (2006) Amniotic membrane transplantation in acute phase of toxic epidermal necrolysis with severe corneal involvement. Ophthalmology 113:126–132

35. Muqit MM, Ellingham RB, Daniel C (2007) Technique of amniotic membrane transplant dressing in the management of acute Stevens-Johnson syndrome. Br J Ophthalmol 91:1536
36. Tandon A, Cackett P, Mulvihill A, Fleck B (2007) Amniotic membrane grafting for conjunctival and lid surface disease in the acute phase of toxic epidermal necrolysis. J AAPOS 11:612–613
37. Shammas MC, Lai EC, Sarkar JS et al (2010) Management of acute Stevens-Johnson syndrome and toxic epidermal necrolysis utilizing amniotic membrane and topical corticosteroids. Am J Ophthalmol 149:203–213
38. Rubinate L, Welch M, Johnson A, DeMartelaere S (2010) New technique for manufacture and placement of enlarged symblepharon ring with amniotic membrane transplantation for ocular surface protection in Stevens-Johnson syndrome and toxic epidermal necrolysis. ARVO Abstr. 1135/D957
39. Gregory DG (2011) Treatment of acute Stevens-Johnson syndrome and toxic epidermal necrolysis using amniotic membrane: a review of 10 consecutive cases. Ophthalmology 118:908–914
40. Hsu M, Jayaram A, Verner R et al (2012) Indications and outcomes of amniotic membrane transplantation in the management of acute Stevens-Johnson syndrome and toxic epidermal necrolysis: a case-control study. Cornea 31:1394–1402
41. Kolomeyer A, Do B, Tu Y, Chu D (2013) Placement of proKera in the management of ocular manifestations of acute Stevens-Johnson syndrome in an outpatient. Eye Contact Lens 39:7–11
42. Tomlins P, Parulekar M, Rauz S (2013) "Triple-TEN" in the treatment of acute ocular complications from toxic epidermal necrolysis. Cornea 32:365–369
43. Im S-K, Lee K-H, Yoon K-C (2010) Combined ethylenediaminetetraacetic acid chelation, phototherapeutic keratectomy and amniotic membrane transplantation for treatment of band keratopathy. Korean J Ophthalmol 24:73–77
44. Anderson D, Prabhasawat P, Alfonso E, Tseng S (2001) Amniotic membrane transplantation after the primary surgical management of band keratopathy. Cornea 20:354–361
45. Kwon YS, Song YS, Kim JC (2004) New treatment for band keratopathy: superficial lamellar keratectomy, EDTA chelation and amniotic membrane transplantation. J Korean Med Sci 19:611–615
46. Esquenazi S, Rand W, Velazquez G, Grunstein L (2008) Novel therapeutic approach in the management of band keratopathy using amniotic membrane transplantation with fibrin glue. Ophthalmic Surg Lasers Imaging 39:418–421
47. Rao A, Sridhar U, Gupta A (2008) Amniotic membrane transplant with superficial keratectomy in superficial corneal degenerations: efficacy in a rural population of north India. Indian J Ophthalmol 56:297–302
48. Li M, Zhu M, Yu Y et al (2012) Comparison of conjunctival autograft transplantation and amniotic membrane transplantation for pterygium: a meta-analysis. Graefes Arch Clin Exp Ophthalmol 250:375–381
49. Kaufman SC, Jacobs DS, Lee WB et al (2013) Options and adjuvants in surgery for pterygium: a report by the American academy of ophthalmology. Ophthalmology 120:201–208
50. Prabhasawat P, Barton K, Burkett G, Tseng SC (1997) Comparison of conjunctival autografts, amniotic membrane grafts and primary closure for pterygium excision. Ophthalmology 104:974–985
51. Ma DH-K, See L-C, Liau S-B, Tsai RJ-F (2000) Amniotic membrane graft for primary pterygium: comparison with conjunctival autograft and topical mitomycin C treatment. Br J Ophthalmol 84:973–978
52. Solomon A, Pires R, Tseng S (2001) Amniotic membrane transplantation after extensive removal of primary and recurrent pterygia. Ophthalmology 108:449–460
53. Tananuvat N, Martin T (2004) The results of amniotic membrane transplantation for primary pterygium compared with conjunctival autograft. Cornea 23:458–463

54. Luanratanakorn P, Ratanapakorn T, Suwan-Apichon O, Chuck R (2006) Randomised controlled study of conjunctival autograft versus amniotic membrane graft in pterygium excision. Br J Ophthalmol 90:1476–1480
55. Küçükerdönmez C, Akova Y, Altinörs D (2007) Comparison of conjunctival autograft with amniotic membrane transplantation for pterygium surgery: surgical and cosmetic outcome. Cornea 26:407–413
56. Ozer A, Yildirim N, Erol N, Yurdakul S (2009) Long-term results of bare sclera, limbal-conjunctival autograft and amniotic membrane graft techniques in primary pterygium excisions. Ophthalmologica 223:269–273
57. Maharajan VS, Shanmuganathan V, Currie A et al (2007) Amniotic membrane transplantation for ocular surface reconstruction: indications and outcomes. Clin Exp Ophthalmol 35:140–147
58. Tseng SC, Di Pascuale M, Liu DDT-S et al (2005) Intraoperative mitomycin C and amniotic membrane transplantation for fornix reconstruction in severe cicatricial ocular surface diseases. Ophthalmology 112:896–903
59. Nava-Castañeda A, Tovila-Canales J, Monroy-Serrano M et al (2005) Comparative study of amniotic membrane transplantation, with and without simultaneous application of mitomycin C in conjunctival fornix reconstruction. Arch Soc Esp Oftalmol 80:345–352
60. Goyal R, Jones SM, Espinosa M et al (2006) Amniotic membrane transplantation in children with symblepharon and massive pannus. Arch Ophthalmol 124:1435–1440
61. Yao Y, Qiu W, Zhang Y, Tseng S (2006) Mitomycin C, amniotic membrane transplantation and limbal conjunctival autograft for treating multirecurrent pterygia with symblepharon and motility restriction. Graefes Arch Clin Exp Ophthalmol 244:232–236
62. Kheirkhah A, Blanco G, Casas V et al (2008) Surgical strategies for fornix reconstruction based on symblepharon severity. Am J Ophthalmol 146:266–275
63. Kheirkhah A, Ghaffari R, Kaghazkanani R et al (2013) A combined approach of amniotic membrane and oral mucosa transplantation for fornix reconstruction in severe symblepharon. Cornea 32:155–160
64. Prabhasawat P, Tesavibul N (2001) Preserved amniotic membrane transplantation for conjunctival surface reconstruction. Cell Tissue Bank 2:31–39
65. Barabino S, Rolando M, Bentivoglio G et al (2003) Role of amniotic membrane transplantation for conjunctival reconstruction in ocular-cicatricial pemphigoid. Ophthalmology 110:474–480
66. Prabhasawat P, Tseng S (1997) Impression cytology study of epithelial phenotype of ocular surface reconstructed by preserved human amniotic membrane. Arch Ophthalmol 115:1360–1367
67. Kim B-H (2011) Surgical treatment of necrotic scleral calcification using combined conjunctival autografting and an amniotic membrane inlay filling technique. Eye 25:1484–1490
68. Kim EC, Kim MS, Kang NY (2012) Excision with corneoscleral lamellar keratoplasty and amniotic membrane transplantation of a corneal displaced recurrent conjunctival melanoma. Korean J Ophthalmol 26:383–387
69. Lee J, Kim Y, Kim M, Kim E (2014) The effect of surgical wide excision and amniotic membrane transplantation with adjuvant topical mitomycin C treatment in recurrent conjunctival-corneal intraepithelial neoplasia. Semin Ophthalmol 29:192–195
70. Pirouzian A (2013) Management of pediatric corneal limbal dermoids. Clin Ophthalmol 7:607–614
71. Sheha H, Kheirkhah A, Taha H (2008) Amniotic membrane transplantation in trabeculectomy with mitomycin C for refractory glaucoma. J Glaucoma 17:303–307
72. Ji Q-S, Qi B, Liu L et al (2013) Comparison of trabeculectomy and trabeculectomy with amniotic membrane transplantation in the same patient with bilateral glaucoma. Int J Ophthalmol 6:448–451

73. Eliezer RN, Kasahara N, Caixeta-Umbelino C et al (2006) Use of amniotic membrane in trabeculectomy for the treatment of glaucoma: a pilot study. Arq Bras Oftalmol 69:309–312
74. Mahdy RA, Nada WM, Almasalamy SM et al (2010) A freeze-dried (lyophilized) amniotic membrane transplantation with mitomycin C and trabeculectomy for pediatric glaucoma. Cutan Ocul Toxicol 29:164–170
75. Stavrakas P, Georgopoulos G, Milia M et al (2012) The use of amniotic membrane in trabeculectomy for the treatment of primary open-angle glaucoma: a prospective study. Clin Ophthalmol 6:205–212
76. Kiuchi Y, Yanagi M, Nakamura T (2010) Efficacy of amniotic membrane-assisted bleb revision for elevated intraocular pressure after filtering surgery. Clin Ophthalmol 4:839–843
77. Drolsum L, Willoch C, Nicolaissen B (2006) Use of amniotic membrane as an adjuvant in refractory glaucoma. Acta Ophthalmol Scand 84:786–789
78. Tawara A, Miyamoto N, Ishibashi S et al (2013) Reconstruction of a nonfunctional trabeculectomy bleb using an amniotic membrane-wrapped silicone sponge to treat refractory glaucoma. Graefes Arch Clin Exp Ophthalmol 251:2013–2018
79. Amini H, Kiarudi MY, Moghimi S et al (2010) Ahmed glaucoma valve with adjunctive amniotic membrane for refractory glaucoma. J Ophthalmic Vis Res 5:158–161
80. Budenz DL, Barton K, Tseng SC (2000) Amniotic membrane transplantation for repair of leaking glaucoma filtering blebs. Am J Ophthalmol 130:580–588
81. Kee C, Hwang J-M (2002) Amniotic membrane graft for late-onset glaucoma filtering leaks. Am J Ophthalmol 133:834–835
82. Rauscher FM, Barton K, Budenz DL et al (2007) Long-term outcomes of amniotic membrane transplantation for repair of leaking glaucoma filtering blebs. Am J Ophthalmol 143:1052–1054
83. Nagai-kusuhara A, Nakamura M, Fujioka M, Negi A (2008) Long-term results of amniotic membrane transplantation-assisted bleb revision for leaking blebs. Graefes Arch Clin Exp Ophthalmol 246:567–571
84. Bochmann F, Azuara-Blanco A (2012) Interventions for late trabeculectomy bleb leak (review). Cochrane Database Syst Rev 9:CD006769
85. Kitagawa K, Yanagisawa S, Watanabe K et al (2009) A hyperdry amniotic membrane patch using a tissue adhesive for corneal perforations and bleb leaks. Am J Ophthalmol 148:383–389
86. Strube YNJ, Conte F, Fariq C et al (2011) Amniotic membrane transplantation for restrictive strabismus. Ophthalmology 118:1175–1179
87. Frangouli O, Adams G (2013) The use of amniotic membrane for the management of fibrosis in complex strabismus surgery. Strabismus 21:13–22
88. Tugcu B, Helvacioglu F, Yuzbasioglu E et al (2013) Amniotic membrane in the management of strabismus reoperations. Jpn J Ophthalmol 57:239–244
89. Poonyathalang A, Preechawat P, Pomsathit J, Mahaisaviriya P (2005) Reconstruction of contracted eye socket with amniotic membrane graft. Ophthalmic Plast Reconstr Surg 21:359–362
90. Ma DDH-K, Wang S-FS SWW-Y, Tsai RJ-FR (2002) Amniotic membrane graft for the management of scleral melting and corneal perforation in recalcitrant infectious scleral and corneoscleral ulcers. Cornea 21:275–283
91. Hanada K, Shimazaki J, Shimmura S, Tsubota K (2001) Multilayered amniotic membrane transplantation for severe ulceration of the cornea and sclera. Am J Ophthalmol 131: 324–331
92. Teping C, Reim M (1989) Tenonplasty as a new surgical principle in the early treatment of the most severe chemical eye burns. Klin Monatsbl Augenheilkd 194:1–5
93. Casas V, Kheirkhah A, Blanco G, Tseng S (2008) Surgical approach for scleral ischemia and melt. Cornea 27:196–201

94. Thatte S, Gupta L (2012) Amniotic membrane transplantation in surgically induced necrotizing scleritis with peripheral ulcerative keratitis. Middle East Afr J Ophthalmol 19:419–421
95. Gris O, Wolley-Dod C, Güell JL et al (2002) Histologic findings after amniotic membrane graft in the human cornea. Ophthalmology 109:508–512
96. Seitz B, Resch M, Schlötzer-Schrehardt U et al (2006) Histopathology and ultrastructure of human corneas after amniotic membrane transplantation. Arch Ophthalmol 124:1487–1490
97. Gregory ME, Spiteri-Cornish K, Hegarty B et al (2011) Combined amniotic membrane transplant and anterior stromal puncture in painful bullous keratopathy: clinical outcome and confocal microscopy. Can J Ophthalmol 46:169–174

Chapter 7
Amniotic Membrane in the Treatment of Burns

Ahmad Sukari Halim, Ehfa Bujang-Safawi, and Arman Zaharil Mat Saad

Abstract Burn wound dressing represents an important aspect of burn care. The various inherent features of amniotic membrane proved to be useful as it contributes towards the attributes of an ideal dressing. This biological dressing is inexpensive, able to enhance re-epithelialization, readily available and stored, hence rendering it suitable for a wide range of burn wound treatment. The different preparations of amniotic membrane provide additional benefit whereby clinicians can have a range of product to choose from. Pain relief, good wound adherence and conformability further increase the importance of amnion as burn wound dressing. The versatility and efficacy of the amnion makes it an attractive option despite of the enormous range commercially available wound management products in the market. This chapter presents a review of the various aspects of amniotic membrane for burn wound treatment.

Keywords Amnion • Burns • Wound healing • Wound dressing • Biological dressing

7.1 Introduction

The skin is the largest organ in the body and plays a crucial role in body homeostasis, fluid and electrolyte balance, thermoregulatory function, vitamin production, providing a physical barrier to external threats, sensory function and personal beauty.

A.S. Halim (✉) • A.Z.M. Saad
Department of Reconstructive Sciences, School of Medical Sciences,
Health Campus, Universiti Sains Malaysia, Kubang Kerian, Kelantan, Malaysia
e-mail: ashalim@usm.my

E. Bujang-Safawi
Department of Surgery, Universiti Malaysia Sarawak, Kuching, Sarawak, Malaysia

A.C. Mamede, M.F. Botelho (eds.), *Amniotic Membrane: Origin, Characterization and Medical Applications*,
DOI 10.1007/978-94-017-9975-1_7

Burn injuries can occur through several mechanisms. Typically, such injuries will affect the skin in a superficial to deep order, except for electrical burns, where the tissue damage internally may be worse than the apparent skin surface damage. Based on this common pattern of injury, burn wounds can be classified according to their depth and according to the total body surface area (TBSA) recorded as percentages.

Superficial burns (first degree) are the mildest form of burns. First-degree burns are characterized by pink discoloration of the skin and dry with no blister formation, but the injury is very painful. This type of burn involves the epidermis only and heals in a matter of a few days.

Partial-thickness burns (second degree) can be further subdivided into superficial dermal and deep dermal burns. Superficial dermal burns may involve the formation of small blisters, pinkness, slight wetness and the presence of capillary refill, along with extreme pain. These burns typically have good potential for spontaneous healing in 5–10 days. With deep dermal burns, large blisters may form, the skin may appear red or pale with fixed staining, thick edema is present, impaired capillary refill may occur and sensation loss may occur. These types of wounds take more than 2 weeks to heal by re-epithelialization of the residual dermal appendages.

Full-thickness burns (third degree) involve the full thickness of the epidermis and dermis. The burn will appear white, pale or black and leathery. There will be an absence of blisters, no capillary refill and no sensation, which may give a false sense of security to the patients, as they do not feel the pain.

When burn injury occurs and disrupts the skin physiological barrier, the injured portions can no longer adequately perform their functions. Therefore, it is extremely important to restore skin integrity as soon as possible. Failure to correct this problem will result in continuous plasma or fluid leakage through the burn wound, burn edema, impaired thermoregulation, potential microbial invasion and potentially fatal consequences such as burn shock and multi-organ dysfunction.

According to the World Health Organization (W.H.O.), almost 200,000 deaths occur annually due to flame burns alone, and more than half of these burns occur in low- and middle-income countries. Non-fatal burns injuries are a major source of morbidity [1]. It is estimated that in 2012 around 450,000 burn injuries received medical attention in the United States, with 3,400 fatalities and 40,000 hospitalizations (30,000 admissions to burn centers). These injuries represent a serious financial and manpower burden to the healthcare system. It is estimated that burn injuries cost healthcare providers around USD2 billion each year, and this amount is estimated to be increasing [2].

7.2 Historical Perspective: Amnion for Burns

Human amnion has been used for centuries as a biological wound dressing. In China and Japan, amnion is considered a potent medication for the treatment of many diseases and is believed to impart the 'magical strength of youth' (quoted by

Tyszkiewicz et al.) [3]. In western medicine, John Staige Davis introduced the use of fetal membrane as a permanent skin transplant at the beginning of the last century. William Thornton, a final-year medical student at Johns Hopkins Hospital in 1910, suggested this concept. However, it soon became clear that amnion could only be used as a temporary biological wound dressing [4].

Human amnion was first used as a biological burn dressing by Sabella and Stem in 1913 [5]. Since then, amnion has been widely used to treat superficial and partial-thickness burns at many centers worldwide [5]. The versatility and reliability of human amnion as a biological dressing resulted in the extension of its application to various clinical conditions in the plastic surgery field.

Ward et al. [6] used this biological dressing in the treatment of chronic ulcers and pressure sores. Moreover, skin graft donor sites can be effectively treated using human amniotic membrane. Recently, radiation-induced ulcers have been effectively treated using human amniotic membrane as shown by Gajiwala and Sharma [7]. Thus, human amnion has the potential to become an attractive biomaterial that is cheap and widely available.

7.3 Types of Amnion Preparation

Amniotic membrane has been used mainly in the developing world due to its low cost and availability [5, 8, 9], and fresh amnion is mostly used in clinical applications. The use of different preparation of amnion such as dried [10], frozen [11], irradiated [10], lyophilized [12] and glycerol-preserved [8] preparations are based on clinical indications and availability in the clinical setting.

Though several preparation and preservation methods are available, dried amnion is the easiest to prepare and store. Gamma-irradiation can be used to further eliminate the risk of microbial contamination from the source [10]. Gamma irradiation at 25 kGy will assure complete bacterial, fungal and viral sterilization (HIV is inactivated at 20 kGy) [13]. Singh and Chacharkar demonstrated using infrared spectral scanning that there is no degradation or changes in the quality and characteristics when using this type of preparation even after long storage periods of up to 5 years. This type of amnion preservation is relatively cheap and does not require special storage. The clinical efficacy of the material is also maintained, as was demonstrated by a series of patient with superficial and partial-thickness burns [14]. The process of lyophilization (freeze drying to only 5 % water content) can also reduce the risk of infection transmission from the amniotic membrane when it is used clinically [13].

Glycerolization is another cost-effective way to preserve and store the amniotic membrane as demonstrated and suggested by Ravishanker et al. This method requires the storage of amnion in the refrigerator at 4 °C and can last for up to 5 years. Ravishanker et al. also suggested that amnion banks be set up for the retrieval, processing and storage of the amniotic membrane [8].

Table 7.1 The biological functions and benefits of amnion as a wound dressing

Reduction of fluid loss
Decrease bacterial count in wound
Stimulation of neovascularization
Decrease in physiological stress for the patient – pain
Promotion of wound healing
Protection of growing epithelium
Tight adherence to the wound surface, increased mobility, and diminished pain
Preparation of skin defects for closure
Immune privileged tissue with no rejection phenomenon
Better scar formation
Ease of application
Readily available
Cost effective

7.4 An Ideal Burn-Dressing Material

There are multitudes of burn wound dressings available on the market. This variety indicates that none of the individual dressings are perfect for all wound conditions at all times. An ideal wound dressing should be readily available, cheap/economical, easy to apply, able to provide good pain relief, able to protect the wound from infection, promote healing, prevent heat and fluid loss, elastic, adhere well to the wound, robust, able to withstand shear force, durable and non-antigenic. Currently, there is no ideal dressing material available that meets all these criteria [15].

A biological dressing may be used temporarily to replace some of the skin function lost due to burn injury until a permanent cover is provided or healing is completed by epithelialization, as a biological dressing may provide a closed resemblance to the skin. Unfortunately, biological dressings may create financial burden for healthcare providers, hence their limited availability. However, a naturally occurring product – the amniotic membrane – is readily available worldwide and is inexpensive. This material is almost always discarded despite its proven efficacy as a true biological dressing [16, 17]. Amnion appears to satisfy most of the criteria for an ideal biological dressing (see Table 7.1). It is thin, elastic, adheres to the wound surface well and, at the same time, it provides an effective vapor barrier, a durable cover for the raw surface, and substantial pain relief with potential bacteriostatic properties. In addition, it facilitates rapid re-epithelization [10, 11, 18, 19].

7.5 Amnion for Burn Wound Care

7.5.1 Superficial (Dermal) Partial-Thickness Burn

Superficial (dermal) partial-thickness burns usually heal within 5–10 days. Therefore, the main principles of management of this type of burn are pain control, infection prevention and rapid healing promotion. In addition, repetitive trauma to the burn surface must be minimized, as this may injure the damaged epithelium and convert the burn to a deeper depth injury [9]. Human amnion is a near ideal temporary biological dressing for these types of burns. It can easily be applied to the surface of the burn wound due to its thinness, pliability and good adherence.

Using aseptic technique, the recipient wound is thoroughly and gently cleansed with normal saline or 0.05 % aqueous chlorhexidine to remove all the debris and dead skin. The amniotic membrane is then applied with the glistening side against the wound surface. All air bubbles and excess fluid should be smoothed out to ensure good contact. Typically, the amniotic membrane adheres immediately, thus preventing excessive loss of fluids and electrolytes. Being semi permeable in nature, it allows exchange of gases and liquids across the membrane.

After application of the amniotic membrane, the recipient wound will epithelialize below the membrane. The amniotic membrane remains attached to the burn wound until the wound has completely healed. Within a week or so, the desiccated membrane will flake off, resulting in a shiny scar. Therefore, if the amniotic membrane remains well adhered to the wound surface, this indicates that the wound has not completely epithelialized. Membrane that is stuck to the burn wound surface should be left alone, and the edge of the amnion that has peeled off should be trimmed using scissors.

However, the amniotic membrane can be changed whenever necessary, especially if the amnion does not 'take.' If there is any sign of infection, if the patient is febrile, or if there is fluid or pus collection below the amniotic membrane, then the burn wound needs to be thoroughly and gently washed and re-dressed. The use of antibiotic cream is not necessary and not advisable. Systemic antibiotics should be given only when indicated.

It is not necessary to apply another dressing overlying the membrane, although a light gauze dressing can be applied to ensure the amniotic membrane remains in place.

7.5.2 *Deep (Dermal) Partial-Thickness and Full-Thickness Burns*

The principles of management for deep (dermal) partial-thickness and full-thickness burns include removing the burn tissue/eschar by tangential excision and replacing the missing skin with autologous skin as soon as possible. In major burns, where there is limited availability of donor autograft skin, the debrided burn wound can be dressed with amniotic membrane. The membrane can be used as a temporary skin replacement until the donor autograft site is ready to be harvested again.

However, in this case, frequent changing of the amniotic membrane is necessary as the membrane may occasionally dissolve. In addition, its biological value has been determined to be inferior to that of skin allografts. The latter is not always an option due to financial constraints, infrastructure limitations, religion or poor societal acceptance. In this modern era, advancements in synthetic dressings with antibacterial properties and bio-engineered skin substitutes have limited the use of amniotic membrane as a temporary skin replacement for excised burn wounds.

However, these modern dressings have the drawbacks of limited availability, limited quantity and cost.

Mohammadi et al. [9], in a randomized controlled trial of 211 patients with varying degree of burn depth, observed that the application of amniotic membrane accelerated the separation of necrotic tissue/burn eschar in deep partial-thickness burns (second degree) and full-thickness burns without the need to surgically excise the burn eschar as compared with conventional silver sulfadiazine dressing. They also noted that the appearance of granulation tissue was faster in the amniotic membrane group and confirmed the findings of other studies that frequent application of amniotic membrane is required for these types of burns. Another interesting finding from their study was that the repeated frequent dressing changes in the silver sulfadiazine group could eventually convert the superficial burns to a deeper wound due to damage and trauma to the fragile surface of the wound. On the other hand, the amniotic membrane adhered well to the wound surface and did not need to be removed. A new membrane could be applied over it when necessary to cover the exposed area.

Therefore, amniotic membrane, with its high tolerability and its ability to reduce pain, remains a useful biological dressing material for the treatment of burn patients, especially in settings where financial resources are limited.

7.5.3 Hand and Facial Burns

The management of hand and facial burns remains one of the greatest challenges in burn care. Historically, superficial (dermal) partial-thickness burns to these areas were treated conservatively by debriding the blisters, daily washing, and the application of new bandages with topical medications two to four times a day. These procedures cause excruciating pain and anxiety in patients.

A number of skin substitutes have been developed in recent years to overcome these problems. However, the contours and continued movement of the face has, in the past, limited the use of skin substitutes. Moreover, burn wounds to the hands covered with skin substitutes are usually bandaged heavily, leading to stiffness. Thus, the standard care of facial burns remains an open technique using topical antibiotics, whereas hand burns are commonly placed in a plastic bag following application of 1 % silver sulfadiazine cream.

As the amniotic membrane is thin and pliable, it conforms well to these special areas. These unique properties allow the simple and easy application of amniotic membranes. No additional dressings are needed, as amniotic membrane adheres well to the wound surface. This open dressing technique allows monitoring of the wound without the need for dressing change and free movements of the fingers, which thus prevents joint stiffness. Amniotic membrane has been reported to be a better treatment option compared to 1 % silver sulfadiazine because the use of the former reduces the length of hospital stay, reduces the number of dressing changes, and promotes earlier wound epithelialization [20].

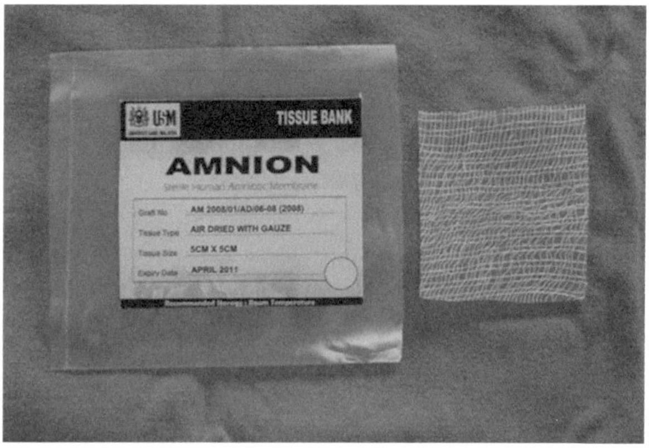

Fig. 7.1 The dried, irradiated amnion sheet is packed in a transparent plastic bag, which minimizes storage space and distribution cost. The amnion is spread on a thin layer of cotton gauze, allowing easy handling and quick application to wounds. (Photo courtesy of USM Tissue Bank)

At our center, superficial partial-thickness facial burns are treated with dried, irradiated human amnion. Application of human amnion to the wound surfaces is easy, simple and quick as the amniotic membranes are supported with cotton gauze (Fig. 7.1). In contrast, glycerol-preserved membranes have to be rinsed thoroughly before spreading over the wound surface. All air bubbles and excess fluid have to be removed. Ravishanker et al. reported that it takes approximately 1 hour for glycerol-preserved amniotic membrane to dry following application to wounds [8].

We conducted a 7-year review (2001–2008) that involved 33 patients with facial burns [21]. None of the patients developed infection, and 85 % of the patients needed only a single application of the dried amnion sheets. Patients with burns confined to the face and with easy access to the hospital were treated on an outpatient basis. The average healing time was 5.4 days (range: 2–14 days). We demonstrated that dried amnion sheets fulfill the qualities of an ideal biological dressing and are effective for the treatment of superficial partial-thickness facial burns.

7.5.4 Amnion as a Graft Overlay

Amniotic membrane can also be used as an overlay for split-thickness skin grafts. Immediate adherence to the grafted area facilitates fixation of the skin graft to the wound bed. This eliminates the need for using sutures or staples to anchor the skin graft thus avoiding painful suture or stapler removal. Traumatic graft loss induced by suture or staple removal can also be prevented [22]. Mohammadi et al. demonstrated that the rate of graft take in extremity burns was higher in a group where the amniotic membrane was used as a dressing to cover the skin graft compared with a conventional graft fixation and dressing group. These authors also noted that the mean duration of graft take was significantly faster in the amnion group than in the control group [23].

7.5.5 Amnion for Split-Skin Graft Donor Site Healing

The ideal split-thickness donor site should heal rapidly with minimal pain and scarring. Hypopigmented, hyperpigmented or hypertrophic resulting scars are dreaded by patients. Various modalities have been developed. Accordingly, the preferred management strategy to achieve the optimal healing of a split-skin graft donor site remains rather controversial. Hydrocolloid, hydrofiber, and silver-coated dressings, along with human recombinant growth hormone, have been demonstrated to produce improved quality scarring [24–26]. However, these modalities are costly and their availability is limited in developing countries. The amniotic membrane is an alternative that provides significant benefits by increasing patient comfort via diminishing the number of dressing changes and facilitating the process of wound healing [27, 28].

7.6 Advantages of Amnion in Burn Treatment

7.6.1 Pain Reduction

Pain reduction in burn wounds is a well-recognized advantage of the amnion that has been observed by many researchers [5, 9, 17, 20, 29–31]. Application of amnion to burn surfaces can protect the exposed nerve endings from external irritants such as clothing and even those as soft as moving air [32]. Amnion has good adherence to wound surfaces and is able to maintain a moist environment, which contributes to this advantage.

Sharma et al., in their observation of patients with partial-thickness burns, demonstrated that amnion application completely alleviated pain [33]. Ravishanker et al. used glycerol-preserved amniotic membrane and noted that it reduced the pain score in adults and quieted a distressed child [8].

7.6.2 Good Wound Adherence

Amnion very firmly adheres to wound beds. The bonding that is established between the amnion and the wound surface is biological, not mechanical, and is considered to be due to the fibrin-elastin formation at the wound-dressing interface [34]. This property, which is shared with other biological dressings, helps to protect the fragile tissue at the burn wound surface, especially the injured cells that have the potential either to recover or to be converted into a deeper burn depth. The strong bond and adherence can also maintain the moisture level over the wound surface, which can accelerate re-epithelialization and protect the newly formed thin epithelium from trauma. The amnion will become dry and readily peel off once the wound is completely healed.

7.6.3 Light, Thin and Transparent

Amniotic membrane is thin and lightweight, which makes it practical to apply. These characteristics, combined with its adherence properties, allow bulky dressings to be avoided, resulting in the patient feeling more comfortable and providing better movement, which can prevent joint stiffness, particularly in the elderly, and avoid respiratory compromise due to restricted chest wall movement, particularly in children. The transparent nature of the membrane also allows visualization of the underlying wound bed without having to remove the membrane. Therefore, any collection or wound surface is easily detectable if present.

7.6.4 Minimizing Fluid/Water Loss

The amnion is made of five distinctive layers: epithelial, basement membrane, connective tissue, fibroblasts, and a spongy layer. This structure resembles the stratum corneum in the human skin epidermis, which is a natural vapor barrier [35]. By acting as a vapor barrier, the amnion can prevent excessive evaporation from the wound surface thus directly and markedly reducing the insensible fluid loss, which, in turn, reduces the amount of replacement fluid required for burn injuries [36]. It is worth to note that the membrane does not influence the fluid shift from the plasma into the interstitial compartment in the extracellular space. An additional advantage of preventing evaporation of the wound is that the temperature regulation mechanism is not overly affected and, in turn, the caloric requirements needed to maintain temperature are also reduced accordingly.

There are multiple lines of evidence in the literature that support the aforementioned properties of the amnion as a burn dressing. Unfortunately, there is still lack of level-I evidence supported by properly conducted randomized controlled trials. The evidence is confined to several poorly designed small studies with a small number of subjects [9, 20, 36, 37] and observational studies [14, 31, 33, 36, 38]. Sharma et al. used amniotic membrane in the treatment of partial-thickness burns and observed that it was associated with minimal oozing and that healing occurred without the need for skin grafting [33]. Pigeon noted that the amniotic membrane was beneficial in the early treatment of burns to prevent fluid or plasma leakage. Its application to third-degree burns and granulated wound beds caused the membrane to disintegrate. Therefore, amnion is less effective for the treatment of third-degree burns as opposed to first- and second-degree burns [31].

7.6.5 Less Frequent Dressing

In a study comparing radiation-sterilized, oven-dried amniotic membrane against topical silver sulfadiazine for the treatment of second-degree burns, Mostaque and

Rahman [20] demonstrated that the amniotic membrane is a better treatment option because its use reduces the length of hospital stay and the number of required dressing changes. Epithelialization of the wound is also quicker. The use of amniotic membrane is painless and odorless. The procedure is easy and comfortable for the doctor, and it is well accepted. Most of the patients remain ambulatory during treatment [8, 20].

7.6.6 Faster Wound Healing

Anecdotally, amnion has been demonstrated to speed up the epithelialization process, which results in faster wound healing in superficial and partial-thickness burns. For example, Miller et al. demonstrated that partial-thickness burn wounds healed faster than controls when protected by immediate allografting. However, the routine use of allografts is limited to special centers as it requires a well-organized skin bank [39]. On the other hand, xenografts (e.g., porcine skin) are inferior to allografts, more expensive and not readily available in many countries [40]. Clinical trials and animal experiments have demonstrated that amnion is superior to both allograft and xenograft materials [11]. Sharma et al. [33] demonstrated that amniotic membrane application to partial-thickness burns is associated with minimal oozing, and healing occurs without the need for autografting. The membrane promotes healing by accelerating the migration of fibroblasts and the development of collagen during the first 6–8 days of repair [33, 41, 42]. The amniotic membrane can also provide a good moisture balance that allows rapid epithelialization to occur. One other suggested mechanism for the observed rapid healing is the inhibition of proteinase activity, which, in turn, reduces inflammatory responses because of the decreased infiltration of polymorphonuclear leukocytes [43].

7.6.7 Neovascularization

Human amnion has inherent angiogenic properties that increase the rate of healing and epithelialization. The vascular response of amnion on the wound bed has been extensively studied using standard histological and immunohistological staining techniques. In a study of the use of amnion in the treatment of chronic leg ulcers, Faulk et al. noted that the most striking effect of amnion was the development of new vessels. This was observed grossly, histopathologically, and immunohistologically [44]. Increased vascularity in the amniotic membrane-treated burn wound was also microscopically noted [30]. Burgos and Sergeant suggested that the proteins produced by amniotic epithelial cells have angiogenic effects, but the precise mode of action remains unknown [12].

7.6.8 Anti-bacterial Properties

Microbial invasion into the burn wound has been recognized as one of the most common obstacles in the recovery and healing process for burn injuries, especially major burns with a deeper burn thickness. The current strategy employed at most institutions is to maintain strict environmental control in the burn unit, early removal/debridement of dead/non-viable burn eschar and the use of topical antimicrobial agents. Human amniotic membrane can provide a mechanical barrier to microbial invasion, and when used, has produced a lower rate of burn wound infection compared with other types of dressings [14, 33, 37]. There are several possible mechanisms for the observed antibacterial effect of amniotic membranes. Many authors agree that amnion is an effective biological dressing, as it adheres intimately onto the contour of the wound and inevitably lies in close opposition to the wound surface [45, 46]. It has been noted that the degree of microbial or bacterial contamination is inversely proportional to the number of leucocytes beneath the membrane [46]. Saymen et al. observed that leukocytes move toward the surface of a covered wound and away from an uncovered wound. Therefore, it is important to have a dressing that can be closely approximated to the wound surface with minimal dead space to gain this advantage. Amnion is better than other biological dressings in this manner as it is thinner and more pliable. It has also been demonstrated that the surface pH falls after amniotic membrane application and that the temperature of the underlying wound bed rises. Both of these changes promote phagocytosis and reduce the exudation of interstitial fluid [47].

Robson and Krizek reported that amniotic membrane is superior to allograft (homograft) and xenograft (heterograft) skin in the prevention of microbial infection. Amniotic membrane was as effective as autograft skin in lowering the bacterial count when applied to 20 % rat scalds topically inoculated with 10^8 Pseudomonas aeruginosa. The authors concluded that no antibacterial substance is produced in this context; however, a substance that enhances leukotaxis or angiogenesis, with subsequent leukotaxis and promotion of host defense mechanism, would be important to the observed effects [40]. At the same time, amnion is an immune privileged tissue with no rejection reaction, which permits continued close surface adherence, providing an additional antibacterial effect [48]. Burleson and Eiseman confirmed that biological dressings that adhere to the wound surface and prevent the accumulation of pus on the granulation surface aid in the sterilization of the wound [34]. The presence of a weak antibiotic-like substance in amniotic fluid has also been demonstrated, and the possibility of its elaboration in the extra-embryonic membrane epithelium remains strong [48–50]. This antibacterial property is suggested to be the result of the lysozyme and progesterone content of the amniotic fluid. The progesterone is bacteriostatic to various Gram-positive organisms.

Singh and Chacharkar demonstrated that dried, irradiated amnion was impermeable to various strains of Gram-negative bacilli strains (E. Coli, Klebsiella, Pseudomonas) and Staphylococcus [14]. In a randomized controlled trial comparing fresh amnion and a conventional antibiotic ointment dressing (nitrofurazone) in the treatment of partial-thickness burns, Ghalambor et al. observed that the amnion group had a significantly lower infection rate (2 % versus 17 %) and less need for surgical debridement of the burn wounds [37].

7.6.9 *No Rejection Phenomenon*

Almost all biological dressings or skin substitutes are predisposed to early graft reaction with the amnion membrane tissue being the only exception. As human amniotic membrane is embryologically derived from fetal ectoderm, it carries the advantages of fetal skin allografts [48]. Therefore, it can fulfill some of the functions of the skin that are destroyed in burn injuries; however, it is different than skin allografts, and the vascularization and rejection of the membrane does not occur when the amnion is directly placed over the wound surface [11]. Faulk associated the feto-maternal relationship of the fetus with the existence of a parasite that successfully avoids host rejection [51]. Later, McIntyre and Faulk demonstrated that a glycoprotein that is found in trophoblasts and amniochorion is very important in suppressing the detection of the fetus as "foreign" to the mother by acting on maternal lymphocytes and preventing rejection throughout pregnancy [52]. The lack of human leukocyte antigen (HLA) A, BC and DR surface antigens and beta-2-microglobulin expression in human amniotic epithelial cells could further contribute to the lower inflammatory response and lack of rejection phenomenon observed in this type of allografting [53].

7.6.10 *Better Scarring*

The use of amnion dressing in the treatment of superficial burns and superficial dermal burns has been observed to produce faster healing and superior cosmetic results for the resultant scars. Pigeon also noted that the use of amnion resulted in an absence of the abnormal pigmentation usually found with healed burn scars [31].

An animal study investigating the resultant scarring from deep dermal burns demonstrated that amniotic membrane produces reduced scar tissue as compared with standard paraffin tulle dressings when assessed histopathologically by significantly inhibiting alpha smooth muscle actin (α-SMA). The amniotic membrane treatment led to a significant reduction in the amount of scar tissue generated in response to an experimental burn compared with moist wound dressings [29].

7.7 Disadvantage of Amniotic Membrane Application to Burns

The risk of disease transmission remains when amniotic membrane is used. An amicable understanding and agreement between the obstetrics and burn units is essential as screening for hepatitis, syphilis and HIV needs to be performed on the donors. Consent needs to be obtained from the potential donors prior to the tests being conducted. Although the processing of amniotic membranes is conducted under aseptic conditions, the possibility of microbial contamination cannot be excluded [54]. Amniotic membranes irradiated at the recommended dose of 25 kGy have a high sterility assurance level and are microbiologically safe for clinical use [55]. However, gamma irradiation requires specialized and expensive equipment.

Amniotic dressings have a slightly unpleasant odor. This is a normal functional response of the amniotic membrane sloughing [37]. No major concerns among patients have been reported. This is not an issue when a dried amnion preparation is used [20].

7.8 Conclusion

Amniotic membrane has the features of an ideal biological burn dressing. The various preparation forms and the inherent properties as well as the outcome of clinical studies provide the evidence to support its use in burn care. Although it is available in various preparation forms, the inherent properties remain. Being thin and pliable, with good adherence properties, amnion is a good choice of dressing for burn wound management. In addition, amnion is inexpensive, readily available, easy to procure and process and offers indefinite shelf life. Established advantages in burns usage include marked pain relief, enhanced wound epithelialization, excellent vapor barrier, lack of antigenicity leading to no rejection, potent neovascularization stimulant, and good antibacterial property. These properties make amnion efficacious in burn wound irrespective of the depth, as well as skin graft donor sites. Therefore it is time for clinicians to consider the usage of amnion in the face of escalating cost in healthcare system, especially in burn management when there is abundance of evidence pointing to the superiority of amnion in burn management.

References

1. WHO (2012) Burns – fact sheet. http://www.who.int/mediacentre/factsheets/fs365/en/
2. ABA (2012) Burn incidence fact sheet. http://www.ameriburn.org/resources_factsheet.php
3. Tyszkiewicz J, Uhrynowska-Tyszkiewicz I, Kaminski A, Dziedzic-Goclawska A (1999) Amnion allografts prepared in the central tissue bank in Warsaw. Ann Transplant 4:85–90
4. Davis J (1910) Skin transplantation with a review of 550 cases at the Johns Hopkins hospital. Johns Hopkins Hosp Rep 15:307–395
5. Subrahmanyam M (1995) Amniotic membrane as a cover for microskin grafts. Br J Plast Surg 48:477–478
6. Ward D, Bennett J, Burgos H, Fabre J (1989) The healing of chronic venous leg ulcers with prepared human amnion. Br J Plast Surg 42:463–467
7. Gajiwala AL, Sharma V (2003) Use of irradiated amnion as a biological dressing in the treatment of radiation induced ulcers. Cell Tissue Bank 4:147–150
8. Ravishanker R, Bath A, Roy R (2003) "Amnion Bank" – the use of long term glycerol preserved amniotic membranes in the management of superficial and superficial partial thickness burns. Burns 29:369–374

9. Mohammadi AA, Sabet B, Riazi H et al (2009) Human amniotic membrane dressing: an excellent method for outpatient management of burn wounds. Iran J Med Sci 34:61–64
10. Rao TV, Chandrasekharam V (1981) Use of dry human bovine amnion as a biological dressing. Arch Surg 116:891–896
11. Colocho G, Graham WP, Greene AE et al (1974) Human amniotic membrane as a physiologic wound dressing. Arch Surg 109:370–373
12. Burgos H, Sergeant R (1983) Lyophilized human amniotic membranes used in reconstruction of the ear. J R Soc Med 76:433
13. Adly O, Moghazy A, Abbas A et al (2010) Assessment of amniotic and polyurethane membrane dressings in the treatment of burns. Burns 36:703–710
14. Singh R, Chacharkar M (2011) Dried gamma-irradiated amniotic membrane as dressing in burn wound care. J Tissue Viability 20:49–54
15. Halim AS, Khoo TL, Yussof SJM (2010) Biologic and synthetic skin substitutes: an overview. Indian J Plast Surg 43:23–28
16. Trelford J, Trelford-Sauder M (1979) The amnion in surgery, past and present. Am J Obstet Gynecol 134:833–845
17. Matthews R, Faulk W, Bennett J (1982) A review of the role of amniotic membranes in surgical practice. Obstet Gynecol Ann 11:31–58
18. Talmi Y, Finkelstein Y, Zohar Y (1990) Use of human amniotic membrane as a biologic dressing. Eur J Plast Surg 13:160–162
19. Unger M, Roberts M (1976) Lyophilized amniotic membranes on graft donor sites. Br J Plast Surg 29:99–101
20. Mostaque A, Rahman K (2011) Comparisons of the effects of biological membrane (amnion) and silver sulfadiazine in the management of burn wounds in children. J Burn Care Res 32:200–209
21. Bujang-Safawi E, Halim A, Khoo T, Dorai A (2010) Dried irradiated human amniotic membrane as a biological dressing for facial burns -a 7-year case series. Burns 36:876–882
22. Mohammadi AA, Mohammadi MK (2011) How does human amniotic membrane help major burn patients who need skin grafting: new experiences. In: Spear M (ed) Ski. grafts – Indic. Appl. Curr. Res. In Tech Europe, Rijeka, Croatia pp 265–276
23. Mohammadi AA, Johari HG, Eskandari S (2013) Effect of amniotic membrane on graft take in extremity burns. Burns 39:1137–1141
24. Barnea Y, Amir A, Leshem D et al (2004) Clinical comparative study of aquacel and paraffin gauze dressing for split-skin donor site treatment. Ann Plast Surg 53:132–136
25. Innes ME, Umraw N, Fish JS et al (2001) The use of silver coated dressings on donor site wounds: a prospective, controlled matched pair study. Burns 27:621–627
26. Losada F, García-Luna PP, Gómez-Cía T et al (2002) Effects of human recombinant growth hormone on donor-site healing in burned adults. World J Surg 26:2–8
27. Singh R, Kumar A, Singh D, Malviya A (2013) Use of gamma-irradiated amniotic membrane for the healing of split skin graft donor site. Tissue Eng Regen Med 10:110–114. doi:10.1007/s13770-013-0004-5
28. Salehi SH, As'adi K, Mousavi SJ, Shoar S (2013) Evaluation of amniotic membrane effectiveness in skin graft donor site dressing in burn patients. Indian J Surg. doi:10.1007/s12262-013-0864-x. http://link.springer.com/search?query=Evaluation+of+Amniotic+Membrane+Effectiveness+in+Skin+Graft+Donor+Site+Dressing+in+Burn+Patients&search-within=Journal&facet-journal-id=12262#page-2. Published online 31 Jan 2013
29. Fraser JF, Cuttle L, Kempf M et al (2009) A randomised controlled trial of amniotic membrane in the treatment of a standardised burn injury in the merino lamb. Burns 35:998–1003
30. Kirschbaum S, Hernandez H (1963) Use of amnion in extensive burns. In: 3rd International Congress in Plastic Surgery. Excerpta Medica, Amsterdam
31. Pigeon J (1960) Treatment of second-degree burns with amniotic membranes. Can Med Assoc J 83:844–845
32. Dino B, Eufemio G, De Villa M (1966) Human amnion: the establishment of an amnion bank and its practical application in surgery. J Philipp Med Assoc 42:357–366

33. Sharma S, Bagree M, Bhat A et al (1985) Amniotic membrane is an affective burn dressing material. Jpn J Surg 15:140–143
34. Burleson R, Eiseman B (1972) Nature of the bond between partial-thickness skin and wound granulations. Surgery 72:315–322
35. Niknejad H, Peirovi H, Jorjani M et al (2008) Properties of the amniotic membrane for potential use in tissue engineering. Eur Cell Mater 15:88–99
36. Gump F, Kinney J (1970) Caloric and fluid losses through the burn wound. Surg Clin N Am 50:1235–1248
37. Ghalambor A, Pipilzadeh MH, Khodadadi A (2000) The amniotic membrane: a suitable biological dressing to prevent infection in thermal burns. Med J Islam Acad Sci 13: 115–118
38. Walker AB, Cooney DR, Allen JE (1977) Use of fresh amnion as a burn dressing. J Pediatr Surg 12:391–395
39. Miller T, Switzer W, Foley F, Moncrief J (1967) Early homografting of second degree burns. Plast Reconstr Surg 40:117–125
40. Robson M, Krizek T, Koss N, Samburg J (1973) Amniotic membranes as a temporary wound dressing. Surg Gynecol Obstet 136:904–906
41. Bose B (1979) Burn wound dressing with human amniotic membrane. Ann R Coll Surg Engl 61:444–447
42. Bapat C, Kothary P (1974) Preliminary report on acceleration of wound healing by amnion membrane graft. Indian J Med Res 62:1342–1346
43. Kim JS, Kim JC, Na BK et al (2000) Amniotic membrane patching promotes healing and inhibits proteinase activity on wound healing following acute corneal alkali burn. Exp Eye Res 70:329–337
44. Faulk WP, Stevens PJ, Burgos H et al (1980) Human amnion as an adjunct in wound healing. Lancet 315:1156–1158
45. Trelford J, Hanson F, Anderson D (1973) Amniotic membrane as a living surgical dressing in human patients. Oncology 28:358–364
46. Gruss JS, Jirsch DW (1978) Human amniotic membrane: a versatile wound dressing. Can Med Assoc J 118:1237–1246
47. Saymen DG, Nathan P, Holder IA et al (1973) Control of surface wound infection: skin versus synthetic grafts. Appl Microbiol 25:921–934
48. Burleson R, Eiseman B (1973) Mechanisms of antibacterial effect of biologic dressings. Ann Surg 177:181–186
49. Galask R, Snyder I (1970) Antimicrobial factors in amniotic fluid. Am J Obstet Gynecol 106:59–65
50. Sachs B, Stern C (1979) Activity and characterization of a low molecular fraction present in human amniotic fluid with broad spectrum antibacterial activity. Br J Obstet Gynaecol 86:81–86
51. Faulk W, Galbraith GM (1979) Trophoblast transferrin and transferrin receptors in the host-parasite relationship of human pregnancy. Proc R Soc B Biol Sci 204:83–97
52. McIntyre JA, Faulk WP (1979) Antigens of human trophoblast: effects of heterologous anti-trophoblast sera on lymphocyte responses in vitro. J Exp Med 149:824–836
53. Akle C, Adinolfi M, Welsh K et al (1981) Immunogenicity of human amniotic epithelial cells after transplantation into volunteers. Lancet 2:1003–1005
54. Dziedzic-Goclawska A (2000) The application of ionizing radiation to sterilise connective tissue allografts. In: Nather A, Yusof N, Hilmy N (eds) Radiation in tissue banking. World Scientific, Singapore, p 586
55. Singh R, Purohit S, Chacharkar M et al (2007) Microbiological safety and clinical efficacy of radiation sterilized amniotic membranes for treatment of second-degree burns. Burns 33:505–510

Chapter 8
Amniotic Membrane in Cancer

Ana Catarina Mamede, Ana Salomé Pires, and Ana Filipa Brito

Abstract Cancer currently represents a public health problem. Despite scientific advances in the area of cancer treatment, particularly in the area of conventional therapies such as chemotherapy and radiotherapy, this disease continues to be responsible for a high rate of morbidity and mortality worldwide. For this reason, the emergence of new anti-cancer therapies in the medical and scientific community has been desired. The application of amniotic membrane in anti-cancer therapy is a recent idea. Since this tissue has anti-angiogenic, pro-apoptotic and immunoregulatory activities, several authors have pointed to a potential benefit resulting from the application of amniotic membrane in cancer therapy. Furthermore, given the well documented stem properties of the cells derived from amniotic membrane, some authors have relied on these knowledge to support its application in cancer therapy. In fact, despite this being a new idea, several papers with promising results have been already published about this topic, all pointing to a potential benefit of the use of amniotic membrane in the treatment of oncological disease.

Keywords Amniotic membrane • Cancer • Pro-apoptosis • Anti-angiogenesis • Immunoregulatory activity

---

A.C. Mamede (✉)
CICS-UBI, Health Sciences Research Center, University of Beira Interior, Covilhã, Portugal

Biophysics Unit, Institute for Biomedical Imaging and Life Sciences (IBILI), Center of Investigation in Environment, Genetics and Oncobiology (CIMAGO), Faculty of Medicine, University of Coimbra, Coimbra, Portugal
e-mail: ana_mamede@hotmail.com

A.S. Pires
Biophysics Unit, Institute for Biomedical Imaging and Life Sciences (IBILI), Faculty of Medicine, University of Coimbra, Coimbra, Portugal

A.F. Brito
Biophysics Unit, Institute for Biomedical Imaging and Life Sciences (IBILI), Center of Investigation in Environment, Genetics and Oncobiology (CIMAGO), Faculty of Medicine, University of Coimbra, Coimbra, Portugal

A.C. Mamede, M.F. Botelho (eds.), *Amniotic Membrane: Origin, Characterization and Medical Applications*,
DOI 10.1007/978-94-017-9975-1_8

8.1 The Cancer

Cancer is the second leading cause of death in developed countries, after cardiovascular diseases. This fact is mainly due to the increase of risk behaviors that enhance the disease, as the adoption of sedentary lifestyles, poor and unbalanced diets, alcoholism and tobacco [1].

In 2008, it is estimated that about 12.7 million of people have developed cancer, and that this disease was the cause of death of about 7.6 million of people worldwide. Among the values presented, it is estimated that 56 % of diagnosed cases and 64 % of deaths occurred in developed countries [2].

Cancer is typically characterized by an uncontrolled cell growth and division. However, cancer is not only a mass of cells with high proliferation index, but rather a complex tissue composed by several distinct types of cells that interact between them [3]. Several genetic mutations are associated with the origin of different types of cancer. These mutations lead to the development of cancer, by a process known as carcinogenesis (Fig. 8.1). Carcinogenesis is a sequence of processes that lead a cell from a healthy state to a precancerous state until the development of cancer [4, 5]. On the other hand, the tumors formation is a process that occurs over the time in several stages involving progressively several metabolic pathways. This development seems to be mainly characterized by a "somatic evolution", which results in the development of new phenotypes that are generated continuously by changes in the expression or in the function of specific genes. These genetic and epigenetic changes are very important constituents in the development of specific characteristics of premalignant lesions [3].

Such intracellular events are only one component of the Darwinian dynamics, which is based on the Darwin's theory, who says that the effect of each phenotypic change on population dynamics is dependent on its interactions with identifiable environmental selection forces. Thus, the loss or maintenance of each genetically encoded phenotypic characteristic is dependent on its contribution to the genome overall constitution, which in turn is dependent on the selection forces present in the microenvironment of the cell in question [3]. Thus, now it is believed that the tumors

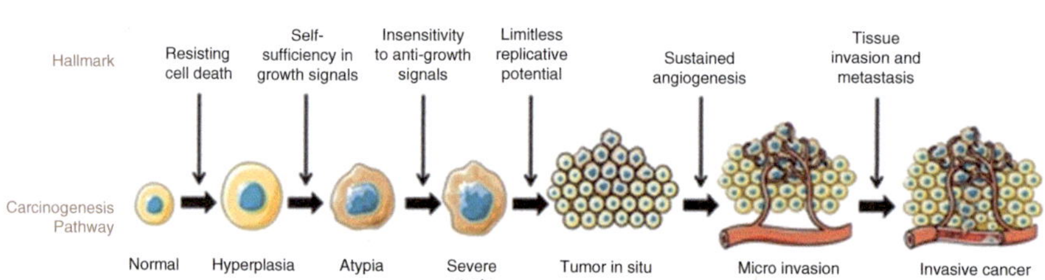

Fig. 8.1 Sequence of processes and respective hallmarks that lead a cell from a healthy state to a precancerous state, until the development of cancer

are more than masses of proliferating cells, and in this context, during the last years, according to Hanahan and Weinberg it has been considered cancer as a consequence of six major changes in cell physiology, which determine the evolution of normal cell to tumor cell. These changes are related with the self-sufficiency to growth factors, the insensitivity to inhibitory factors, the evasion of apoptosis, the metastasis and tissue invasion, the unlimited replicative potential and the sustained angiogenesis [6]. Subsequently, more recently, the same authors complemented this theory, and it is now known that tumors are complex systems with a wide variety of different cell types. Thus, cells recruited by the tumor to form the tumor stroma will become active participants in tumorigenesis, rather than passive spectators. These stromal cells contribute to the development and expression of certain cancer characteristics. Thus, and based on these considerations, the cancer biology can no longer be understood by the simply enumerating of tumor cells characteristics, but should include also the tumor microenvironment contributions in the carcinogenesis process [7].

Therefore, the six cancer characteristics initially described, distinct but complementary, that provide a solid basis for understanding the cancer biology, have resulted in more characteristics to be considered such as: the tumors capacity to modify or reprogram its metabolism in order to better withstand the tumor proliferation and the ability to evade to immune destruction. Furthermore, also the genomic instability associated to the subsequent mutagenicity allows the tumor genetic alterations that facilitate their proliferation. The associated inflammatory process may also enhance the other characteristics already enumerated, and moreover the immune system plays an important role in the tumor development and progression [7].

Thus, according to Hanahan and Weinberg, ten hallmarks of cancer must be considered: resisting cell death, inducing angiogenesis, enabling replicative immortality, activating invasion and metastasis, evading growth supressors, sustaining proliferative signaling, avoiding immune destruction, tumor-promoting inflammation, genome instability and mutation and deregulated cellular energetics [7].

Given the tumors multiple characteristics, the majority of the anticancer therapies currently used are not successful in many types of cancer. Solid tumors surgery, when applicable, is the most effective treatment. Complementarily to surgery or as a first-line therapy, there are also commonly used chemotherapy and radiotherapy. However, these therapies are not always effective and have many adverse effects. For a long time, systemic therapy has been dominated by the use of cytotoxic agents, and most of these drugs are deoxyribonucleic acid (DNA)-damaging agents that are designed only to kill or inhibit cells with high proliferation rates [8, 9].

In order to make the anti-cancer therapies more personalized, in the last few years new therapeutic approaches have been developed, that take into account all signaling networks that regulate cell proliferation and survival, as well as several specific features of each type of tumor. These new therapies are specific to new targets which include growth factors, molecules that promote angiogenesis, cell cycle proteins, signaling molecules and modulators of apoptosis [9].

The hypothesis that inhibition of angiogenesis (anti-angiogenesis) would be an effective strategy to treat human cancer is one of the oldest, but it is still a valid theory, especially for highly vascularized tumors [9, 10]. The growth of new vessels from the pre-existing vasculature, called angiogenesis is a crucial event in solid tumors development. In particular, tumor expansion is dependent on this phenomenon allowing them to overcome starvation and hypoxia [9, 11, 12]. The angiogenic vasculature is also important in metastasis because cancer cells metastasize to the distant organs through this vasculature [9, 13].

The resistance to cell death, and more specifically resistance to apoptosis (programmed cell death), is also one of the most known characteristics of tumor cells, and a major obstacle to the therapies commonly used for cancers. It is known, for example, that the majority of tumors have mutations in the TP53 tumor suppressor gene and also exhibit a deregulated expression of proteins belonging to the BCL-2 family, namely an increased expression of anti-apoptotic proteins such as BAX and a decrease of anti-apoptotic proteins like BCL-2. These characteristics confer to the tumor cells an intrinsic resistance to therapies which often leads to disappointing clinical results [6, 7].

The relationship between cancer and the immune system has been more recently discovered. However, it is a highly important topic. It is known that cells and tissues are constantly monitored by an ever-alert immune system, being the immune surveillance responsible for the recognition and elimination of the vast majority of incipient tumor cells. However, some data indicate that tumors can in some way prevent detection by the immune system by limiting the immunological killing, thereby evading eradication [7].

8.2 The Cancer and Amniotic Membrane: The Hypothesis

One of the promising ideas that emerged in recent years is the potential application of amniotic membrane in cancer therapy. In fact, the publication of scientific papers in this area has grown. Currently, there are some data that prove, in vitro and in vivo, that this biological tissue can be useful in this area. However, there is still little information about the molecular mechanisms in which the anti-cancer effect of the amniotic membrane is based. In fact, researchers have been harvested the know-how obtained in other application areas, e.g. ophthalmology [14–19] or dermatology [20–22], and tried to explore the application of amniotic membrane in cancer therapy. Seo et al. [23] were the first to introduce in the scientific community the idea that amniotic membrane could possess anti-carcinogenic properties. The authors suggested that the anti-angiogenic, pro-apoptotic and immunoregulatory activity of amniotic membrane can contribute to the anti-cancer properties of this tissue (Fig. 8.2). These ideas were later reinforced by Niknejad and his team [24–26].

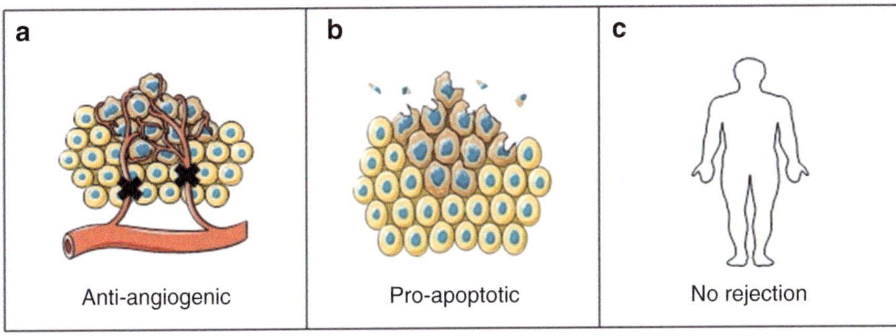

Fig. 8.2 Amniotic membrane suggested properties that determine its anti-cancer potential: (**a**) anti-angiogenic, (**b**) pro-apoptotic, (**c**) immunoregulatory activity

8.2.1 Anti-angiogenic Activity of Amniotic Membrane and the Tumor Biology

Concept

The formation of new blood vessels, or angiogenesis, promotes the influx of nutrients and oxygen to the primary tumor. Consequently, the tumor will grow and cells will migrate from the primary tumor site to other organs, leading to the process of metastasis. Therefore, angiogenesis is an important control point in the cancer progression [27, 28]. In order to reduce local vascularization, and consequently tumor aggression and metastatic potential, new therapeutic strategies to inhibit angiogenesis has been investigated. Since amniotic membrane has anti-angiogenic features and acts as a potential physical barrier, this tissue can be potentially used in cancer therapy.

Basis of the Concept

The amniotic membrane expresses interleukin (IL)-1 antagonist receptor, all members of the family of tissue inhibitors of metalloproteinases, collagen XVIII, IL-10, endostatin and thrombospondin-1, as well as several cytokines with anti-angiogenic effects [29]. The amniotic membrane also contains various matrix proteins, such as collagen α2 (IV), fibronectin and collagen type VII, that are involved in the suppression of corneal vascularization [30]. Human amnion contains a large quantity of laminin isoforms (-2, -4, -5, -6, -7, -10, and -11) [31]. Through amniotic membrane proteomic analysis the expression of several proteins involved in the anti-angiogenic activity of this tissue were also confirmed [32]. Shao et al. [33] also reported that inhibition of endothelial cell growth by the amniotic membrane and the consequent inhibition of corneal neovascularization is due to the pigment epithelium derived factor (PEDF). Kobayashi et al. [34] found that the supernatant from the cell culture of cells derived from the amniotic membrane also contains effective inhibitors of neovascularization. This author also argues that inhibition of growth and migration of vascular endothelial cells is due to the action of amniotic membrane as a physical barrier, preventing the diffusion of promoters of vascularization.

8.2.2 Pro-apoptotic and Immunoregulatory Activity of Amniotic Membrane and the Tumor Biology

Concept

Apoptosis is closely involved in a variety of disorders, including cancer. Indeed, tumor cells have a high resistance to apoptosis, a characteristic that influences tumor aggressiveness, as well as the efficacy of the treatments [35–38]. Although the apoptotic mechanism is complex, vast literature suggests that it is possible to model apoptosis in cancer [39]. Once the amniotic membrane transplantation is possible without any signs of rejection being observed, and since this tissue has the capacity to secrete factors with pro-apoptotic activity, it is expected that amniotic membrane can induce apoptosis in tumor cells.

Basis of the Concept

Li et al. [40] found that epithelial cells derived from amniotic membrane secrete soluble factors that inhibit the innate and adaptive system. In fact, the amniotic membrane is an immunologically privileged tissue as it contains an innumerable amount of immunoregulatory factors, such as human leukocyte antigen (HLA)-G and Fas ligand [41]. After transplantation of amniotic membrane in volunteers, any clinical sign of acute rejection has been shown [42]. On the other hand, the pro-apoptotic activity of the amniotic membrane is little explored but it is known that this tissue has the ability to secrete factors that has pro-apoptotic activity, such as interferon-γ. It is also known that apoptosis of interferon-γ activated macrophages promoted by amniotic membrane is not mediated directly by nitric oxide (NO) and tumor necrosis factor alpha (TNF-α). This effect is actually caused by interruption of survival pathways mediated by both nuclear factor kappa-light-chain-enhancer of activated B cells (NFκB) and Akt-FKHR. The amniotic membrane is also capable of stimulating polymorphonuclear neutrophil apoptosis [23, 26, 43–45].

8.2.3 Stem Characteristics of Amniotic Membrane Cells: What Is Its Role in Cancer Therapy?

When delivered systemically, stem cells exhibit the capability to migrate and engraft into tumor sites. Until now, the mechanism behind this event is not totally understood. The more acceptable theory is that damaged tissue expresses specific receptors, or ligands, which enables adhesion and extravasation of stem cells to the site of damage and recruitment to inflammation sites due to the release of chemotactic gradients from the tumors (Fig. 8.3). In fact, stem cells can home and modulate to tumor microenvironment, therefore representing a potential delivery vehicle for anti-cancer agents [46].

Although a percentage of the scientific community believes in the potential anti-cancer effect of stem cells, there is no real consensus about the potential beneficial or mutually antagonistic relationship between cancer cells and stem cells. Indeed, therapy with stem cells as an alternative therapy in regenerative medicine and treatment of various diseases, such as cancer, has been increasingly investigated [47–49].

Once it has been already proved that the cells derived from the amniotic membrane have stem characteristics, research on the potential use of these cells in anti-cancer therapy has recently been initiated, showing positive results [26, 50–52]. In fact, many of these articles were inspired by numerous previously published articles which proved that stem cells from different sources may have an anti-carcinogenic effect. Khakoo et al. [53] proved that human mesenchymal stem cells are able to exert anti-carcinogenic effects in a model of Kaposi's sarcoma. Le Blanc et al. [54] also proved that mesenchymal stem cells derived from bone marrow exert a potent immunosuppressive effect in vivo. Ehtesham et al. [55] also proved that neural stem cells can serve as carriers for targeted delivery of drugs for the treatment of intracranial glioma. It was also found that transplantation of nonmyeloablative allogeneic stem-cell induces regression of metastatic renal-cell carcinoma in patients who did not respond to conventional immunotherapy [56]. Belmar-Lopez et al. [46] revealed that mesenchymal stem cells from different sources, including amniotic membrane, are efficient in cell-based anti-tumor therapy. In fact, there are numerous articles in the literature with results that indicate a potential application of stem cells in cancer disease. It thus becomes important to determine whether the cells derived from the amniotic membrane, also with stem characteristics, may or may not present a potential anti-cancer effect.

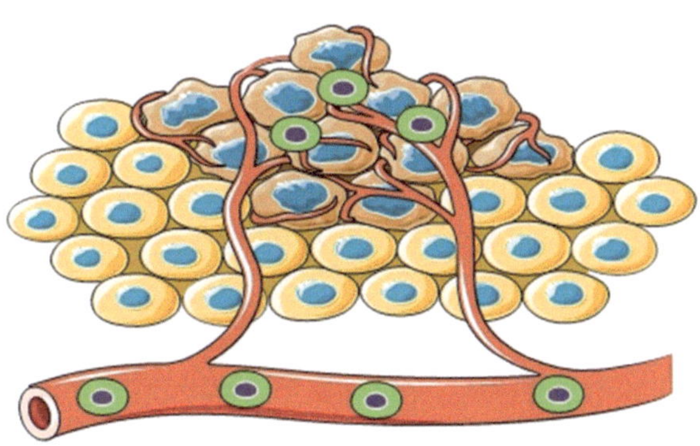

Fig. 8.3 Stem cells tumor-homing. Tumoral tissue can express specific receptors that enable adhesion of stem cells, which are able to modulate tumor microenvironment

8.3 Hypothesis and Proof: What We Know So Far?

While in 2008 the application of amniotic membrane in cancer therapy was no more than a hypothesis, in 2012 the first evidence provided from results obtained in cellular and animal models begin to emerge. In fact, several studies using amniotic membrane isolated-cells have been recently conducted. Some of them support the idea that the administration of human amniotic membrane derived stem cells represents a potential route for the tumor growth reduction and apoptosis induction.

Jiao et al. evaluated the therapeutic effect of the human amniotic membrane derived-mesenchymal cells (hAMCs) on the glioma growth [51]. After isolation and propagation of hAMCs, the authors verified that the administration of intratumoral single doses of hAMCs in glioma xenografts reduced significantly the tumor volume in 30.9 %, being also obtained a reduction of 49.5 % of the tumor for animals receiving multiple doses. The data obtained after analyzing glioma cells of hAMCs-treated animals also revealed that hAMCs induced apoptosis by a caspase-3-mediated mechanism. The susceptibility of tumor cells to the induction of apoptosis can be controlled by the ratio between pro-apoptotic proteins, like BAX and BAD, and anti-apoptotic proteins, like BCL-2 [57]. Measurement of BAX and BCL-2 expression corroborated the apoptosis results, in that a single intratumoral dose of hAMCs induced an increase in BAX expression accompanied by a decrease in BCL-2 expression in glioma cells [51]. Furthermore, it is also known that the principal players in apoptosis pathways are the caspases, a group of intracellular proteases accountable for the deliberated desintegration of cell, originating apoptotic bodies. Once activated, caspase-8 propagate the apoptotic signal, inducing several morphological and biochemical changes that will lead to cell death [58, 59]. In glioma tumors treated with hAMCs, Jiao et al. verified that active caspase-8 and total and active caspase-3 expression was substantially increased. These results, in addition to the above presented, suggest that hAMCs could, in vivo, induce apoptosis of glioma cells, revealing its potential as a new anti-cancer therapeutic agent [51].

The effect of hAMCs was also studied in vitro in different haematopoietic and non-haematopoietic tumor cell lines. Magatti et al. demonstrated that hAMCs have inhibitory effects on cell proliferation, in both direct and indirect co-culture conditions, suggesting a possible release of unknown inhibitory soluble factors by these cells. In this case, cell proliferation was not mediated by apoptosis but it was related with cell cycle arrest [52]. Some authors advocate that cell cycle machinery is a valid target and an alternative approach for cancer diagnostic and therapeutic applications, in that its deregulation plays a central role in the aberrant cell proliferation that is a guidemark of cancer and it also acts as an integration point for information transduced through oncogenic signaling networks [60]. Seven different cancer cell lines were cultured alone and co-cultured with hAMCs in two conditions: direct contact or with a physical separation with transwell chambers. Magatti et al. then evaluated cell proliferation and concluded that hAMCs reduce cell proliferative activity in a cell-dose-dependent way, being this effect verified in all cell lines tested and not only when cells are cultured in cell-cell contact but also in transwell co-cultures. However, when cancer cells are co-cultured with bone marrow mesenchymal stem cells (BM-MSCs) an anti-proliferative effect is observed just in direct co-cultures, suggesting that the inhibitory effects induced by hAMCs on different cancer cell lines can be due to the release of yet-unknown soluble factors. Besides, the authors ascertained that hAMCs-induced cell proliferation reduction was not mediated by promotion of cancer cell apoptosis but by cell cycle arrest. Experiments conducted in leukaemia and lymphoma cells cultured alone and co-cultured with hAMCs revealed a cell cycle block in G0/G1 phase accompanied by a decrease of the percentage of cells in the S phase. A subsequent analysis of the expression of a panel of genes related to cell cycle was consistent with the increase of cells in phase G0/G1 without progression to phase S. In the presence of hAMCs, several cyclins (cyclin D2, cyclin E1, cyclin H) and cyclin-dependent kinases (CDK4, CDK6 and CDK2) responsible for cell cycle progression were down-regulated, while negative regulators of cell cycle (CDKN1A or P21 and CDKN2B or P15) were up-regulated [52]. The authors clearly provide evidences of the beneficial anti-cancer effects of amniotic membrane-derived cells.

Meanwhile, also in 2012, a new strategy using human amniotic membrane-derived epithelial stem cells (hAECs) transplantation against breast cancer was proposed. Kang et al. [50] studied the effect of hAECs in an in vitro model using MDA-MB-231 cells and in a female BALB/c nude mouse xenograft model. First of all, they concluded that hAECs have an anti-proliferative effect in breast cancer cells, whereas these cells did not affect normal bovine fibroblast cells. It is known that cellular perturbations like inflammation, infection and immunity induce the release of cytokines that can function to inhibit tumor development and progression [61]. So, hAECs inhibitory effect can be related with the expression and release of cytotoxic cytokines that were collected from the hAECs cultures, such as transforming growth factor beta (TGF-β), tumor necrosis factor beta (TNF-β) and alpha (TNF-α), interferon gamma (IFN-γ), IL-1 and -8, and others [50]. The anti-cancer effect of hAECs was also obtained in vivo, since the circumtumoral injections of hAECs in a MDA-MB-231 cell xenograft model significantly inhibited tumor growth and increased animal survival; however, it was not dependent on the number of injected cells. Interestingly, hAECs were found to be localized within the tumor tissue, possibly benefiting hAECs transplantation therapy. Likewise, the authors performed a histological analysis of tumor and breast adjacent tissues, concluding that hAECs treatment not only suppress breast cancer infiltration, but also help to maintain the original breast tissue structure and tumor targeting, while 5-FU provoked serious side-effects in adjacent tissues [50]. Although further studies have to be performed to determine the mechanism by which hAECs have their anti-cancer effect, it became clear that hAECs transplantation may be a new and effective approach to the treatment of breast cancer.

Aiming to understand a little more about the anti-tumor mechanism of amniotic membrane, in 2013, Niknejad et al. carried out a study using amniotic membrane as whole tissue and amniotic epithelial cells. They hypothesized that amniotic membrane has anti-tumor effects through the induction of apoptosis and that hAECs are the key element of the amniotic membrane anti-tumor properties. Firstly, both supernatant of amniotic membrane tissue as hAECs decreased cell viability in a dose-dependent way, being this anti-cancer effect associated with the increase of caspase-3 and -8 expressions, as well as the increase of cell degradation and the reduction of cancer cell motility. The fact that results obtained for amniotic membrane tissue and hAECs were similar, led the authors to consider that hAECs could really be the source of the anti-cancer effect of amniotic membrane. So, they performed the aorta ring assay in order to assess the capillary formation with or without hAECs and/or hAMCs cells. In fact, they observed signs of angiogenesis even in the presence of the hAMCs, while in the presence of hAECs no capillary formation was observed. It was previously described that both hAECs and hAMCs express anti-angiogenic and anti-inflammatory proteins [29]. Otherwise, hAMCs also revealed markedly up-regulated expression of angiogenic factors [like epidermal growth factor (EGF), IL-8, EGF-1] and pro-angiogenic genes, vascular endothelial growth factor (VEGF) -A, angiopoietin 1 (Ang-1), hepatocyte growth factor (HGF), and fibroblast growth factor (FGF)-2 [62, 63]. These reports corroborate the author's hypothesis that hAECs confer to amniotic membrane its anti-angiogenic properties. Moreover, authors also defend that a possible mechanism through which amniotic membrane could induce apoptosis and inhibit angiogenesis and cell cycle arrest can be the heat shock protein 90 (HSP90) modulation. HSP90 is an adenosine triphosphate (ATP)-dependent molecular chaperone that is required for the function of some tumor-promoting client proteins (like signaling kinases and P53). Thus, HSP90 inhibitors could provide a new and targeted approach in the treatment of cancer [64, 65]. Niknejad et al. verified that condition medium of amniotic membrane decrease the levels of some HSP90 client proteins in cancer cells, as well as, the amount of ATP-bound HSP90, suggesting that amniotic membrane interferes with the ability of HSP90 to bind to ATP [26]. Although further studies are needed, the issue that unknown substances released by amniotic membrane inhibit HSP90, thereafter revealing anti-proliferative, pro-apoptotic, anti-angiogenic properties remains unsolved.

In different studies conducted in different types of cancer, the anti-tumor effect of amniotic membrane begins to be documented (Table 8.1), pointing essentially to its anti-angiogenic and immunoregulatory properties as the source of this therapeutic effect. However, many questions regarding the use of amniotic membrane in cancer therapy remain unanswered, underlying the need for further research.

Table 8.1 Anti-cancer effects of amniotic membrane or amniotic membrane-derived cells in cultured cells and animal models

Reference	Cell line/animal model	Treatment	Effects
Jiao et al. [51]	Glioma xenograft	Intratumoral administration of single and multiple doses of hAMCs	Inhibited tumor growth Increased BAX, caspase-8 and -3 expression Decreased BCL-2 expression Increased apoptosis
Magatti et al. [52]	Jurkat cells (human T-cell leukaemia) U937 cells (human histiocytic lymphoma)	Co-culture in both direct contact or in transwell setting with hAMCs	Inhibited cancer cell proliferation Cell cycle arrest in G0/G1 phase Down-regulated expression of genes involved in cell cycle promotion Up-regulated expression of negative regulators of cell cycle No induced apoptosis Inhibited cancer cell proliferation
Kang et al. [50]	MDA-MB-231 cells (human breast carcinoma) Breast cancer xenograft	Direct co-culture with hAECs Circumtumoral injection of hAECs	Expression of multiple cytotoxic cytokines and interleukins Inhibited tumor growth without side effects Increased mouse survival rate hAECs intratumorally localized
Niknejad et al. [26]	Hela cells (human cervical cancer) MDA-MB-231 cells (human breast carcinoma)	Co-culture with condition medium of either amniotic membrane tissue or hAECs	Decreased cell viability and proliferation Increased caspase-8 and -3 expression Induced apoptosis Decreased cell motility

References

1. Murphy SL, Xu J, Kochanek KD (2013) National vital statistics reports – deaths: final data for 2010. Natl Vital Stat Rep 61:1–52
2. Jemal A, Bray F, Ferlay J et al (2011) Global cancer statistics. CA Cancer J Clin 61:69–90
3. Gatenby RA, Gillies RJ (2007) Glycolysis in cancer: a potential target for therapy. Int J Biochem Cell Biol 39:1358–1366
4. Valko M, Rhodes C, Moncol J et al (2006) Free radicals, metals and antioxidants in oxidative stress-induced cancer. Chem Biol Interact 160:1–40

5. Verrax J, Taper H, Calderon PB (2008) Targeting cancer cells by an oxidant-based therapy. Curr Mol Pharmacol 1:80–92
6. Hanahan D, Weinberg RA (2000) The hallmarks of cancer. Cell 100:57–70
7. Hanahan D, Weinberg RA (2011) Hallmarks of cancer: the next generation. Cell 144:646–674
8. Kerbel RS, Kamen BA (2004) The anti-angiogenic basis of metronomic chemotherapy. Nat Rev Cancer 4:423–436
9. Urruticoechea A, Alemany R, Balart J et al (2010) Recent advances in cancer therapy: an overview. Curr Pharm Des 16:3–10
10. Folkman J (1971) Tumor angiogenesis: therapeutic implications. N Engl J Med 285:1182–1186
11. Bergers G, Benjamin LE (2003) Tumorigenesis and the angiogenic switch. Nat Rev Cancer 3:401–410
12. Hanahan D, Folkman J (1996) Patterns and emerging mechanisms of the angiogenic switch during tumorigenesis. Cell 86:353–364
13. Kato H, Ichinose Y, Ohta M et al (2004) A randomized trial of adjuvant chemotherapy with Uracil–Tegafur for adenocarcinoma of the lung. N Engl J Med 350:1713–1721
14. Fernandes M, Sridhar MS, Sangwan VS, Rao GN (2005) Amniotic membrane transplantation for ocular surface reconstruction. Cornea 24:643–653
15. Baradaran-Rafii A, Aghayan H-R, Arjmand B, Javadi M-A (2007) Amniotic membrane transplantation. Iran J Ophthalmic Res 2:58–75
16. Hill SJ (2008) Placental amniotic membrane: the pathway to ocular transplantation. AORN 88:731–746
17. Rahman I, Said D, Maharajan V, Dua H (2009) Amniotic membrane in ophthalmology: indications and limitations. Eye 23:1954–1961
18. Dua HS, Gomes JA, King AJ, Maharajan VS (2004) The amniotic membrane in ophthalmology. Surv Ophthalmol 49:51–77
19. Murube J (2006) Early clinical use of amniotic membrane in medicine and ophthalmology. Ocul Surf 4:114–119
20. Singh R, Chacharkar M (2011) Dried gamma-irradiated amniotic membrane as dressing in burn wound care. J Tissue Viability 20:49–54
21. Meller D, Pires RT, Mack RJ et al (2000) Amniotic membrane transplantation for acute chemical or thermal burns. Ophthalmology 107:980–990
22. Gajiwala K, Gajiwala AL (2004) Evaluation of lyophilised, gamma-irradiated amnion as a biological dressing. Cell Tissue Bank 5:73–80
23. Seo JH, Kim YH, Kim JS (2008) Properties of the amniotic membrane may be applicable in cancer therapy. Med Hypotheses 70:812–814
24. Niknejad H, Khayat-Khoei M, Peirovi H (2012) Inhibition of MMPs might increase anticancer properties of amniotic epithelial cells. Med Hypotheses 78:690–691
25. Niknejad H, Yazdanpanah G, Mirmasoumi M et al (2013) Inhibition of HSP90 could be possible mechanism for anti-cancer property of amniotic membrane. Med Hypotheses 81:862–865
26. Niknejad H, Khayat-Khoei M, Peirovi H, Abolghasemi H (2014) Human amniotic epithelial cells induce apoptosis of cancer cells: a new antitumor therapeutic strategy. Cytotherapy 16:33–40
27. Kerbel RS (2000) Tumor angiogenesis: past, present and the near future. Carcinogenesis 21:505–515
28. Folkman J (2002) Role of angiogenesis in tumor growth and metastasis. Semin Oncol 29:15–18
29. Hao Y, Ma DH-K, Hwang DG et al (2000) Identification of antiangiogenic and antiinflammatory proteins in human amniotic membrane. Cornea 19:348–352
30. Fukuda K, Chikama T, Nakamura M, Nishida T (1999) Differential distribution of subchains of the basement membrane components type IV collagen and laminin among the amniotic membrane, cornea and conjunctiva. Cornea 18:73–79

31. Takashima S, Yasuo M, Sanzen N et al (2008) Characterization of laminin isoforms in human amnion. Tissue Cell 40:75–81
32. Hopkinson A, McIntosh RS, Shanmuganathan V et al (2006) Proteomic analysis of amniotic membrane prepared for human transplantation: characterization of proteins and clinical implications. J Proteome Res 5:2226–2235
33. Shao C, Sima J, Zhang SX et al (2004) Suppression of corneal neovascularization by PEDF release from human amniotic membranes. Investig Ophthalmol Vis Sci 45:1758–1762
34. Kobayashi N, Kabuyama Y, Sasaki S et al (2002) Suppression of corneal neovascularization by culture supernatant of human amniotic cells. Cornea 21:62–67
35. Bold RJ, Termuhlen PM, McConkey DJ (1997) Apoptosis, cancer and cancer therapy. Surg Oncol 6:133–142
36. Evan GI, Vousden KH (2001) Proliferation, cell cycle and apoptosis in cancer. Nature 411:342–348
37. Kerr JF, Winterford CM, Harmon BV (1994) Apoptosis: its significance in cancer and cancer therapy. Cancer 73:2013–2026
38. Lowe SW, Lin AW (2000) Apoptosis in cancer. Carcinogenesis 21:485–495
39. Wong RS (2011) Apoptosis in cancer: from pathogenesis to treatment. J Exp Clin Cancer Res 30:87
40. Li H, Niederkorn JY, Neelam S et al (2005) Immunosuppressive factors secreted by human amniotic epithelial cells. Invest Ophthalmol Vis Sci 46:900–907
41. Kubo M, Sonoda Y, Muramatsu R, Usui M (2001) Immunogenicity of human amniotic membrane in experimental xenotransplantation. Invest Ophthalmol Vis Sci 42:1539–1546
42. Toda A, Okabe M, Yoshida T, Nikaido T (2007) The potential of amniotic membrane/amnion-derived cells for regeneration of various tissues. J Pharmacol Sci 105:215–228
43. Li W, He H, Kawakita T et al (2006) Amniotic membrane induces apoptosis of interferon-gamma activated macrophages in vitro. Exp Eye Res 82:282–292
44. He H, Li W, Chen S-Y et al (2008) Suppression of activation and induction of apoptosis in RAW264.7 cells by amniotic membrane extract. Invest Ophthalmol Vis Sci 49:4468–4475. doi:10.1167/iovs.08-1781
45. Mamede AC, Carvalho MJ, Abrantes AM et al (2012) Amniotic membrane: from structure and functions to clinical applications. Cell Tissue Res 349:447–458
46. Belmar-Lopez C, Mendoza G, Oberg D et al (2013) Tissue-derived mesenchymal stromal cells used as vehicles for anti-tumor therapy exert different in vivo effects on migration capacity and tumor growth. BMC Med 11:1–16
47. Klopp AH, Gupta A, Spaeth E et al (2011) Dissecting a discrepancy in the literature: do mesenchymal stem cells support or suppress tumor growth? Stem Cells 29:11–19
48. Strauer B (2003) Stem cell therapy in perspective. Circulation 107:929–934
49. Watt FM, Hogan BL (2000) Out of Eden: stem cells and their niches. Science 287:1427–1430, 80-
50. Kang N-H, Yi B-R, Lim SY et al (2012) Human amniotic membrane-derived epithelial stem cells display anticancer activity in BALB/c female nude mice bearing disseminated breast cancer xenografts. Int J Oncol 40:2022–2028
51. Jiao H, Guan F, Yang B et al (2012) Human amniotic membrane derived-mesenchymal stem cells induce C6 glioma apoptosis in vivo through the Bcl-2/caspase pathways. Mol Biol Rep 39:467–473
52. Magatti M, De Munari S, Vertua E, Parolini O (2012) Amniotic membrane-derived cells inhibit proliferation of cancer cell lines by inducing cell cycle arrest. J Cell Mol Med 16:2208–2218
53. Khakoo AY, Pati S, Anderson SA et al (2006) Human mesenchymal stem cells exert potent antitumorigenic effects in a model of Kaposi's sarcoma. J Exp Med 203:1235–1247
54. Le Blanc K, Rasmusson I, Sundberg B et al (2004) Treatment of severe acute graft-versus-host disease with third party haploidentical mesenchymal stem cells. Lancet 363:1439–1441
55. Ehtesham M, Kabos P, Gutierrez MA et al (2002) Induction of glioblastoma apoptosis using neural stem cell-mediated delivery of tumor necrosis factor-related apoptosis-inducing ligand. Cancer Res 62:7170–7174

56. Childs R, Chernoff A, Contentin N et al (2000) Regression of metastatic renal-cell carcinoma after nonmyeloablative allogeneic peripheral-blood stem-cell transplantation. N Engl J Med 343:750–758
57. Yip K, Reed J (2008) Bcl-2 family proteins and cancer. Oncogene 27:6398–6406
58. Sun Y, Peng Z-L (2009) Programmed cell death and cancer. Postgrad Med J 85:134–140
59. Ghobrial IM, Witzig TE, Adjei AA (2005) Targeting apoptosis pathways in cancer therapy. CA Cancer J Clin 55:178–194
60. Williams G, Stoeber K (2012) The cell cycle and cancer. J Pathol 226:352–364
61. Dranoff G (2004) Cytokines in cancer pathogenesis and cancer therapy. Nat Rev Cancer 4:11–22
62. Kim S-W, Zhang H-Z, Kim CE et al (2012) Amniotic mesenchymal stem cells have robust angiogenic properties and are effective in treating hindlimb ischaemia. Cardiovasc Res 93:525–534
63. Kim S-W, Zhang H-Z, Guo L et al (2012) Amniotic mesenchymal stem cells enhance wound healing in diabetic NOD/SCID mice through high angiogenic and engraftment capabilities. PLoS One 7:e41105
64. Miyata Y, Nakamoto H, Neckers L (2013) The therapeutic target Hsp90 and cancer hallmarks. Curr Pharm Des 19:347–365
65. Trepel J, Mollapour M, Giaccone G, Neckers L (2010) Targeting the dynamic HSP90 complex in cancer. Nat Rev Cancer 10:537–549

Chapter 9
Amniotic Membrane in Oral Medicine

Manuel Marques Ferreira and Eunice Virgínia Palmeirão Carrilho

Abstract It has been discovered that human amniotic membrane, have several properties which allow them to be applied in many medical fields, namely in the treatment of superficial wounds and burns. In dentistry, some surgical procedures results the exposure of bone surfaces to the oral cavity that is prone not only to infection but also to scar formation during secondary healing. The proper covering of the exposed periostium or bone surface is often needed to prevent these complications. Therefore, the use of amniotic membrane transplantation, to accelerate oral wound healing in humans, regardless of the genetic background, is promising. Regenerative endodontic procedure aims to provide a continuation of root development. In this field the use of amniotic membranes inside the root canals, after disinfection of the immature nonvital permanent teeth, is a promising method. This chapter will discuss the use of amniotic membranes in several fields of oral medicine.

Keywords Amniotic membrane • Oroantral fistula • Pulp regeneration • Periapical surgery • Oral surgery

9.1 Introduction

Surgical procedures in the oral cavity that include vestibuloplasty for preprosthetic treatments and resection of broad mucosal lesions in the gingival and alveolar areas result in the exposure of bone surfaces to the oral cavity. These are prone not only to infection but also to scar formation during secondary healing.

M.M. Ferreira (✉) • E.V.P. Carrilho
Faculty of Medicine, University of Coimbra, Coimbra, Portugal
e-mail: mmferreira@fmed.uc.pt

A.C. Mamede, M.F. Botelho (eds.), *Amniotic Membrane: Origin, Characterization and Medical Applications*,
DOI 10.1007/978-94-017-9975-1_9

A proper covering of the exposed periostium or bone surface is often needed to prevent these complications.

Mucosal and skin autografts have been used for this purpose and seem biologically ideal. However, these grafts require a separate surgical procedure and have other disadvantages, such as limited size of the donor mucosal graft and patient discomfort resulting from mismatching between the mucosa and the skin graft.

For these reasons, other biologic materials, such as bovine-derived collagen, chitin membranes, and human skin allografts, have been used to cover surgical defects, and the biocompatibility and availability of these materials have been investigated.

The human amniotic membrane, the innermost layer of the placenta, is a suitable tissue for allografts because of its low immunogenicity [1]. It also possesses anti-inflammatory, wound-protecting, and scar reducing properties [2].

9.2 Embryogenesis of the Mouth

The mouth is a cavity occupied by the tongue and teeth. It is limited in the front by the lips, behind by the jaws, above by the palate, below by the floor of the mouth and the tongue and laterally by the cheeks. At about the fourth week of gestation, the first branquial arch establishes the maxillary and mandibular process of the oral cavity. Thus the oral cavity is formed after development of the secondary palate, in the seventh and 8 weeks and is completed at the third month of gestation [3].

The mandible is formed from the Merckel's cartilage after condensation of ectomesenchyme by endochondral and intramembranous ossification. The maxilla develops from a center of ossification of the maxillary process of the first branchial arch by intramembranous ossification [3]. Covering the bone tissue is the oral mucosa, which is divided into three categories: lining mucosa, masticatory mucosa and specialized mucosa [4]. The attached gingiva, bonded firmly to the bone, is continuous with alveolar mucosa. In the oral cavity the teeth are inside the bone with the periodontal ligament joining the tissues (Fig. 9.1). The tissues of the teeth are enamel, cementum and dentin recovering the pulp cavity (Fig. 9.2). The alveolar bone is a connective tissue with cells and extracellular matrix with 60 % mineral components, 20 % water and 20 % organic. The organic material components are 80 % hydroxiapatite and cells which are mainly osteoblasts and osteoclasts [5].

9.3 Regeneration in Periapical Surgery

Endodontic retreatment surgery is a therapeutic option in teeth with apical periodontitis and may be indicated when there is failure in primary endodontic treatment or nonsurgical endodontic retreatment. Another indication is the

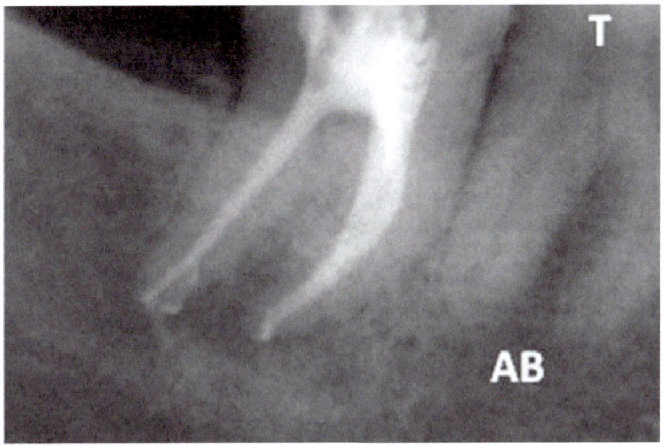

Fig. 9.1 Clinical case of the teeth (*T*) inside the alveolar bone (*AB*)

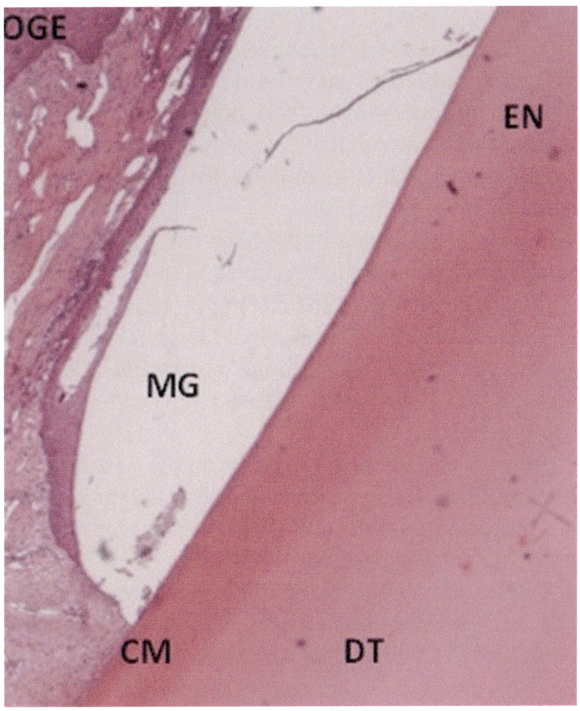

Fig. 9.2 Detail of the marginal gingiva with oral gengival epithelium (*OGE*), marginal groove (*MG*), enamel (*EN*), cementum (*CM*) and dentin (*DT*). Hematoxylin and eosin stain (×40)

inability to perform nonsurgical retreatment or for excision of lesions for histological analysis [6, 7].

Currently, this surgery is performed with the aid of a surgical microscope and with the use of ultrasound for preparation of a cavity in the root apex, which is sealed with cement-based mineral trióxide (MTA) [8]. The purpose of this treatment is to prevent the microleakage of bacteria and fluids into the dental canal and the periapical tissue, so that the residual bacteria don not feed and perpetuate the injury. This technique, known as microsurgery, now has a success rate of around 90 %, with values similar to those of primary endodontic treatment [9].

These surgical procedures include several phases and concepts:

(a) Need for deep anesthesia and perfect hemostasis;
(b) Manipulation of the soft tissues;
(c) Manipulation of hard tissue, bone and tooth root;
(d) Correct surgical access;
(e) Surgical access to the apical region of the tooth;
(f) Periradicular curettage, resection of the root apex of the tooth;
(g) Preparation of the apical root recess;
(h) Filling the cavity created in the apical portion of the root;
(i) Repositioning and suturing of soft tissue;
(j) Post-operative care.

After incision and detachment of the mucoperiosteum, the endodontist accesses the periapical lesion, and needs to gain surgical access with rotary instruments, through the cortical bone. This, however, may sometimes already destroyed by the injury, with fenestration of root dental organ. As a consequence of the cystic lesion, there is a significant loss of bone structure, which needs local conditions to be created leading to its healing.

This wound healing in the bone and dental tissues of the periodontal ligament can be made by repair or regeneration of lost tissue.

Repair occurs when the healing process with new tissue is formed in different cells and structures of the original fabric.

Regeneration occurs with the destruction of the original tissue disease process and it is substituted by a cellular fabric with a similar composition structure and ability to react [10].

Repair or regeneration depends on the type of healing, the availability of undifferentiated mesenchymal cells (stem cells), growth factors and cell differentiation and the local environment, such as the presence of adhesion molecules, extracellular matrix proteins and not collagens [11].

Endodontic surgery is a clinical procedure of great dynamism and development and the – healing after periapical surgery, includes the regeneration of alveolar bone, periodontal ligament and cementum [11].

However, it must be supported by quality scientific research and a constant update of new techniques and materials. It is obvious that the root end that was removed cannot be restored.

The regeneration of tissues after apical surgery means: (I) regeneration of alveolar bone and peri-radicular tissues; (II) restoring periodontal ligament, to surround the surgically exposed root surface, and (III) the formation of new cementum on the sectioned root surface.

However, the nature of the periapical tissues regenerated after use in tissue regeneration of apical surgery remains unknown, although periapical radiographs reveal some evidence of the formation of new periodontal space (PDL).

The techniques for tissue regeneration are based on cell differentiation in cell proliferation and the induction of tissue formation.

These effects are obtained with the various protocols used in clinical practice such as bone substitutes, barrier membranes, growth factors, or a combination of these agents and materials.

Guided tissue regeneration (GTR) is a therapeutic procedure which has been recommended by some authors in periapical surgery in order to promote healing of bone injury [12].

GTR was introduced and used with great success in apical margin defects for periodontal regeneration. The placing of a physical barrier over the defect is to prevent bone growth and rapid proliferation of oral epithelium and gingival connective tissue to bone lesion, allowing the stem cells of the periodontal ligament and bone to colonize the blood clot and thus permit regeneration of lost tissue [13].

However, the use of these materials has some disadvantages such as high costs, the possibility of contamination and infection, as well as the difficulty of bringing together the edges of the wound and complete coverage of the membrane [14].

Some studies also report the risk of ankylosis that may occur with the use of membrane and tissue regeneration [15, 16].

There are several local conditions that may be encountered during apical surgery. These include: (I) large apical cystic lesions, (II) lesions in bone tunnel and (III) apical margin lesions.

In large cystic lesions or lesions in bone tunnel, where there is destruction of bone from cortical bone to the palate, that it can interfere with bone growth, resulting in the formation of a fibrous scar tissue.

For didactic, surgical and therapeutic aspects, periapical lesions are classified according to various authors [8, 17, 18]:

Type I – Lesion is limited to the apical area.

Type II – Lesion with cortical bone destruction lingual/palatal (with or without erosion of cortical bone), resulting in a failure of the tunnel.

Type III – Injury apical margin with complete exposure of the buccal surface of the root teeth.

Experimental animal studies and assessments of clinical cases are not unanimous as regards the use of membranes in periapical surgery.

Bernabe et al. [19] evaluated the healing of periapical tissues after the use of bone grafts, membrane, or combinations in guided tissue regeneration in apical surgery, carried out on dogs. Histological evaluation and histomorphometry showed that the repair process was similar in all groups.

Taschieri et al. [20] assessed the outcome of large surgery apical lesions (>10 mm), with or without guided tissue regeneration.

After 1 year of follow-up, the success rates of the test sites (75.0 %) and control sites (61.5 %) did not differ significantly.

In recent years, platelet-rich plasma (PRP) combined with graft materials has been used for the purpose of periodontal regeneration [21].

Platelet-rich plasma is rich in growth factors and target-specific polypeptides that play a role in cell proliferation and differentiation and can thus encourage wound repair.

PRP is a volume of autologous plasma that has a higher platelet concentration than baseline.

One histologic study comparing autologous platelet concentrate with a bioabsorbable membrane in periodontal defects found similar results between the two groups, suggesting that autologous platelet concentrate could be used in lieu of a membrane for periodontal GTR applications [22].

However, in endodontics, no reports have been found comparing the efficacy and equivalence of PRP with GTR membrane during the treatment of apical margin defects.

The application of PRP to apical margin defects, although the improvement registered over GTR is marginal, has emerged as an alternate route of treatment to the established practice of GTR membrane.

In conclusion and in resemblance to periodontally intrabony defects, it appears that GTR principles rather than the use of an osteoconductive biomaterial would be justified to enhance the surgical endodontic outcome compared with the traditional approach [23].

Published work suggests that human amniotic membrane, provides a source of undifferentiated cells with great potential that may be used in regenerative medicine [24].

As the cells are derived from epiblastic amniotic membrane before the phase of gastrulation, they are considered a source of stem cells [25].

Pre-clinical and clinical studies have shown the use of amniotic membrane stem cells for repairing tissues such as the cornea [26] or spinal cord injury [27], cerebral ischemia [28] and Parkinson's disease [29].

Despite amniotic membrane containing a large amount of undifferentiated cells that can be cryopreserved and stored as a useful potential source, fetal tissues are routinely rejected after childbirth [30–33].

A major advantage of cells isolated from the human amniotic membrane is that they are harvested after birth and can be cryogenically stored to be available in a timely manner for patient therapy after being thawed and expanded for use in tissue engineering, cell transplantation and gene therapy.

Therefore, the human amniotic membrane represents a very useful source of progenitor cells for a variety of applications.

The ability of amniotic membrane to stimulate rapid production of collagen fibres in the gingival wound has also suggested. This membrane contains transforming growth factor beta (TGF-β) and tissue inhibitors of metalloproteinase polypeptides (TIMPs), which up-regulate the production of collagen fibres by fibroblasts in wound healing [34, 35].

Amniotic membrane contains growth factors, necessary for epithelialization, and provides substrates such as laminins for rapid epithelial cell attachment which, in turn, up-regulates the expression of growth factor receptors such as epidermal growth factor (EGF) receptor on epithelial cells [35, 36].

Therefore, one may assume that these growth factors might induce rapid gingival epithelial cell migration and attachment to the wound area, thereby stimulating rapid cell proliferation and differentiation.

Therefore, the use of amniotic membrane transplantation to accelerate gingival wound healing in humans, regardless of genetic background, is promising.

9.4 Oroantral Fistula

The jaw bone is a pneumatic structure, which has inside the maxillary sinus and the paranasal sinus, bilaterally.

Because of the size of the sinus and because the proximity to the roots of some upper posterior teeth makes it easier, in some circumstances during dental extractions direct access is formed between this and the oral cavity, being termed or antral communication.

When this communication access between the oral cavity and sinus is lined with epithelial tissue, originating from the proliferation of tissues surrounding the communication, it is designated and or antral fistula [37].

The etiology of this pathology are extractions of teeth with long roots that protrude into the maxillary sinus.

A diagnosis of oroantral fistulae involves clinical and radiographic procedures. Clinical signs reported by patients are continued passing of fluid into the nose, nasal timbre, disorders in swallowing liquids and food, halitosis, runny nose, altered taste, unilateral nasal obstruction, facial pain or frontal headache (when acute maxillary sinusitis), unilateral nasal discharge and nocturnal cough due to drainage of exudate into the pharynx. The exudate swallowed may produce anorexia in the morning, as well as the patient suffering from epistaxis on the affected side and inability to blow or smoke.

One of the main complications of oroantral communications is acute or chronic maxillary sinusitis due to contamination of the maxillary sinus by the bacterial flora of the oral cavity.

Communications may be evidenced through periapical radiography where one observes the discontinuity of radiopaque line that delimits the maxillary sinus floor. Extraoral radiographs (panoramic and Incidence of Waters) are also limited with respect to small communications but may have great importance in the observation of the maxillary sinus involved, which may present a diffuse radiopacity (sinus opacification) , compared with the opposite.

The patient with acute maxillary sinusitis may have swelling and redness in the area of the sinus and malar eminence, as well as pain in the eye. Palpation of the maxilla increases pain, and teeth with roots adjacent to the sinus will often present pain or be sensitive to percussion. The treatment of oroantral communications should be done immediately, so that the opening is created, or later, in cases of failed attempts at primary closure [38].

The closure of the fistula can be made by a free tissue graft taken from the palate or via buccal flap.

The amniotic membrane possesses antibacterial, anti-tumor, anti-inflammatory properties and promotes tissue healing. It also reduces pain, inhibits scarring, and shows little or no immunogenicity. These properties enable surgeons to apply the human amniotic membrane graft on various tissue surfaces without suturing. The low cost of human amniotic membrane graft preparation and the very good clinical results in different applications, have been relevant in proposing the amnion as an alternative to other natural and synthetic wound dressings. This application may be used in the closure of the oroantral fistula.

9.5 Apexogenesis

In developing teeth, root formation starts with the epithelial cells from the cervical loop that proliferate apically and influence the differentiation of odontoblasts from undifferentiated mesenchymal cells and cementoblasts from -mesenchymal follicles [39].

This apically extending two-layered epithelial wall (merging of the inner and outer enamel epithelium) forms Hertwig's epithelial root sheath (HERS), which is responsible for determining the shape of the root(s) and forms cementum through epithelial-mesenchymal transition [40].

It is known that dental papilla contributes to tooth formation and eventually converts to pulp tissue [39, 41, 42].

Cessation of root development caused by trauma or pulpal disease presents both an endodontic and restorative challenge. The divergent apical architecture makes complete debridement and control of obturation material nearly impossible [43]. Fragile root canal walls may be too weak to withstand the normal forces of mastication, becoming more prone to fracture [44].

Pulp regeneration can be a form of therapy for resolving teeth with open apex and thin walls resulting from trauma, decay and iatrogenic procedures, leading to necrosis of the pulp and stopping tooth development.

Several published works refer to the use of paste consisting of three antibiotics (metronidazole, ciprofloxacin and minocycline), in order to eradicate all bacteria in the pulp space. After this procedure, the channel is irrigated with a solution of sodium hypochlorite (at a concentration of 2.5–5.25 %) to remove debris and organic waste [45]. It has been reported that the apical papilla harbours multipotent cells (MSCs) that express various MSC markers and that they are capable of forming odontoblast-like cells, produce dentin in vivo, and are likely to be the cell source of primary odontoblasts for root dentin formation [46].

In order for these cells to enter the pulp space, bleeding is induced in the periapical tissues vessel, so as to enter the canal and be a source of cells to complete root teeth development.

After the formation of the clot within the pulp chamber, a bioactive cement (MTA) is placed over the clot, which will serve as a matrix for the filling of the root canal [45].

Amniotic membrane contains growth factors, necessary for epithelialization, and provides substrates such as laminins for rapid epithelial cell attachment which, up-regulates the expression of growth factor receptors such as EGF receptor on epithelial cells [35, 36].

Therefore, one may assume that these growth factors might induce rapid cell migration and attachment to the wound area, by this means stimulating rapid cell proliferation and differentiation.

So, the use of amniotic membrane transplantation in the root canal may be promising as a scaffold to be used in apexogenesis therapy, instead of the use of a biomaterial as MTA.

References

1. Kubo M, Sonoda Y, Muramatsu R, Usui M (2001) Immunogenicity of human amniotic membrane in experimental xenotransplantation. Invest Ophthalmol Vis Sci 42:1539–1546
2. Faulk WP, Stevens PJ, Burgos H et al (1980) Human amnion as an adjunct in wound healing. Lancet 315:1156–1158
3. Nanci A (2008) Embriology of the head, face and oral cavity. Ten cate's oral. Histol. Dev. Struct. Funct. Elsevier Health Sciences, p 411
4. Roed-Petersen B, Renstrup G (1969) A topographical classification of the oral mucosa suitable for electronic data processing: its application to 560 leukoplakias. Acta Odontol Scand 27: 681–695
5. Garant PR (2003) Oral cells and tissues. Quintessence Publishing Company, Chicago, 430
6. Gutmann J, Harrison J (1985) Posterior endodontic surgery: anatomical considerations and clinical techniques. Int Endod J 18:8–34
7. Tsesis I, Faivishevsky V, Kfir A, Rosen E (2009) Outcome of surgical endodontic treatment performed by a modern technique: a meta-analysis of literature. J Endod 35:1505–1511
8. Kim S, Kratchman S (2006) Modern endodontic surgery concepts and practice: a review. J Endod 32:601–623
9. Tsesis I, Rosen E, Schwartz-Arad D, Fuss Z (2006) Retrospective evaluation of surgical endodontic treatment: traditional versus modern technique. J Endod 32:412–416
10. Pecora G, De Leonardis D, Ibrahim N et al (2001) The use of calcium sulphate in the surgical treatment of a "through and through" periradicular lesion. Int Endod J 34:189–197
11. Lin L, Chen MY-H, Ricucci D, Rosenberg P a (2010) Guided tissue regeneration in periapical surgery. J Endod 36:618–625
12. Tobón S, Arismendi J, Marín M et al (2002) Comparison between a conventional technique and two bone regeneration techniques in periradicular surgery. Int Endod J 35:635–641
13. Dahlin C, Linde A, Gottlow J, Nyman S (1988) Healing of bone defects by guided tissue regeneration. Plast Reconstr Surg 81:672–676
14. Pecora G, Baek S, Rethnam S, Kim S (1997) Barrier membrane techniques in endodontic microsurgery. Dent Clin N Am 41:585–602
15. Karring T, Nyman S, Lindhe J, Sirirat M (1984) Potentials for root resorption during periodontal wound healing. J Clin Periodontol 11:41–52
16. Cury PR, Furuse C, Trierveiler M et al (2005) Root resorption and ankylosis associated with guided tissue regeneration. J Am Dent Assoc 136:337–341

17. von Arx T, Cochran DL (2001) Rationale for the application of the GTR principle using a barrier membrane in endodontic surgery: a proposal of classification and literature review. Int J Periodontics Restorative Dent 21:127–139
18. Dietrich T, Zunker P, Dietrich D, Bernimoulin J-P (2002) Apicomarginal defects in periradicular surgery: classification and diagnostic aspects. Oral Surg Oral Med Oral Pathol Oral Radiol Endod 94:233–239. doi:10.1067/moe.2002.123864
19. Bernabé PFE, Gomes-Filho JE, Cintra LTÂ et al (2010) Histologic evaluation of the use of membrane, bone graft and MTA in apical surgery. Oral Surg Oral Med Oral Pathol Oral Radiol Endod 109:309–314
20. Taschieri S, Del Fabbro M, Testori T, Weinstein R (2007) Efficacy of xenogeneic bone grafting with guided tissue regeneration in the management of bone defects after surgical endodontics. J Oral Maxillofac Surg 65:1121–1127
21. Plachokova AS, Nikolidakis D, Mulder J et al (2008) Effect of platelet-rich plasma on bone regeneration in dentistry: a systematic review. Clin Oral Implants Res 19:539–545
22. Artzi Z, Wasersprung N, Weinreb M et al (2012) Effect of guided tissue regeneration on newly formed bone and cementum in periapical tissue healing after endodontic surgery: an in vivo study in the cat. J Endod 38:163–169
23. Nyman S, Gottlow J, Lindhe J et al (1987) New attachment formation by guided tissue regeneration. J Periodontal Res 22:252–254
24. Toda A, Okabe M, Yoshida T, Nikaido T (2007) The potential of amniotic membrane/amnion-derived cells for regeneration of various tissues. J Pharmacol Sci 105:215–228
25. Ilancheran S, Michalska A, Peh G et al (2007) Stem cells derived from human fetal membranes display multilineage differentiation potential. Biol Reprod 77:577–588
26. Shimmura S, Tsubota K (2002) Ocular surface reconstruction update. Curr Opin Ophthalmol 13:213–219
27. Sankar V, Muthusamy R (2003) Role of human amniotic epithelial cell transplantation in spinal cord injury repair research. Neuroscience 118:11–17
28. Sakuragawa N, Misawa H, Ohsugi K et al (1997) Evidence for active acetylcholine metabolism in human amniotic epithelial cells: applicable to intracerebral allografting for neurologic disease. Neurosci Lett 232:53–56
29. Kakishita K, Nakao N, Sakuragawa N, Itakura T (2003) Implantation of human amniotic epithelial cells prevents the degeneration of nigral dopamine neurons in rats with 6-hydroxydopamine lesions. Brain Res 980:48–56
30. Chang Y, Hwang S, Tseng C et al (2010) Isolation of mesenchymal stem cells with neurogenic potential from the mesoderm of the amniotic membrane. Cells Tissues Organs 192:93–105
31. Niknejad H, Peirovi H, Jorjani M et al (2008) Properties of the amniotic membrane for potential use in tissue engineering. Eur Cell Mater 15:88–99
32. Hennerbichler S, Reichl B, Pleiner D et al (2007) The influence of various storage conditions on cell viability in amniotic membrane. Cell Tissue Bank 8:1–8. doi:10.1007/s10561-006-9002-3
33. Wilshaw SS-P, Kearney JNJ, Fisher J, Ingham E (2006) Production of an acellular amniotic membrane matrix for use in tissue engineering. Tissue Eng 12:2117–2129
34. Hao Y, Ma DH-K, Hwang DG et al (2000) Identification of antiangiogenic and antiinflammatory proteins in human amniotic membrane. Cornea 19:348–352
35. Yam H-F, Pang C-P, Fan DS-P et al (2002) Growth factor changes in ex vivo expansion of human limbal epithelial cells on human amniotic membrane. Cornea 21:101
36. Kurpakus M, Daneshvar C, Davenport J, Kim A (1999) Human corneal epithelial cell adhesion to laminins. Curr Eye Res 19:106–114
37. Visscher SH, van Minnen B, Bos RR (2010) Closure of oroantral communications: a review of the literature. J Oral Maxillofac Surg 68:1384–1391
38. Lazow SK (1999) Surgical management of the oroantral fistula: flap procedures. Oper Tech Otolaryngol Head Neck Surg 10:148–152
39. D'Souza R (2002) Development of the pulpodentin complex. In: Hargreaves KM, Goodi HE (eds) Seltzer Bender's Dent. pulp, 3rd edn. Quintessence Publishing, pp 13–40

40. Sonoyama W, Seo B-M, Yamaza T, Shi S (2007) Human hertwig's epithelial root sheath cells play crucial roles in cementum formation. J Dent Res 86:594–599
41. Linde A, Goldberg M (1993) Dentinogenesis. Crit Rev Oral Biol Med 4:679–728
42. Ruch JV, Lesot H, Begue-Kirn C (1995) Odontoblast differentiation. Int J Dev Biol 39:51–68
43. Rafter M (2005) Apexification: a review. Dent Traumatol 21:1–8
44. Rabie G, Trope M, Garcia C, Tronstad L (1985) Strengthening and restoration of immature teeth with an acid-etch resin technique. Dent Traumatol 1:246–256
45. Banchs F, Trope M (2004) Revascularization of immature permanent teeth with apical periodontitis: new treatment protocol? J Endod 30:196–200
46. Ferreira MM, Botelho MF, Abrantes M et al (2014) Quantitative scintigraphic analysis of pulp revascularization in autotransplanted teeth in dogs. Arch Oral Biol 55:825–829

Chapter 10
Amniotic Membrane in Gynaecology

Maria João Carvalho

Abstract The application of amniotic membrane (AM) in several areas of Medicine has been developed in recent decades. It harbors characteristics that promote a microenvironment favorable to tissue repair and regeneration possibly due to stem cell-*like* properties. Considering gynaecological applications, there are in fact two fields in which it has been used, the Asherman Syndrome (intrauterine adhesions – IUA) and vaginoplasties. The principle for the application of AM on IUA is the use of a biologically active mechanical separator after hysteroscopic adhesiolysis. The reports are limited but point that amnion graft ameliorates the clinical outcome and the type of graft, fresh or dried, does not seem to influence the outcomes. The Mayer-Rokitansky-Küster-Hauser Syndrome (MRKH) can be categorized in type I, with isolated utero-vaginal aplasia and type II, an incomplete aplasia and/or association with other malformation. The application of AM in based on a metaplasia of amniotic epithelium into vaginal epithelium. It has been applied in techniques of vaginoplasty, is a safe and simple procedure with satisfactory functional results.

Keywords Amniotic membrane • Histeroscopia adhesiolysis • Vaginoplasty

10.1 Asherman Syndrome

Intrauterine adhesions were first described in 1894 by Heinrich Fritsch but the complete description was performed by Joseph Asherman, an Israeli gynecologist, and nowadays it is still recognized as Asherman syndrome (AS). The first reports concerned stenosis of cervical ostium and latter were published intrauterine adhesions

M.J. Carvalho
Gynaecology A Service, Hospitalar and University Center of Coimbra, Biophysics Unit, Faculty of Medicine, Center of Investigation of Genetics and Oncobiology (CIMAGO), University of Coimbra, Coimbra, Portugal
e-mail: mariajoaosflcarvalho@gmail.com

A.C. Mamede, M.F. Botelho (eds.), *Amniotic Membrane: Origin, Characterization and Medical Applications*,
DOI 10.1007/978-94-017-9975-1_10

(IUA) that involved the cavity [1]. IUA are permanent adhesions to the uterine walls that obliterates partial or totally the uterine cavity after a trauma to the basalis endometrial layer. The consequences of these adhesions include dysfunctional endometrium that can lead to infertility and altered menstrual pattern (amenorrhea and hypomenorrhea), depending on the degree and location in the uterine cavity. The origin of this adhesion include any type of uterine trauma either following a less invasive procedure or an aggressive curettage following a pregnancy complication. Clinically, the AS can be associated to infertility, miscarriage, impaired implantation after in-vitro fertilization and abnormal placentation [2]. The prevalence of IUA is around 1.5 % in infertile women and 7 % in women with secondary amenorrhea [3]. But it is recognizable that the presence of IUA can be associated with no symptoms. The introduction of hysteroscopy revealed a much more common disease. The main risk factor include abortions, genital tuberculosis and any other intervention to the uterine cavity. From the physiological point of view, some studies have pointed a modification in endometrial vascularization that reduce endometrial receptivity and regeneration [4]. Probably there are a group of molecules involved in adhesion formation, as β-fibroblast growth factor, platelet derived growth factor and transforming growth factor type 1 [5]. The central role for this pathogenesis seems to be inflammation, which produces release of factors that lead to fibrotic tissue after trauma combined with ischemia induced by a traumatic procedure. The most frequent risk factors include iatrogenic trauma to the endometrium, listed in Table 10.1. In clinical practice there are cases that develop adhesions without any traumatic event associated.

The hysterotomy approach is an open-surgery with a transfundal separation of the uterine cavity. It is a rare intervention due to the poor results and technical complexity. The D&C is a blind intervention with risk of perforation and lacking possibility of definitive diagnosis. Finally, the hysteroscopy is the gold standard. It is a diagnostic and surgical procedure, possible in an office setting, sometimes with just a tip of the hysteroscope and uterine distention. Surgical hysteroscopy is used for firm and dense adhesions, either with cold-knife or electric or laser surgery. Over the years were described several methods including myometrial scoring that consists of incisions in the cavity that enlarge the uterus and recover unfunctional endometrium [6]. The results after hysteroscopic treatment depend on the severity of adhesions and recurrence. Usually the menstrual pattern recover but the fertility success is more variable.

For the prevention of the adhesions were described several strategies. The recurrence rate can range 3.1–23.5 % [1]. The purely cervicoisthmic adhesions have a good prognosis. The intrauterine device (IUD) was one of the first procedures described, helping to separate the uterine walls. The results in literature are not precise, some reports refer that the size if the IUD is important and the usual T-shaped are too tiny. It was also mentioned that the inflammation of the IUD could have a deleterious impact [7]. Others methods to prevent adhesion recurrence are intrauterine balloon stent, Foley catheter, hyaluronic acid (HA) and anti-adhesion

Table 10.1 Risk factors for intrauterine adhesions

Risk factors	Frequency
Miscarriage curettage	66.7 %
Postpartum curettage	21.5 %
Caesarean section	2 %
Trophobalstic disease evacuation	0.6 %
Mullerian duct malformation	16 %
Infection (genital tuberculosis)	4 %
Diagnostic curettage	1.6 %
Abdominal myomectomy	1.3 %
Uterine artery embolization	14 %
Hysteroscopic surgery	
Metroplasty	6 %
Myomectomy (single myoma)	31.3 %
Myomectomy (multiple myomas)	45.5 %
Endometrial ablation	36.4 %
Insertion of intrauterine device (IUD)	0.2 %
Uterine compressive sutures for post-partum haemorrhage	18.5 %

Adapted from Conforti et al. [1]

barriers [8, 9]. No randomized controlled trials are available to assure the best practice considering reduction of adhesion recurrence.

The medical treatment to restore functional endometrium is a consensual approach but the precise regimen is not defined. The aim is to promote a fast growth to prevent scar formation of the normal endometrium. The majority of the regimens include oestrogen-progestin treatment [1]. Sildenafil citrate is also used to restore endometrial thickness in several clinical contexts due to its vasodilator effect [10]. Recently was reported the application of endometrial stem cells from autologous stem cells isolated from bone marrow in a patient with AS [11].

The post-operative evaluation of the uterine cavity is recommended in order to access the restoration of endometrium and possible recurrences. The hysteroscopy is the most common used method in clinical practice.

10.1.1 Rational for Amniotic Membrane Application

The principle for the application of amniotic membrane on IUA is the use of a biologically active mechanical separator after hysteroscopic adhesiolysis. In one hand it suppresses adhesion formation and on the other hand promotes epithelial healing. The human amnion cells (hAC) produces factors that create a microenvironment favorable for tissue repair and regeneration. One of the proposed mechanism is the stimulation of endogenous stem cells, a small population of quiescent cells with regenerative properties. The stroma expresses growth factors, natural inhibitors of proteases, antiangiogenic factors (inhibitors of metalloproteinases and nitric oxide synthase) and anti-inflammatory proteins, interleukin (IL)-10 and IL-1-receptor antagonist that points it as an epithelialization promotion, inflammation and fibrosis inhibition. The amniotic membrane can function as an anatomic barrier and have additional benefits as inhibition of infection, fibrosis and re-adhesions. The adhesiolysis may expose endometrial cavity to infection and amniotic membrane can suppress bacterial proliferation and the elimination of the dead space between graft and wound eliminates the exudates and anchorage for bacterial growth [12]. This mechanism can be explained by the trap of the bacteria between the thin fibrin matrix that links the collagen fibers of the graft and the collagen of the wound. This fibrin matrix is the substrate for phagocytes migration. Amniotic membrane also downregulates the transforming growth factor-β pathway of myofiblroblastic cells that deposit collagen and leads to fibrosis [13]. The amniotic membrane also promotes epithelialization as it facilitates the migration of epithelial cells, reinforces the adhesion of epithelial cells and its differentiation and prevents apoptosis [12]. Concerning immunological benefits, the epithelial surface of the membrane lacks human leukocyte antigen (HLA) A, B, C, DR and β2-microglobulin. This permits the utilization of different donor placenta. The risk of cross-infection is decrease using a membrane derived from elective cesarean section without meconium from a seronegative mother considering hepatite B, C, human immunodeficiency virus (HIV) and syphilis. The application of preserved membrane decreases the risk of cross-infection but the impact on epithelial cells viability and growth factors can be a question of debate. The lyophilized membrane can be convenient for transport, storage and sterilization [3]. Preservation of fresh membrane has been demonstrated for 21 days. There are reports that emphasize a limit of 7 days to keep the surface of the membrane apart [3].

10.1.2 Reports Considering Amniotic Membrane Application

The reports of the application of amniotic membrane after intrauterine adhesyolisis were performed after hysteroscopic procedures [3, 12], Table 10.2. The hysteroscopy was performed with a rigid hysteroscope and simultaneously a laparoscopic visualization. The fresh membranes were obtained 1–6 days before de surgery from elective cesarean sections. The amniotic membrane with the corion were separated from the placenta, washed with a saline solution and stored at 4 °C with a 50,000 IU of crystalline penicillin. The membrane was then peeled free from the corion, obtaining the epithelial layer on the basement membrane with an underlying collagen matrix containing few fibroblasts. In one work the use of fresh membrane was compared with sterilized freeze-dried (lyophilized)

Table 10.2 Summary of publication using amniotic membrane in hysteroscopic adhesiolysis

Author	Year	Number of cases	Objectives	Material and methods	Results and conclusions
Amer et al. [12]	2006	25	Evaluate efficacy of amnion graft after hysteroscopic lysis	Hysteroscopic adhesiolysis was followed by intrauterine application of a fresh amnion graft over an inflated balloon of a Foley's catheter for 2 week	No infections Adhesion reformation after 4 months 48 %, all in the group of severe adhesions
Amer et al. [3]	2010	45	Evaluate efficacy of fresh and dried amnion graft after hysteroscopic lysis of severe intrauterine adhesions in decreasing its recurrence and encouraging endometrial regeneration	Prospective randomized study, using intrauterine balloon only (group 1) or either fresh amnion graft (group 2) or dried amnion graft (group 3) for 2 weeks	Improvement in adhesion grade with amnion graft vs intrauterine balloon alone. Improvement was greater with fresh amnion than with dried amnion

amnion grafts obtained from Atomic Energy Organization of Egypt [3]. The amniotic membrane was hydrated with a saline solution for 10 min before the procedure.

The amnion was placed with the corion surface outside on the surface of a pediatric Folly's catheter number 10F. The catheter was then placed inside the uterine cavity and insufflate with 3.5–5 mL of saline solution. The balloon was removed 2 weeks after the intervention. The patients took a hormonal treatment with estrogens or sequential administration f natural estrogens and progestogens. The uterine cavity was evaluated 2–4 months after intervention to assess the possible recurrence of adhesions and the endometrial regeneration.

The primary outcomes were menstrual restoration, uterine length and sounding, improvement of adhesions, need to repeat treatment and complications rate.

The first report of Amer et al. [12] applied fresh membrane after the hysteroscopic adhesiolysis. The restoration of normal menstrual flow was achieved in 80 % of the patients (N = 25). The improvement in IUA was reported in 100 % of patients with moderate disease but the group with severe disease, 92 % had adhesions inside the uterine cavity but were minimal in the majority of the cases. In 2010, the investigators published a pilot randomized comparative study with blinded independent evaluations. There were three groups compared, group one received amniotic balloon without amnion graft; group two received fresh amnion graft and group three received dried amnion. The increase in uterine length and menstrual cycle was reported in all groups, without statistical differences. The adhesion grade was reduced in the amnion graft groups, reaching statistical significance ($p = 0.003$). The improvement of adhesion grade, menstruation and uterine length was greater with fresh amnion than with dried amnion but the only parameter with statistical significance was adhesion grade ($p = 0.01$). The pregnancy rate was twice in the graft groups comparing with intrauterine balloon alone but without significance. These studies are composed by a restrict number of patients that limit valid conclusions. The results point that amnion graft ameliorates the results and the type of graft, fresh or dried, does not seem to influence the endpoints. Due to the limited number of cases in a single center, a multicenter study may improve re-enforce the conclusions.

10.2 Congenital Absence of the Vagina

The syndrome was described for the first time by Mayer in 1829 and Rokitansky published a similar malformations, in 1910 Küster performed the first review about the syndrome. Hauser and Schreiner called for the first time Mayer-Rokitansky-Küster-Hauser Syndrome (MRKH) in 1961 [14]. The syndrome characterizes a normal female phenotype with Wolffian (mesonefric) ducts regression and Müllerian (paramesonefric) ducts absent, due to failure of fusion before 12th week fetal development. It represents an aplasia of the uterus, cervix and upper two-thirds of the vagina in a women with normal karyotype (46,XX). The incidence estimated is 1 in 4,500 female births [15].

The MRKH syndrome was considered a sporadic disease, the first reports mention a possible association with drugs that were after rejected [16]. The latest reports point a manifestation of a variably expressed genetic defects closely related to the embryogenesis. The genetic defect seems to be transmitted as an autosomal dominant trait with incomplete penetrance and variable expressivity. The hypothesis of polygenic/multifactorial causes of the syndrome is gaining more conviction with the identification of interstitial deletion in chromosome 22 and terminal deletion in chromosome 4 [17].

The syndrome can be categorized in type I, with isolated utero-vaginal aplasia. The incomplete aplasia and/or association with other malformation is type II or MURCS association (Mülerian duct aplasia, renal dysplasia and cervical somite anomalies). The associated malformations include renal (unilateral agenesis, ectopic kidney or horse-shoe kidney), skeletal and vertebral (Klippel-Feil anomaly, fused vertebrae mainly cervical, scoliosis), hearing defects and rare cardiac and digital anomalies (syndactyly, polydactyly) [14].

Usually the clinical presentation is primary amenorrhea in a normal female phenotype with normal ovarian function and without androgen excess. The secondary sexual characteristics developed normally as well as external genitaliae. The vaginal dimple has a variable deep (2–7 cm). The gynaecological examination should exclude transverse vaginal septum and imperforate hymen. The anatomic evaluation is performed to evaluate complete uterine aplasia with two rudimentary horns linked by peritoneal fold, that correspond to type I; or an hypoplasic uterus symmetric or assymetric, type II [18]. The imaging technique used are commonly ultrasound, Magnetic resonance imaging, celioscopy. The differential diagnosis includes other causes of primary amenorrhea with normal secondary sexual characteristics. The main exclusion diagnosis are gonadal dysgenesis, isolated vaginal atresia and androgen insensitivity.

The management should include an adequate counseling and support as it is a disorders that bring high anxiety for the patient and family. The creation of a neovagina must be offered to a patient who wishes to star sexual activity. There is no consensus regarding the ideal procedure. The differences between the procedures are access (laparoscopic, transabdominal, vaginal), type of tissue to cover neovaginal cavity (skin graft, peritoneal, intestinal tissue, amniotic membrane and artificial material). The treatment options include the creation of a new cavity or a vaginal replacement. The nonsurgical creation of a neovagina is known as the Franck's dilator method and is based on the application of vaginal dilators (Hegar candles). The surgical technique described is Abbe-McIndoe operation, that consists on the placement of a mold covered with skin, peritoneal, minora labia amniotic membranes or synthetic graft in vesicorectal space and subsequent dilatation [19]; Veccchietti operation, the neovagina is created with the association of dilation and traction using a device placed subperitoneally by abdominal approach (laparotomic or laparoscopic) [20]. Finally the sigmoidal colpoplasty uses a graft of a long sigmoid segment, with good results but prolonged care [21].

10.2.1 Rational for Amniotic Membrane Application

The first description of vaginal reconstruction with amniotic membrane was in 1934 in the French-language literature. The application of amniotic membrane is limited [22–28]. The mature squamous epithelium indicates the metaplasia of amniotic epithelium into vaginal epithelium. The layers described in histology are similar to normal vagina, with a superficial, intermediate and deep layers [22]. Amnion is available and does not require additional incisions as with skin grafts. Again, there are no problems with immune rejection, due to the lake of histocompatible antigens (HLA-A, -B, -C and DR or β2-microglobulin).

10.2.2 Reports Considering Amniotic Membrane Application

The technique reported used processed and sterilized freeze-dried human amniotic membranes from term gestations of healthy women. The surgery consists of a transverse incision in the vaginal dimple and was performed a dissection of vesicorectal cavity of 12 cm at the level of peritoneum [29]. After, a mold of glass covered by amniotic membranes was inserted and the labia minora were fixed around the stent to prevent expulsion. The vagina is then tamponed in order to fill the neovaginal cavity with a sterile gauze with estriol creme (1 %) to promote epithelialization. The patients remained on bed rest during 5 days and then a soft silicon vaginal dilator replaced the mold with estriol creme. The dilator was used during 3–5 months to prevent vaginal retraction. The patients were encouraged to engage sexual activity 6–8 weeks after vaginoplasty. There are 75 reported patients with neovagina reconstruction with amniotic membrane. Only in four cases were reported rectal injury and the anatomic and functional results are satisfactory.

The latest report [29] using this technique encompasses seven patients with 17–26 years old. The mean operating time ranged 20–33 min, with 8–14 days of hospitalization. In only one case was reported a urinary tract infection in the first postoperative week, solved with antibiotic therapy. There was a case of infection an ulceration and was diagnosed a rectovaginal fistula that required surgical repair and diverting loop ileostomy. The infection lead to a stricture and shortening vagina, leading to reconstruction in 4 months with sigmoid loop, complicated by peritonitis but finally with satisfying functional and anatomic results. In all patients, was described a neovagina length of 9 cm (range 9–12 cm) and the vaginal cavity had a diameter of two fingers. At 6 months, the neovagina was epithelized without stricture formation or shortening. The mean time from surgery to vaginal intercourse was 5.28 months. There were no reports of dyspareunia or bleeding during intercourses. The functional satisfaction was similar to the matched healthy women considering Rosen Female Sexual Function Index (FSFI) questionnaire (around 30 points) (Table 10.3).

The ideal vaginoplasty technique depends on several factor, including patients' preference, expectations and mainly surgeons' experience. Patients should be aware of potential complications, long-term results and sexual function. The reconstruction with amniotic membrane is safe and simple, without major complications and satisfactory functional results.

Table 10.3 Summary of literature published considering the application of amniotic membrane in vaginoplasty in cases of congenital absence of vagina, vaginal atresia, partial vaginal atresia and androgen insensibility syndrome

Author	Year	Number of cases	Material and methods	Results and conclusions
Tancer et al. [26]	1979	4	Reconstruction of the vagina in 2 patients with vaginectomy for diffuse carcinoma in situ, 1 patient following severe, corrosive, vulvovaginal burns, 1 patient with müllerian agenesis	Epithelialization was complete within 8 weeks
Dhall et al. [27]	1984	5	Patients were selected for surgical treatment due to congenital absence of vagina, amnion's mesenchymal surface came in contact with the host tissue	Vagina was found to be well formed and was of normal depth and caliber. Vaginal biopsies taken 4–6 weeks after operation showed early epithelialization, mature epithelium on 8–10 weeks
Morton et al. [24]	1986	27	Used human amnion for various forms of vaginoplasty due to complete and partial vaginal agenesis	Excellent epithelialization occurred with good functional results. No long-term complications
Ashworth et al. [23]	1986	15	Human amnion was applied over a mold placed in a new vagina or a strictured one	Improvement in all of the patients with vaginal strictures
Nisolle et al. [25]	1992	10	Amniotic membranes as a graft on vaginoplasties. Amnion was not stripped from the chorion	Vagina was well formed and of normal depth and caliber
Bleggi-Torres et al. [22]	1997	10	Patients with congenital absence of the vagina were surgically treated with application of an amniotic membrane graft using the modified McIndoe and Bannister technique	The cells were arranged in layers as in the normal vaginal epithelium. The human amniotic membrane is able to complete metaplasia into squamous cells
Zafar et al. [28]	2007	Case report	A woman with vaginal atresia, secondary to a mismanaged spontaneous vaginal delivery, underwent vaginoplasty using human amnion kept in its place using a 60 cc syringe as a mold	Restoration of normal coital function
Fotopoulou et al. [29]	2010	7	Young women with MRKH were submitted to McIndoe procedure modified by the application of human freeze-dried amniotic membranes	Satisfying anatomic and functional outcome with low perioperative morbidity

References

1. Conforti A, Alviggi C, Mollo A et al (2013) The management of Asherman syndrome: a review of literature. Reprod Biol Endocrinol 11:118
2. Yu D, Wong Y-M, Cheong Y et al (2008) Asherman syndrome: one century later. Fertil Steril 89:759–779
3. Amer MI, Abd-El-Maeboud KHI, Abdelfatah I et al (2010) Human amnion as a temporary biologic barrier after hysteroscopic lysis of severe intrauterine adhesions: pilot study. J Minim Invasive Gynecol 17:605–611
4. Chen Y, Chang Y, Yao S (2013) Role of angiogenesis in endometrial repair of patients with severe intrauterine adhesion. Int J Clin Exp Pathol 6:1343–1350
5. Tao Z, Duan H (2012) Expression of adhesion-related cytokines in the uterine fluid after transcervical resection of adhesion. Zhonghua Fu Chan Ke Za Zhi Chin J Obstet Gynecol 47:734–737
6. Protopapas A, Shushan A, Magos A (1998) Myometrial scoring: a new technique for the management of severe Asherman's syndrome. Fertil Steril 69:860–864
7. March C (1995) Intrauterine adhesions. Obstet Gynecol Clin North Am 22:491–505
8. Lin X, Wei M, Li T et al (2013) A comparison of intrauterine balloon, intrauterine contraceptive device and hyaluronic acid gel in the prevention of adhesion reformation following hysteroscopic surgery for Asherman syndrome: a cohort study. Eur J Obstet Gynecol Reprod Biol 170:512–516
9. Reijnen MM, Falk P, van Goor H, Holmdahl L (2000) The antiadhesive agent sodium hyaluronate increases the proliferation rate of human peritoneal mesothelial cells. Fertil Steril 74:146–151
10. Takasaki A, Tamura H, Miwa I et al (2010) Endometrial growth and uterine blood flow: a pilot study for improving endometrial thickness in the patients with a thin endometrium. Fertil Steril 93:1851–1858
11. Nagori CB, Panchal SY, Patel H (2011) Endometrial regeneration using autologous adult stem cells followed by conception by in vitro fertilization in a patient of severe Asherman's syndrome. J Hum Reprod Sci 4:43–48
12. Amer MI, Abd-El-Maeboud KH (2006) Amnion graft following hysteroscopic lysis of intrauterine adhesions. J Obstet Gynaecol Res 32:559–566
13. Mamede AC, Carvalho MJ, Abrantes AM et al (2012) Amniotic membrane: from structure and functions to clinical applications. Cell Tissue Res 349:447–458
14. Oppelt PG, Lermann J, Strick R et al (2012) Malformations in a cohort of 284 women with Mayer-Rokitansky-Küster-Hauser syndrome (MRKH). Reprod Biol Endocrinol 10:57
15. Carson S, Simpson J, Malinak L et al (1983) Heritable aspects of uterine anomalies: II – genetic analysis of Müllerian aplasia. Fertil Steril 40:86–90
16. Morcel K, Camborieux L, Guerrier D (2007) Mayer-Rokitansky-Küster-Hauser (MRKH) syndrome. Orphanet J Rare Dis 2:13
17. Cheroki C, Krepischi-Santos AC, Rosenberg C et al (2006) Report of a del22q11 in a patient with Mayer-Rokitansky-Küster-Hauser (MRKH) anomaly and exclusion of WNT-4, RAR-gamma, and RXR-alpha as major genes determining MRKH anomaly in a study of 25 affected women. Am J Med Genet A 140A:1339–1342
18. Strübbe E, Willemsen W, Lemmens J et al (1993) Mayer-Rokitansky-Küster-Hauser syndrome: distinction between two forms based on excretory urographic, sonographic and laparoscopic findings. Am J Roentgenol 160:331–334
19. Jackson N, Rosenblatt P (1994) Use of interceed absorbable adhesion barrier for vaginoplasty. Obstet Gynecol 84:1048–1050
20. Borruto F, Chasen S, Chervenak F, Fedele L (1999) The Vecchietti procedure for surgical treatment of vaginal agenesis: comparison of laparoscopy and laparotomy. Int J Gynecol Obstet 64:153–158

21. Louis-Sylvestre C, Haddad B, Paniel B-J (1997) Creation of a sigmoid neovagina: technique and results in 16 cases. Eur J Obstet Gynecol Reprod Biol 75:225–229
22. Bleggi-Torres L, Werner B, Piazza M (1997) Ultrastructural study of the neovagina following the utilization of human amniotic membrane for treatment of congenital absence of the vagina. Braz J Med Biol Res 30:861–864
23. Ashworth M, Morton K, Dewhurst J et al (1986) Vaginoplasty using amnion. Obstet Gynecol 67:443–446
24. Morton KKE, Dewhurst C (1986) Human amnion in the treatment of vaginal malformations. Br J Obstet Gynaecol 93:50–54
25. Nisolle M, Donnez J (1992) Vaginoplasty using amniotic membranes in cases of vaginal agenesis or after vaginectomy. J Gynecol Surg 8:25–30
26. Tancer M, Katz M, Veridiano N (1979) Vaginal epithelialization with human amnion. Obstet Gynecol 54:345–349
27. Dhall K (1984) Amnion graft for treatment of congenital absence of the vagina. Br J Obstet Gynaecol 91:279–282
28. Zafar M, Saeed S, Kant B et al (2007) Use of amnion in vaginoplasty for vaginal atresia. J Coll Physicians Surg Pak 17:107–109
29. Fotopoulou C, Sehouli J, Gehrmann N et al (2010) Functional and anatomic results of amnion vaginoplasty in young women with Mayer-Rokitansky-Küster-Hauser syndrome. Fertil Steril 94:317–323

Chapter 11
Amniotic Membrane and the Controlled Drug Release

Miklós D. Resch, László Marsovszky, Erzsébet Csányi, Mária Budai-Szűcs, and János Németh

Abstract In this chapter we give an overview on the factors affecting drug permeability in the normal eye, and demonstrate the effect of amniotic membrane on the ocular surface on drug release. Furthermore we focus mainly on topical ocular drug administration with regards to ocular surface and amniotic membrane. Amniotic membrane can have various effects on drug permeability and can significantly alter bioavailability of the drugs used. Beyond several biochemical, biological and biophysical benefits of amniotic membrane transplantation, amniotic membrane serves a layer with a more stable drug concentration on the ocular surface. Amniotic membrane has dual effect on drug penetration. First it acts as a barrier against drug penetration, and it has also a drug release activity.

Keywords Pharmacology • Permeability • Drug penetration • Amniotic membrane • Ocular surface

11.1 Routes of Drug Administration After Amniotic Membrane Transplantation

In ophthalmic pharmacotherapy, different drugs are being used. The delivery of these drugs to the eye can be made through topical, local or systemic administration. The most frequently exercised method of administration is through drops however subconjunctival, subtenon and intraocular administrations are still in use [1]. The rationale of topical route of application is that it maximizes anterior segment concentration and minimizing systemic toxicity. Topical administration is a relatively

M.D. Resch (✉) • L. Marsovszky • J. Németh
Department of Ophthalmology, Semmelweis University, Budapest, Hungary
e-mail: miklosresch@gmail.com

E. Csányi • M. Budai-Szűcs
Department of Pharmaceutical Technology, University of Szeged, Szeged, Hungary

A.C. Mamede, M.F. Botelho (eds.), *Amniotic Membrane: Origin, Characterization and Medical Applications*,
DOI 10.1007/978-94-017-9975-1_11

easy way of drug delivery and it can be repeated several times. The most challenging feature of topical administration however, that the drug is in contact with the ocular surface for a relatively short time limiting its efficacy. As an additional layer, amniotic membrane alters the bioavailability of the drug being applied through topical administration.

In this chapter we give an overview on the factors affecting drug permeability in the normal eye, and then demonstrate the effect of amniotic membrane on the ocular surface on drug release. Furthermore we are going to focus mainly on topical administration with regards to ocular surface and amniotic membrane.

Most of the topically administered ocular therapeutics are simple aqueous solution eyedrops. The main drawbacks of aqueous eyedrops are their inability to deliver lipophilic and insoluble molecules, their low retention time, and their limited ability to resist the washout effect of blinking and tears turnover. Amniotic membrane can potentially serve as a drug reservoir in addition to its other beneficial biochemical effects.

11.2 Barriers Affecting Drug Permeability in the Normal Eye

11.2.1 Tear Film

The very first contact in the interaction between drug and ocular surface is the tear film. The drug needs to penetrate the tear film layers in order to be absorbed. The capability of this transmission depends on the physiologic properties of the carrier molecules which subsequently alters the bioavailability of the drug used. One eyedrop from a commercial dispenser comprises 50 μL. This volume is delivered to the ocular surface during drop application and retains mainly in the lower conjunctival sac afterwards. The lower conjunctival sac is only 7–13 μL in volume in a blinking eye of an upright patient, thus about 20 % of the administered drug remains in the conjunctival sac. This is further diluted by tear turnover and the effect of the tear drainage system due to blinking giving less than 10 % of the administered drug to be present after a few minutes of instillation. This available drug then has to pass through the ocular permeation barriers such as the conjunctiva and cornea. Altogether it is estimated that only 1 % or less of the administered dose can penetrate the ocular surface [2, 3].

King-Smith et al. [4] and Azartash et al. [5] in a recent work evaluated the tear film thickness and found it to be approximately 3.0 μm in the healthy human. The tear film structurally comprises three different layers, however novel studies indicate that those layers are in dynamic balance in normal eyes rather than expressing a static noncommunicating nature. The tear film is a constantly changing layer which needs 5–7 s to build up and reach its highest regularity and optical quality [6]. The drug availability on the ocular surface greatly depends on the tear turnover which can be highly elevated after drop instillation. Nonetheless, blinking rate has also a significant impact on drug availability, as it increases tear turnover by moving the tear toward the canaliculi systems.

11.2.2 Conjunctiva

The conjunctiva is covering the eyeball (bulbar conjunctiva) and the inner surface of the eyelids (tarsal conjunctiva), and these parts are connected by the upper and lower fornices. The conjunctiva comprises 95 % of the ocular surface [7]. The most superficial layer of the conjunctiva is composed of squamous nonkeratinized cells of two to four layers, which similarly to the cornea act as an obstacle against drug penetration. The conjunctival stroma contains elastic collagen fibres, which has low barrier effect. Conjunctiva covers the largest tear reservoir in the lower fornix, and in itself can serve as a drug reservoir after eyedrop administration.

Conjunctiva plays many roles including protection of ocular surface by acting as a passive physical barrier, production of lipids by its goblet cells and maintenance of tear film. It also acts as an active barrier in the interaction between the environment and the ocular surface rendering it a compliant frontliner of ocular surface immunology. The conjunctiva, which is a moderately tight epithelium, endowed with various transport mechanism for different molecules including ions, solutes, and water to maintain homeostasis. As the conjunctiva expresses much larger surface area than that of cornea coupled with the expression of several key transport processes, this renders it an attractive route for drug delivery toward the posterior portion of the eye. Not only the larger surface but the readily available reservoir function of the subconjunctival space makes this area feasible for subconjunctival injection of microparticles and matrix materials. This allows sustained release to maintain reasonable intraocular levels of various drugs, perhaps attesting to the fact that conjunctiva per se may contribute as a part of multiple transport barrier(s) in ocular drug delivery.

11.2.3 Cornea

The cornea comprises only 5 % of the ocular surface, and its structure is more compound compared to the conjunctiva. Cornea has five layers: an external hydrophobic epithelium about 50 μm thick, the Bowman's layer which is the basement membrane of the inner cuboid like epithelial cells, the hydrophilic stroma, Descemet's membrane (basal membrane of the endothelium) and the single layer of endothelium. However cornea is devoid of blood and lymphatic vessels, – along with conjunctiva – it is also endowed with resident immune cells regulating the local immune responses to external and internal stimuli in certain conditions [8]. The outermost two-to three layers of the corneal epithelial cells form a barrier by their tight intercellular boundings called zonulae occludentes. The epithelium contains more lipid than the stroma, which provides a tissue reservoir for lipofil drugs. Corneal stroma comprised compact collagen fibers/lamellae organized to perfection in order to maintain the corneal transparency. The stroma represents the 80 % of the thickness of the cornea and acts as a sceleton to provide corneal shape. The stroma is hydrophilic and also called the corneal collegenous shell. The endothelial cells are not connected by zonulae occludentes subsequently exorting minimal barrier function, however by controlling the water content of the cornea, it has an indirect effect on the transcorneal penetration.

11.2.4 Sclera

When drug penetrates through the conjunctiva, it enters the subconjunctival space. To reach the intraocular receptors it needs to get through the sclera. Scleral stroma consists of bundles of irregularly organized collagen which makes the skeleton of the eyeball. The permeability of the sclera is approximately the same as the corneal stroma and sclera has been shown to be permeable to solutes up to 70 kDa in molecular weight [9]. In conclusion, the epithelium of the conjunctiva and cornea have mainly barrier function against drug penetration. The stroma of the cornea and conjunctiva have dual effect, barrier on one hand and drug reservoir on the other hand. The subconjunctival space is supposed to have drug reservoir function.

11.3 Factors Affecting Drug Permeability of the Eye After Amniotic Membrane Transplantation

Beyond the properties of the normal human eye found in different drug permeations and reservoir, one should calculate with the altering effect of amniotic membrane when choosing the dosing regime. Different aspects of the drug transport modifying properties of amniotic membrane are summarized in the following sections.

11.3.1 Ocular Surface Disease

Amniotic membrane transplantation (AMT) is indicated in several ocular surface diseases [10, 11] (see also Chap. 6), in a wide range of pathologies that could have altering effects on drug penetration. The average corneal thickness varies between 540 and 570 μm. Corneal thickness in ocular surface disease is rather variable. In case of ulceration, significant thinning of the cornea can be present, whilst significant corneal thickening is usually encountered in ophthalmic inflammatory conditions. Alterations in corneal thickness like in persistent corneal epithelial defects may give route to an increased permeation of drugs compared to normal corneas.

On the other hand, excessive presence of fibrotic tissues in cicatricial conditions like ocular pemphigoid or in Stevens-Johnson syndrome may reduce the drug conduit toward deeper ocular tissues.

As a surgical procedure, AMT requires antibiotic treatment during the first weeks of postoperative period. Kim et al. [12] evaluated the effect of amniotic membrane on the permeation of ofloxacin into the cornea, its concentration in aqueous humor, and tears in vivo on the rabbit cornea. They concluded that amniotic membrane transplantation interferes with the ocular penetration of topical ofloxacin in normal rabbit corneas but enhances ofloxacin penetration in corneas with epithelial defects.

Several factors can have impact on transamniotic drug penetration. O'Brien et al. [13] reported that inflammation after corneal deepithelialization enhances the ocular penetration of topical antibiotics. Healy et al. [14] and Robert and Adenis [15] showed that transcorneal penetration of most drugs, including the fluoroquinolones, occur primarily by passive diffusion and is correlated in a positive manner with the drug's aqueous solubility and degree of lipophilicity. The average corneal thickness varies between 540 and 570 μm. Corneal thickness in ocular surface disease is rather variable. In case of ulceration, significant thinning of the cornea can be present, whilst significant corneal thickening is usually encountered in ophthalmic inflammatory conditions.

Vascularization of the cornea may significantly alter drug absorption. Corneal neovascular vessels are usually fenestrated, thus more permeable to any agent, including drug molecules. In addition, neovascularization usually changes epithelial structure, giving easier access for drug molecules to pass through into deeper parts of the eye. Angiogenesis and lymphangiogenesis of the cornea was demonstrated by Cursiefen et al. [16] in certain ophthalmic inflammatory disorders. If a neovascularized cornea is covered by amniotic membrane, increased blood flow with elevated concentration of drug in the blood serum is expected. Furthermore, not only blood vessels, but lymphatic vessels are present in ocular surface diseases, rendering the process of drug penetration more complex.

11.3.2 Surgical Factor

The choice of operating technique for AMT is determined not only by the ocular surface pathology, but localization and etiology [17, 18]. Amniotic membrane promotes the healing of the cornea, and conjunctival lesions not only by its natural remedial properties, but via exploiting the drug release capacity and barrier function of the amniotic membrane. Different techniques have been proposed over the years and there have been contradictory reports concerning the proper placing of the amniotic membrane on the ocular surface. The dimensions of the amniotic membrane used in surgery renders the surgery to amniotic membrane grafting or patch technique procedure. Corneal amniotic membrane graft is covering the cornea only partially while corneal patch is covering the cornea and limbus completely. One could argue that the dimension of amniotic membrane used to cover the ocular surface would interfere with the drug penetration in a negatively correlated fashion. Giving the complex nature of the amniotic membrane, the drug dynamics does not follow this role. In this chapter our intension is to examine this phenomen further and give an overview on the mechanisms operating in it. In case of sandwich technique, a combination of graft and patch techniques is used, meaning that several layers of amniotic membrane are being transplanted to the ocular surface. One could assume that the pharmacological effect of AMT could be different, but no quantitative evidence has been published yet (but comparative studies are yet to be carried out).

AMT is also used to substitute conjunctiva, when conjunctiva is partially missing, or degenerated and amniotic membrane is covering Tenon's fascia or sclera, or the complete fornix needs reconstruction in case of ocular pemhigoid [19].

The orientation of the amniotic membrane graft and patch can also have potential role in drug penetration. As described in Chap. 1, amniotic membrane consists of a single layer cuboid epithelium and a thick stroma. Amniotic membrane is usually transplanted with the epithelial side up, but in some cases epithelial side down technique is recommended especially in cases of patch technique in persistent epithelial defects of the cornea.

Amniotic membrane can be fixed to the ocular surface in several ways. Fixation of amniotic membrane is usually performed by multiple interrupted sutures, but tissue adhesives can also be used, or the combination of these techniques might also be reasonable [20, 21]. Sutures have angiogenetic effects, and mechanical irritation can result in local enhancement of the inflammatory reaction and hyperaemia of the conjunctiva and sclera. Tightness of sutures plays and important role in the fixation of grafts, and in preventing the corneal epithelium from growing underneath the amniotic membrane stroma. In multilayer graft transplantation the stromal defect is filled with multiple layers of amniotic membrane, but only the most superficial layer or/the uppermost layer is fixed by sutures. The number of applied layers is also very important, because corneal epithelium can grow between the amniotic membrane layers [22]. As shown above, epithelium acts as a barrier for drug penetration, thus if corneal epithelium grows between multi-layered amniotic membrane in the cornea it can decrease its permeability. One interesting area that awaits for exploration is the properties of deepithelialized amniotic membrane to drug permeation. It has been shown by Zhang et al. [23] that a de-epithelialized amniotic membrane could preserve the basement membrane integrity, extracellular matrix (ECM), and growth factor composition if treated with urea. Theoretically the denuded amniotic membrane would allow faster penetration into the cornea while preserving its drug reservoir properties. This might have/bear a clinical significance on the wound healing of ulcerative corneal pathologies.

In order to quantify trans amniotic drug penetration, an in vitro model was developed [24, 25]. Experiments were performed with ofloxacin, a widely used topical antibiotic eyedrop. Amniotic membrane pieces were mounted in a vertical Franz-diffusion cell system (Fig. 11.1) equipped with an autosampler. In vitro release of 300 mg of 3 % commercially available ofloxacin ophthalmic solution was determined by quantitative absorbance measurement carried out with an ultraviolet (UV) spectrophotometer (wavelength, 287 nm). We found, that ofloxacin was detectable in the acceptor phase, 1 min after instillation, and a gradual increase of concentration was detected in a period

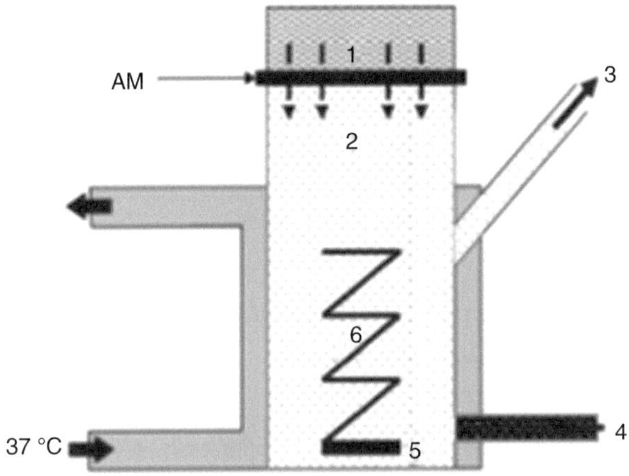

Fig. 11.1 Schematic drawing of amniotic membrane mounted in the Franz cell. The donor compartment (*1*) above contains the tested drug solution. The compartment below is the acceptor phase (*2*), from which samples are taken through the sampling port (*3*), to the acceptor phase replacing port (*4*). The acceptor compartment is surrounded with a water jacket kept at 37 °C. At the bottom of the acceptor phase, a stir-bar (*5*) and a helix mixer (*6*) are rotated magnetically (With the permission of ARVO, originally published in Resch et al. [24])

of 90 min. The in vitro model of the Franz-diffusion cell system was found to be applicable for drug permeability studies of human amniotic membranes to water-based solutions. The filter membrane and amniotic membrane were permeable to a water-based solution of ofloxacin. Significant barrier function of the amniotic membrane could be measured in ofloxacin permeability.

The same method was used to demonstrate drug reservoir function of amniotic membrane. In those experiments, antibiotic solution was not added to the donor compartment, but the amniotic membrane itself was soaked in 3 % ofloxacin ophthalmic solution. Ofloxacin was detectable in the acceptor phase 1 min after mounting in all groups. Until 120 min, rapid increase of released ofloxacin could be observed. From 120 to 450 min amount of released ofloxacin showed a slower increasing pattern. Significant ofloxacin reservoir capacity of a single human amniotic layer could be demonstrated in vitro. Amniotic membrane acted as an ofloxacin slow release device for up to 7 h in vitro, depending on the duration of pretreatment of amniotic membrane. Individual pretreatment of amniotic membrane could increase beneficial effects of amniotic membrane transplantation especially in infectious keratitis.

Carrier molecules play an important role in drug delivery. Drug needs to be in contact with the amniotic membrane surface to be able to penetrate. This bioavailability of the drug has been discussed above. Those molecules that are ready to travel through the barriers must face further difficulties given the lipophilic property of amniotic membrane. Hydrophilic molecules are less capable to penetrate unless disguise themselves in a lipophilic coat. This draw our attention to examine the drug penetration of diclofenac through amniotic membrane with and without cyclodextrin

(CD). We found the latter in greater concentration at the acceptor phase which should be taken into consideration on treatment choice and regime [24, 25].

Mayer and coauthors soaked amniotic membrane in solution of bevacizumab and in their in-vitro setting effective vascular endothelial growth factor (VEGF)-blockade could be demonstrated for up to 1 week, and stated, that amniotic membrane might be potentially used as a carrier for drugs delivered to the cornea [26].

11.3.3 Microstructure of the Transplanted Amniotic Membrane

Biomicroscopically amniotic membrane remains on the ocular surface for shorter or longer time depending on the surgical technique, pathology, comorbidities and treatment [22]. Patch usually acts like a temporary contact lens and disappears within 2 weeks. Grafts however can be integrated into the corneal tissue up to several years [27].

Amniotic membrane can integrate into the host corneal tissue after AMT in intraepithelial, subepithelial, or intrastromal patterns, or may be localized on the corneal surface. The thicknesses of corneal epithelium and amniotic membrane are extremely variable. The morphology of the individual pattern seems to depend on the ocular surface disorder, and the AMT technique.

The application technique of AMT seems to have a great impact on the morphology of amniotic membrane integration. The surgical technique is dependent on the type and severity of ocular surface disease. In general, a patch disappears most often during the first 1 or 2 weeks after AMT without remnants. Using the sandwich technique the patch is typically lost very early after AMT (similar to single patch), but all layers of grafts may be integrated into the corneal stroma and stay there for many months [22].

Type of preservation is certainly affecting the structure of amniotic membrane. Preservation of amniotic membrane is detailed in Chap. 13. Freshly prepared amniotic membrane (rarely used in ophthalmology) is supposed to have more intact amniotic epithelial cells, than cryopreserved amniotic membrane, since cryopreservation usually interferes with the survival rate of amniotic epithelial cells [28]. There are commercially available or dried amniotic membrane on the market, which do not contain any viable amniotic epithelial cell or amniotic fibroblast (Ambiodry) [29, 30]. It has not been measured before, but we suppose, that a pure stroma without cells can serve as a good device for sustained drug release, because it lacks the barrier function of the epithelial cells.

In case of integrated amniotic membrane, the type of the epithelium covering amniotic membrane can play a role in drug penetration. Amniotic epithelium is usually missing or non-viable after cryopreservation [31]. Tissue engineering enables us to modify the epithelial cells on the superficial layer of the amniotic membrane. Amniotic membrane transplantation effectively expands the remaining limbal epithelial stem cells in patients with partial limbal stem cell deficiency.

Corneal epithelial cells or limbal stem cells can be cultured on amniotic membrane ex vivo [32]. AMT is performed after ex vivo stem cell expansion, recovering not only the ocular surface, but the epithelial cell layer [33–37]. In contrast by denuding amniotic membrane we can use the pure stroma of the amniotic membrane [38]. There is no sufficient information, but it is postulated, that the ultrastructure of the amniotic membrane can modify its drug penetration features [39]. Ultrastructural and immunohistochemical studies showed, that the basement membrane of the amniotic membrane is very similar to the corneal epithelial basement membrane, which makes it a good niche to expand the corneal cells on [40].

11.3.4 Pharmacological Factors

Patients after AMT usually need prolonged and multiple topical pharmacotherapy. Several pharmacological factors are involved in the transcorneal and transconjunctival drug penetration.

Effect of an ophthalmic solution can be described in several phases:

1. Contact with the eye surface
2. Wetting and hydration of the dosage form (in case of solid or partially hydrated formulations),
3. Blending with tear film and spreading on it,
4. Drug release from the dosage form and diffusion into lacrimal fluid (liberation),
5. Drug permeation trough the cornea to the anterior ocular tissues (absorption)

Evolution of the first three steps strongly depends on ingredients, formulation and the dosage form, while in case of the latter steps physico-chemical characteristics of the active substance come to the front. Beside the drainage of the formulation, the tear turnover and the conjunctival absorption (mentioned earlier), there is some other important factor in the precorneal area, which can decrease the amount of the active ingredient before the penetration. They are the protein binding and the metabolism, such as hydrolysis, oxidation or reduction. For the determination of the required amount of the drug these phenomena must also be taken into account. The most important factors of the trans-corneal absortion are the contact time between the formulations and the eye surface, and the drug permeability in the cornea. The contact time can be improved by special additives and formulation; these possibilities will be discussed in the next chapter. Drug absorption can take place either by the way of passive diffusion, facilitated diffusion or by active transport. Passive diffusion do not require transporter proteins, but it is influenced by the physico-chemical characteristics of the molecule such as lipophilicity, solubility, molecular size and shape, charge and degree of ionization. Formulation of an eyedrop depends on its solubility: a drug with adequate solubility can be formulated in an aqueous solution, but drugs with poor solubility may need to be provided in a suspension. Suspensions may be more irritative to the ocular surface than solutions, which can promote reflex tearing and further dilution of the drug.

Viscosity is important, since it can prolong contact time of the drug with the ocular surface. Adding high viscosity substances to the solution increases drug retention in the inferior cul-de-sac, promoting drug penetration. Methylcellulose and polyvinyl alcohol are frequently used substances with high viscosity.

To traverse the cornea, a drug must pass sequentially through the lipophilic environment of the epithelial cell membranes through the hydrophilic environment of the stroma, and finally through the lipid barrier at the endothelium. Drugs with greater relative lipid solubility can better penetrate cellular membrane, however are poorly soluble in tear. Hydrophilic molecules are simple aqueous eyedrops, their ocular penetration is critically limited by the intercellular zonulae occludentes and the hydrophobic nature of the epithelium cell membranes of the ocular surface. In contrast, lipophilic molecules can penetrate through cell membranes of the ocular surface epithelium but their formulation for topical ocular administration is complex because they are insoluble in simple aqueous solutions.

Small (molecular weight (MW) <500) non ionizable hydrophilic molecules are able to penetrate through intercellular spaces in the corneal epithelium, while these types of lipophilic molecules can go directly through the cells. Contrarily molecule with MW of 5×105 or over can diffuse easily through the stroma. The barrier function of the endothelium is not as remarkable as that of epithelium, because its permeability is 200-fold weaker than that of the latter. Both the epithelium and the stroma have a reservoir function; the hydrophilic stroma serves as a depot for water-soluble molecules whereas the epithelium is the main depot for lipophilic ones [41].

For the optimal corneal penetration it was found that the ideal lipophilicity corresponds to logp (lipid/water partition coefficient) values of 2–3. The drugs with higher lipophilicity (logp>3) have showed lower permeability, because they cumulate and move slower from the lipophilic epithelial to the hydrophilic stroma [42]. For ionisable drugs, the unionized form is preferred to penetrate the epithelium and endothelium, while the ionized form is desirable to stromal transition. On basis of the Henderson-Hasselbach equation:

$$pH = pKa + log\left(\frac{\left[A^{-}\right]}{\left[HA\right]}\right) \tag{11.1}$$

where, [HA] is the molar concentration of the undissociated weak acid, [A^-] is the molar concentration of this acid's conjugate base and pKa is – log(Ka) where Ka is the acid dissociation constant – in case of weak acid or a weak base the pH of the ophthalmic drug delivery will determine the ratios of the ionized and un-ionized species of the molecule [43]. In the case of ionized molecules, over the degree of ionization the charge of the molecule also affect the cornea penetration [44]. The cornea epithelium is negatively charged above its isoelectric point, from this reason hydrophilic charged cationic molecules permeate easier through the cornea than anionic molecules. Below the isoelectric point, the cornea is permeable to negatively charged molecules, however, this acidic field strongly irritates the eye and triggers an increased lacrimation [45].

Not only drugs, but additional compounds of the eyedrops affect drug penetration. Preservatives (benzalkonium chloride, tiomersal) are applied frequently in commercially available eyedrops, which prevent bacterial contamination, alter structure of the corneal and conjunctival epithelial cell membranes. Disruption of the epithelial cell wall reduce the barrier effect of the corneal surface and increase drug permeability. Application of ointments increase the contact time of ocular medications. Ingredients of the ointments are lipophilic thus most water-soluble medications are insoluble in the ointments and are present as microcrystals. In this formulations slow release can be provided on the ocular surface, but penetration is usually very limited.

Tear film alterations are not rare comorbidities in cases after AMT. Excessive tearing is encountered in the first few days after AMT causing significantly reduced bioavailability of the drugs on the ocular surface. On the other hand local or systemic pathologies resulting in dry eye also bear a considerable impact on drug availability and penetration. Either dry eye or excessive tearing can be present in patients after AMT. Systemic autoimmune disease in the background of ocular surface disease usually reduce tear volume and ingredients of the tear film also change. Meibomian gland dysfunction determines the lipid content of the tear film. The synchronized movement of the eyelids can be limited in ocular surface disease needing AMT, thus the spreading of the precorneal tearfilm is damaged. In ocular pemhigoid not only the lipid content, but the mucin layer of the tear film is pathologic.

11.4 Enhancements in Topical Ocular Drug Delivery Systems (DDS)

As the eye is an organ with special anatomy, physiology and biochemistry, the improvement of bioavailability and extension of the residence time of drugs administrated on eye is one of the most interesting and challenging problems in the past decades. Despite that many different ophthalmic drug delivery strategies have been published, the conventional ocular dosage forms dominate the clinical therapy [46]. The strategies to improve the bioavailability of ocular drug can be divided into two groups:

- Improving residence time on the eye surface
- Enhancing corneal permeability
- Enhancing site specificity [47]

11.4.1 Improved Corneal Permeability

For enhancement of corneal drug permeability, there are two main possibilities. The first one is to modify the physico-chemical characteristics of the drug and the second approach is a temporary modification of the integrity of the corneal epithelium [48]. Prodrugs are derivatives of drugs which are chemically or enzymatically converted to their active parent drugs. These prodrugs have adequate lipophilicity, solubility and pKa for the corneal penetration and modify to parent drug in the tissue of the eye. Recently the concept of double prodrug (prodrug of prodrug) is also gaining importance [47].

Cyclodextrins (CDs) are cyclic oligosacharide and can improve the physico-chemical characteristics of drugs forming complexes with them. They are often used to increase the solubility and the dissolution rate of the active agent. Moreover CDs can adhere to the biological membrane [49].

In the early studies chelating agents, preservatives, surfactants and bile salts were used as penetration enhancers. Nowadays these substances are less favourable due to the local toxicity associated with them and new types of penetration enhancer molecule are on the focus of the innovators, such as chitosan, poly-L-arginine, aminated gelatin and many types of thiolated polymers. The permeation enhancing activity of these polymers can be explained by opening the thigh junction by inhibiting the activity of P-gp on the epithelial cells.

Drug loaded nanoparticles are particles with 1–1,000 (500) nm size. They contribute to the drugs attaching to a nanoparticle matrix, or dissolved, encapsulated and entrapped in that. The nanoparticle are able to carry the drug through the barriers (enhancing the corneal penetration of drugs) or target them to the specific site of action [50]. Beyond that -depending on the formulation and ingredients- modified drug release can also be achieved.

11.4.2 Improvement of Retention of the DDS

The most commonly used ocular formulations are solutions and aqueous suspensions. They are well tolerated and effective systems but their bioavailabilities are limited. have low bioavailability. To overcome these mishaps, viscous formulations (ointments) or viscosity enhancing agents (polymers) are mainly used in the preparations. Since increased viscosity can causes the sticky foreign body sensation and blurred vision, it is important to determine an optimal viscosity range. The topically applied hydrogels can be divided into preformed (classical) and in situ forming gel. Whereas performing hydrogels are highly viscous systems with complicated administration, the in situ forming gels are low viscous liquids which upon exposure to physiological conditions of the ocular surface/exposure to ocular surface Ph. transforms to a gel phase [51, 52]. From this reason these systems can be easily installed as a drop overcoming the problem of repeated administration. The sol-gel transition can be induced by the ion strength (gellan gum) [53], alginates [54], pH (latex) and the temperature (poloxamers) [55] of the biological environment.

The last decade has seen a move forward in the formulation of topically administrated DDSs bioadhesive preparations. They gained interest lately and have come into highlights. Another approach to improve the residence time of the ocular dosage form was the implementation of the mucoadhesive concept. In many papers interactions of different polymers with mucins were evaluated. Cellulose derivatives, acrylates, alginates, hyaluronan, polysaccharides, chitosan and the new type of mucoadhesives such as thiomers were defined as good bioadhesive polymers.

There are several potencial techniques that can enhance topical drug delivery. One group of these techniques is the polymeric drug delivery system. Nanoparticles and nanocapsules offer more possibilities for controlled drug release. Particulate systems are more stable than other colloidal systems, and long term storage is possible after freeze drying. Nowadays, carriers are nontoxic, nonimmunogenic, biocompatible, uniform and biodegradable. Nanoparticles are able to protect the delivered molecules while interacting with the ocular surface. Nanocapsules are colloid systems with an oil core and polymeric coating, which is planned to further improve ocular bioavailability. Nanocapsules present an advantage over nanoparticles as carriers for lipophilic drugs, as they can incorporate within their core oil phase marked amounts of lipophilic drugs.

Micelles are amphiphilic molecules, thus surfactants, which form nanoscopic supramolecular structures with a hydrophobic core and a hydrophilic shell. Liposomes are vesicles consisting of an aqueous compartment core surrounded by a lipid bilayer that mimics a cell membrane. The topical ophthalmic use of lipophilic or macromolecular therapeutic substances has been limited clinically because of their poor ocular bioavailability.

Several pharmaceutical developments have been made in delivery systems, but a very simple method can be applied: amniotic membrane extraction solutions [56]. Amniotic membrane extract is produced by lyophilized preparation. In amniotic membrane extract all biologically active components are readily available (growth factors, neutrophins, interleukins, receptors, fibronectins and different types of collagen) for the treatment of corneal surface defects [57]. It is also possible to add drugs to amniotic membrane extract in order to combine biological and pharmacological effects. Ha [58] and coworkers further demonstrated, that amniotic membrane extract is beneficial in alkali burns whereas Jiang et al. [59] provided evidence on the antiangiogenic effects of amniotic membrane extract.

Mencucci et al. [60] soaked amniotic membrane in acyclovir or trifluridine and put on cell cultures. They have noted, that a sustained release of antiviral drug was possible by amniotic membrane, since it inhibited the reproduction of herpes simplex virus significantly in cell cultures.

Further evidence have been presented by Singh et al. [61] who developed the silver-impregnated amniotic membranes, which provided an effective barrier to bacterial penetration, and they have demonstrated the ability of silver-impregnated amniotic membrane to combat microbial infection and its ideal physical characteristics for clinical use as a burn wound dressing.

11.5 Conclusions

Amniotic membrane can have various effects on drug permeability and can significantly alter bioavailability of the drugs used. Beyond several biochemical, biological and biophysical benefits of amniotic membrane transplantation, amniotic membrane serves a layer with a more stable drug concentration on the ocular surface. If we could achieve exact knowledge of the pharmakokinetics of amniotic membrane, an individual treatment of patients needing AMT could be the future therapeutical option.

References

1. American Academy of Ophthalmology (2005) Ocular pharmacology. Fundam Princ Ophthalmol (Basic Clin Sci Course) 2:319–332
2. Burstein NL, Anderson JA (1985) Corneal penetration and ocular bioavailability of drugs. J Ocul Pharmacol 1:309–326
3. Shell JW (1985) Ophthalmic drug delivery systems. Drug Dev Res 6:245–261
4. King-Smith PE, Fink BA, Fogt N et al (2000) The thickness of the human precorneal tear film: evidence from reflection spectra. Invest Ophthalmol Vis Sci 41:3348–3359
5. Azartash K, Kwan J, Paugh JR et al (2011) Pre-corneal tear film thickness in humans measured with a novel technique. Mol Vis 17:756–767
6. Németh J, Erdélyi B, Csákány B et al (2002) High-speed videotopographic measurement of tear film build-up time. Invest Ophthalmol Vis Sci 43:1783–1790
7. Ahmed I (2003) The noncorneal route in ocular drug delivery. In: Mitra AK (ed) Ophthalmic drug delivery system, 2nd edn. Marcel Dekker, New York, pp 335–363
8. Marsovszky L, Resch MD, Németh J et al (2012) In vivo confocal microscopic examination of corneal Langerhans cell density, distribution and evaluation of dry eye in rheumatoid arthritis. Innate Immun 19:348–354
9. Berman ER (1991) Biochemistry of the eye. Plenum press, New York, NY, USA p 476
10. Azuara-Blanco A, Pillai C, Dua HS (1999) Amniotic membrane transplantation for ocular surface reconstruction. Br J Ophthalmol 83:399–402
11. Fish R, Davidson RS (2010) Management of ocular thermal and chemical injuries, including amniotic membrane therapy. Curr Opin Ophthalmol 21:317–321
12. Kim H-S, Sah W-J, Kim Y-J et al (2001) Amniotic membrane, tear film, corneal and aqueous levels of ofloxacin in rabbit eyes after amniotic membrane transplantation. Cornea 20:628–634
13. O'Brien T, Sawusch M, Dick J, Gottsch J (1988) Topical ciprofloxacin treatment of pseudomonas keratitis in rabbits. Arch Ophthalmol 106:1444–1446
14. Healy DP, Holland EJ, Nordlund ML et al (2004) Concentrations of levofloxacin, ofloxacin and ciprofloxacin in human corneal stromal tissue and aqueous humor after topical administration. Cornea 23:255–263
15. Robert P-Y, Adenis J-P (2001) Comparative review of topical ophthalmic antibacterial preparations. Drugs 61:175–185
16. Cursiefen C, Chen L, Dana MR, Streilein JW (2003) Corneal lymphangiogenesis: evidence, mechanisms and implications for corneal transplant immunology. Cornea 22:273–281
17. Seitz B, Das S, Sauer R et al (2011) Simultaneous amniotic membrane patch in high-risk keratoplasty. Cornea 30:269–272
18. Dietrich T, Sauer R, Hofmann-Rummelt C et al (2011) Simultaneous amniotic membrane transplantation in emergency penetrating keratoplasty: a therapeutic option for severe corneal ulcerations and melting disorders. Br J Ophthalmol 95:1034–1035

19. Tseng SC, Di Pascuale M, Liu DDT-S et al (2005) Intraoperative mitomycin C and amniotic membrane transplantation for fornix reconstruction in severe cicatricial ocular surface diseases. Ophthalmology 112:896–903
20. Liu B-Q, Wang Z-C, Liu L-M et al (2008) Sutureless fixation of amniotic membrane patch as a therapeutic contact lens by using a polymethyl methacrylate ring and fibrin sealant in a rabbit model. Cornea 27:74–79
21. Hick S, Demers PE, Brunette I et al (2005) Amniotic membrane transplantation and fibrin glue in the management of corneal ulcers and perforations: a review of 33 cases. Cornea 24:369–377
22. Seitz B, Das S, Sauer R et al (2009) Amniotic membrane transplantation for persistent corneal epithelial defects in eyes after penetrating keratoplasty. Eye 23:840–848
23. Zhang T, Yam GH-F, Riau AK et al (2013) The effect of amniotic membrane de-epithelialization method on its biological properties and ability to promote limbal epithelial cell culture. Invest Ophthalmol Vis Sci 54:3072–3081
24. Resch M, Resch B, Csizmazia E et al (2010) Permeability of human amniotic membrane to ofloxacin in vitro. Invest Ophthalmol Vis Sci 51:1024–1027
25. Resch M, Resch B, Csizmazi E et al (2011) Drug reservoir function of human amniotic membrane. J Ocul Pharmacol Ther 27:323–326
26. Mayer WJ, Grüterich M, Kook D et al (2013) Modification of amniotic membrane as a depot carrier for bevacizumab: an in-vitro model for a slow release mechanism. Curr Eye Res 38:445–450
27. Resch M, Schlötzer-Schrehardt U, Hofmann-Rummelt C et al (2006) Integration patterns of cryopreserved amniotic membranes into the human cornea. Ophthalmology 113:1927–1935
28. Hahn A, Thanos M, Reinhard T et al (2010) Procedural guidelines: good practice procedures for acquisition and preparation of cryopreserved human amniotic membranes from donor placentas. Ophthalmologe 107:1020–1031
29. Mahdy RA, Nada WM, Almasalamy SM et al (2010) A freeze-dried (lyophilized) amniotic membrane transplantation with mitomycin C and trabeculectomy for pediatric glaucoma. Cutan Ocul Toxicol 29:164–170
30. Nakamura T, Inatomi T, Sekiyama E et al (2006) Novel clinical application of sterilized, freeze-dried amniotic membrane to treat patients with pterygium. Acta Ophthalmol Scand 84:401–405
31. Kruse FE, Joussen AM, Rohrschneider K et al (2000) Cryopreserved human amniotic membrane for ocular surface reconstruction. Graefe's Arch Clin Exp Ophthalmol 238:68–75
32. Schwab IR, Isseroff RR (2002) Ex vivo stem cell expansion. In: Holland E, Mannis M (eds) Ocular surface disease: medical and surgical management. Springer, New York, pp 232–242
33. Meller D, Pires R, Tseng S (2002) Ex vivo preservation and expansion of human limbal epithelial stem cells on amniotic membrane cultures. Br J Ophthalmol 86:463–471
34. Meller D, Pauklin M, Westekemper H, Steuhl K-P (2010) Autologe transplantation von kultiviertem Limbusepithel. Ophthalmologe 107:1133–1138
35. Grueterich M, Espana EM, Tseng SC (2003) Ex vivo expansion of limbal epithelial stem cells: amniotic membrane serving as a stem cell niche. Surv Ophthalmol 48:631–646
36. Shortt AJ, Secker GA, Rajan MS et al (2008) Ex vivo expansion and transplantation of limbal epithelial stem cells. Ophthalmology 115:1989–1997
37. Ti S-E, Anderson D, Touhami A et al (2002) Factors affecting outcome following transplantation of ex vivo expanded limbal epithelium on amniotic membrane for total limbal deficiency in rabbits. Invest Ophthalmol Vis Sci 43:2584–2592
38. Hopkinson A, Shanmuganathan VA, Gray T et al (2008) Optimization of amniotic membrane (AM) denuding for tissue engineering. Tissue Eng C 14:371–381
39. Deolinda de Oliveira Pena J, de Melo GB, Gomes JÁP et al (2007) Ultrastructural and growth factor analysis of amniotic membrane preserved by different methods for ocular surgery. Arq Bras Oftalmol 70:756–762
40. Dietrich-Ntoukas T, Hofmann-Rummelt C, Kruse FE, Schlötzer-Schrehardt U (2012) Comparative analysis of the basement membrane composition of the human limbus epithelium and amniotic membrane epithelium. Cornea 31:564–569

41. Mindel JS, Smith H, Jacobs M et al (1984) Drug reservoirs in topical therapy. Invest Ophthalmol Vis Sci 25:346–350
42. Huang H-S, Schoenwald RD, Lach JL (1983) Corneal penetration behavior of β-blocking agents: II – assessment of barrier contributions. J Pharm Sci 72:1272–1279
43. Greaves J, Olejnik O, Wilson C (1992) Polymers and the precorneal tear film. STP Pharma Sci 2:13–33
44. Liaw J, Rojanasakul Y, Robinson JR (1992) The effect of drug charge type and charge density on corneal transport. Int J Pharm 88:111–124
45. Järvinen K, Järvinen T, Urtti A (1995) Ocular absorption following topical delivery. Adv Drug Deliv Rev 16:3–19
46. Touitou E, Barry BW (2007) Enhancement in drug delivery. CRC Press, Boca Raton, FL, USA 656
47. Thakur RR, Kashiv M (2011) Modern delivery systems for ocular drug formulations: a comparative overview W.R.T conventional dosage form. Int J Res Pharm Biomed Sci 2:8–18
48. Sasaki H, Yamamura K, Nishida K et al (1996) Delivery of drugs to the eye by topical application. Prog Retin Eye Res 15:583–620
49. Loftssona T, Järvinen T (1999) Cyclodextrins in ophthalmic drug delivery. Adv Drug Deliv Rev 36:59–79
50. Xie S, Zhu L, Dong Z et al (2011) Preparation and evaluation of ofloxacin-loaded palmitic acid solid lipid nanoparticles. Int J Nanomedicine 6:547–555
51. Kumar S, Haglund BO, Himmelstein KJ (1994) In situ-forming gels for ophtalmic drug delivery. J Ocul Pharmacol 10:47–56
52. Zignani M, Tabatabay C, Gurny R (1995) Topical semi-solid drug delivery: kinetics and tolerance of ophthalmic hydrogels. Adv Drug Deliv Rev 16:51–60
53. Mazuel C, Friteyre M-C (1989) Ophthalmological composition of the type which undergoes liquid-gel phase transition. US 4861760 A.
54. Cohen S, Lobel E, Trevgoda A, Peled Y (1997) A novel in situ-forming ophthalmic drug delivery system from alginates undergoing gelation in the eye. J Control Release 44:201–208
55. Greaves JL, Wilson CG (1993) Treatment of diseases of the eye with mucoadhesive delivery systems. Adv Drug Deliv Rev 11:349–383
56. Liang L, Li W, Ling S et al (2009) Amniotic membrane extraction solution for ocular chemical burns. Clin Exp Ophthalmol 37:855–863
57. Kordić R, Suić S, Jandroković S et al (2013) Application of the amniotic membrane extract (AMX) for the persistent epithelial defect (PED) of the cornea. Coll Antropol 37:161–164
58. Ha SW, Kim JS, Cheong TB, Kim JC (2001) Therapeutic effect of amniotic membrane extract on keratitis following corneal alkali burn. J Korean Ophthalmol Soc 42:1555–1561
59. Jiang A, Li C, Gao Y et al (2006) In vivo and in vitro inhibitory effect of amniotic extraction on neovascularization. Cornea 25:S36–S40
60. Mencucci R, Paladini I, Menchini U et al (2011) Inhibition of viral replication in vitro by antiviral-treated amniotic membrane: possible use of amniotic membrane as drug-delivering tool. Br J Ophthalmol 95:28–31
61. Singh R, Kumar D, Kumar P, Chacharkar MP (2008) Development and evaluation of silver-impregnated amniotic membrane as an antimicrobial burn dressing. J Burn Care Res 29:64–72

Part III
Technical Issues About Amniotic Membrane

Chapter 12
Isolation and Characterization of Mesenchymal Stem Cells from Amniotic Membrane

Ana Lopes, Mário Grãos, and Artur Paiva

Abstract Placenta and amniotic membrane emerge as important sources of stem cells, since it is possible to perform their study with non-invasive methods. Also, is possible to obtain higher numbers of stem cells from these tissues compared with other sources. Several studies have reported that multipotent adult stem cells represent an attractive stem cell source for regenerative medicine and transplantation therapy, since these cells have a high degree of plasticity and multi-lineage differentiation potential. Among adult stem cells, mesenchymal stem cells (MSC) arise since they are able of both supporting hematopoiesis and differentiating into mesoderm, endoderm and ectoderm cells and are known to be weakly immunogenic and to exhibit immunomodulatory properties, which is important to escape to the immunological defence mechanisms and to suppress several functions. Diverse studies associated the MSC heterogeneity with their differential potential. MSC have different characteristics related to the source of isolation, therefore they can differ, not only, in terms of phenotype, but also morphology, ultrastructure and even function. Although MSC phenotype is well described in the literature, the absence of a single marker expressed only by this cell type could be a problem for the isolation of a homogeneous stem cell population and for the identification of their therapeutic potential. In this sense, recent efforts have been made in order to isolate more homogeneous cell populations, avoiding contamination by other

A. Lopes • A. Paiva (✉)
Flow Cytometry and Cellular Immunology Lab – Blood and Transplantation Center of Coimbra, Portuguese Institute of Blood and Transplantation, Coimbra, Portugal
e-mail: Artur.Paiva@ipst.min-saude.pt

M. Grãos
Cell Biology Unit, Biocant, Center of Innovation in Biotechnology, Coimbra, Portugal

A.C. Mamede, M.F. Botelho (eds.), *Amniotic Membrane: Origin, Characterization and Medical Applications*,
DOI 10.1007/978-94-017-9975-1_12

cellular types, and to develop a set of markers for the characterization of this stem cell type. In this context, this chapter gives an overview of protocols for isolation, a proposal for their characterization, and differentiation of mesenchymal stem cells from human amniotic membrane (hAMSC) in order to potentiate its application to clinical practice.

Keywords Amniotic membrane • Mesenchymal stem cell • Immunophenotype • Cell sorting • Gene expression • Differentiation

12.1 Amniotic Membrane-Derived Cells

In the search for other potential sources of adult stem cells, placenta and amniotic fluid have clearly been in focus. Cells from the mesenchymal and epithelial regions of the amnion can be isolated by mechanical separation. Epithelial cells can be isolated after digestion with trypsin or dispase and mesenchymal stem cells from human amniotic membrane (hAMSC) after digestion with collagenase. MSC have important applications in the field of regenerative medicine for therapeutic purposes due to their high degree of plasticity and also, based on their immunomodulatory properties, may contribute to the maintenance of feto-maternal tolerance, to reverse acute graft *versus* host disease (GVHD) refracted to conventional therapy as well as lupus nephritis in systemic lupus erithematosous. Another advantage for the use of placental derived cells is related to the ethical problems, which could be avoided since these tissues are generally discarded [1]. Accordingly, they can be useful for understanding general cellular processes like embryogenesis, organogenesis, cancer or ageing [2]. Regarding their multipotent properties, these cells are an attractive source in the transplantation field [3].

Regardless of MSC type has not a specific phenotypic marker that allows direct identification, diverse laboratorial procedures and strategies have been developed for MSC processing [4]. For example, several groups still perform isolation of this cellular type only by plastic adherence, although this method does not provide a purified cell population, once monocytes also have adherence for the plastic [5]. Moreover, the culture conditions are not consensual too. Therefore there is no standard protocol for MSC proceedings and the comparisons between the results can be compromised. To overcome these limitations, several strategies have been developed to facilitate the prospective isolation and characterization of MSC based on the selective expression or absence of surface markers [6].

In this chapter, we aim to summarize different methods for the isolation of MSC from the amniotic membrane after caesarean; to establish a proposal for an extensive immunophenotypic characterization of MSC, and also for their differentiation into diverse cell lineages under specific culture conditions.

12.2 Technical Proceedings/Methods and Technologies

12.2.1 hAMSC Isolation and Culture

Placentas are harvested within 4 h of delivery after cesarean section. The fresh amnion should be mechanically separated from the chorion and washed extensively in phosphate-buffered saline (PBS) containing 100 U/ml penicillin and 100 μg/mL streptomycin in order to remove blood and cellular debris. The amnion should be cut into 2×2 cm pieces and incubated for 7 min at 37 °C in PBS containing 2.4 U/mL dispase. Then, the pieces should be placed in Roswell Park Memorial Institute medium (RPMI) 1 640 medium, supplemented with 10 % heat-inactivated fetal bovine serum (FBS) and 2 mM L-glutamine at room temperature [7, 8].

After a resting period (5–10 min) the fragments should be digested with 0.75 mg/mL collagenase and 20 μg/mL DNAse for approximately 3 h at 37 °C. Amnion fragments should be then removed and the cell suspension filtrated through a 100 μm cell strainer and the cells collected by centrifugation at 200× g for 10 min. The cells are cultured in mesenchymal stem cell growth medium (MSCGM) and, when plated at low density, formed distinct colonies that could be isolated using cloning rings and expanded for several passages. The culture medium should be changed once a week until fibroblast-like cells appeared and twice a week thereafter [7, 8].

Several protocols have been reported to be useful for the isolation of different cell types from the amniotic membrane. In Table 12.1 are summarized some proceedings and different reagents used

12.2.2 hAMSC Characterization

Regardless of the source of MSC, they express several phenotypic markers, however none are specific for this cell population. It is generally accepted that MSC do not express the hematopoietic cell markers CD45, CD34, CD14, or CD11b. They also do not express the costimulatory molecules CD80, CD86, or CD40 or the adhesion molecules CD31 [platelet/endothelial cell adhesion molecule (PECAM)-1], CD18 [leukocyte function-associated antigen- 1 (LFA-1)], or CD56 (neuronal cell adhesion molecule-1), but they express CD105 (endoglin), CD73 (ecto-5′-nucleotidase), CD90 (Thy-1), CD13 (aminopeptidase N) and CD29 (integrin β1). Also, there are conditions that MSC can express CD271 [nerve growth factor receptor (NGFR)], CD44 (glycoprotein), CD71 [transferrin receptor protein 1 (TfR1)] and Stro-1, as well as the adhesion molecules CD106 [vascular cell adhesion molecule (VCAM])-1], CD166 [activated leukocyte cell adhesion molecule (ALCAM)], CD54 [intercellular adhesion molecule (ICAM)-1] [13–15].

Table 12.1 Different proceedings for the mesenchymal stem cells from human amniotic membrane isolation

	Soncini et al. (2007) [7]	Kim et al. (2007) [9]	Alviano et al. (2007) [10]	Koo et al. (2012) [11]	Leyva-Leyva et al. (2013) [12]
Mechanical separation	Yes	Yes	Yes	Yes	Yes
Amnion washing solution	PBS with 100 U/ml penicillin + 100 μg/mL streptomycin	PBS	PBS with 200 U/mL penicillin + 200 μg/mL streptomycin	DPBS	PBS with 100 U/ml penicillin + 100 μg/mL streptomycin + 0.25 μg/mL amphotericin B
Size of amnion pieces	2×2 cm	5×5 cm	Condition NM	2×2 cm	10×10 cm
1st stage of enzymatic digestion	PBS with 2.4 U/mL dispase	DMEM with 0.25 % trypsin[a]	EDTA with 0.25 % trypsin	EDTA with 0.5 % trypsin	EDTA (0.5 mM) with 0.125 % trypsin
Time of incubation	7 min at 37 °C[b]	30 min at 37 °C stirring[a]	15 min at room temperature	5 min at room temperature	30 min at 37 °C
2nd stage of enzymatic digestion	RPMI with 0.75 mg/mL collagenase and 20 μg/mL DNAse	PBS + 32 mg/mL collagenase A + 0.05 mg/mL DNase	EDTA with 0.25 % trypsin + 10 U/mL DNase I + 0.1 % collagenase IV in DMEM	0.3 % collagenase I	DMEM with 100 U/ml collagenase II + 3 mM calcium chloride
Time of incubation	3 h at 37 °C	2 h at 37 ° C stirring	5 min at 37 °C[b]	20–30 min at 37 °C	2 h at 37 °C
Filtrate	100 μm cell strainer	Condition NM	Condition NM	100 μm cell strainer	Yes (conditions NM)
Centrifugation	200×g for 10 min	500×g	1500 rpm for 10 min	2500 rpm for 5 min	Yes (conditions NM)
Culture medium	Mesenchymal stem cell growth medium (MSCGM)	DMEM + 10 % FBS + 100 U/mL penicillin + 0.1 mg/ml streptomycin + 3.7 mg/mL sodium bicarbonate	DMEM + 20 % FBS + 100 U/mL penicillin + 100 μg/ml streptomycin	α-MEM + 10 % FBS + 1 % penicillin-streptomycin	DMEM + 10 % FBS + 100 U/ml penicillin + 100 μg/mL streptomycin + 0.25 μg/mL amphotericin B
Seeding area	162 cm^2 flaks	75 cm^2 flaks	25 cm^2 flaks	25 cm^2 flaks	25 cm^2 flaks

NM not mentioned, *DMEM* Dulbecco's Modified Eagle's Medium, *EDTA* Tripotassium ethylene diamine tetraacetic acid, *α-MEM* Minimum Essential Medium Alpha

[a]Repeat twice

[b]Inactivation stage

Immunophenotyping

The International Society of Cellular Therapy has identified a set of criteria to define human MSC that included the phenotypic analyses, simultaneous positively expression of CD105, CD73 and CD90 in the absence of hematopoietic markers expression as previously referred.

Multiparameter flow cytometry arises as a powerful technology that represents a great technical advantage, since it allows the individual measurement of physical and chemical characteristics using an extended panel of markers as cells pass one by one through a laser [16]. This technique is capable of rapid and highly quantitative identification of individual cells for the presence or absence of a wide range of fluorescence and light scattering signals that correlate to cell morphology, surface and intracellular protein expression, gene expression, and cellular physiology [17]. The increased sensitivity of flow cytometry allows a direct identification and measurement of multiple antigens at a single cell level that is useful for accurate MSC identification.

The composition of the following proposed antibody panel was drawn based on the International Society for Cellular Therapy criteria [4] and improved by testing the various combinations of fluorochrome-conjugated antibody clones according to our experience and based on EuroFlow methodology [18].

The application of this eight-color staining protocol for MSC characterization enables the analyses of different proteins (17) in one single file; applying the merge and calculate tools it is possible to integrate information from different tubes or conjugated with the same fluorochrome into the same data file with a new statistical approach that may be used for the automated generation of flow cytometry data files containing information on single events about a virtually infinite number of parameters [19, 20].

The immunophenotypic characterization of hAMSC can be performed directly (e.g. in amniotic fluid) and also after enzymatic treatment (e.g. with tryptase and accutase) for detachment and isolation, however a wash procedure using PBS and centrifugation is required. These cells are stained for surface cell markers using a stain-wash direct immunofluorescence procedure. 2×10^5 hAMSC are distributed in different tubes and stained with the following combinations of monoclonal antibodies (Table 12.2).

Cells can be incubated for 15 min at room temperature in the dark. The isotype control for each antibody/fluorochrome should be used to exclude nonspecific binding and a non-stained sample also should be used as negative control. After the incubation period, a wash protocol should be followed: 2 mL of PBS was added, mixed and then centrifuged at 1500 rpm for 5 min. Cells were resuspended in 0.5 mL of PBS before acquisition in the flow cytometer [21].

Dead cells and cell debris could be excluded by forward scatter area (FSC-A) and side scatter area (SSC-A), while doublets are excluded by FSC-A versus forward scatter height (FSC-H). The analysis might be performed in Infinicyt Software (Cytognos, Salamanca, Spain) (Fig. 12.1).

Table 12.2 Panel of monoclonal antibodies for MSC characterization [21]

	FITC	PE	$PerCP_{Cy}5.5$	$PE_{Cy}7$	APC	APCH7	PB	PO
Tube 1	**CD49e**	**CD73**	**CD34**	**CD13**	**CD90**	**HLA-DR**	**CD11b**	**CD45**
Clone	SAM1	AD2	8G12	Immu103.44	5E10	L243	ICRF44	HI30
Company	Beckman Coulter	BD Pharmingen	BDB	Beckman Coulter	BD Pharmingen	BDB	BD Pharmingen	Invitrogen
Tube 2	**CD31**	**NGFR**	**CD14**	**CD13**	**CD133**	–	**CD11b**	**CD45**
Clone	WM59	C40-1457	M5E2		293C3			
Company	BD Pharmingen	BD Pharmingen	BD Pharmingen		Miltenyi Biotec			
Tube 3	**CD15**	**CD146**	**CD24**	**CD13**	**CD90**	**CD29**	**CD11b**	**CD45**
Clone	HI98	P1H12	ALB9			TS2/16		
Company	BDB	BD Pharmingen	Beckman Coulter			BioLegend		
Tube 4	**CD106**	**CD105**	–	**CD13**	**HLA-A, B, C**	–	**CD11b**	**CD45**
Clone	51-10C9	1G2			G46-2.6			
Company	BD Pharmingen	Beckman Coulter			BD Pharmingen			
Tube 5	–	**CD73**	**CD24**	**CD13**	**CD90**	–	**CD11b**	**CD45**

FITC fluorescein isothiocyanate, *PE* phycoerythrin, *PerCPcy5.5* peridinin chlorophyll protein cyanine 5.5, *PEcy7* phycoerythrin cyanine 7, *APC* allophycocyanin, *APCH7* allophycocyanin hilite7, *PB* pacific blue, *PO* pacific orange

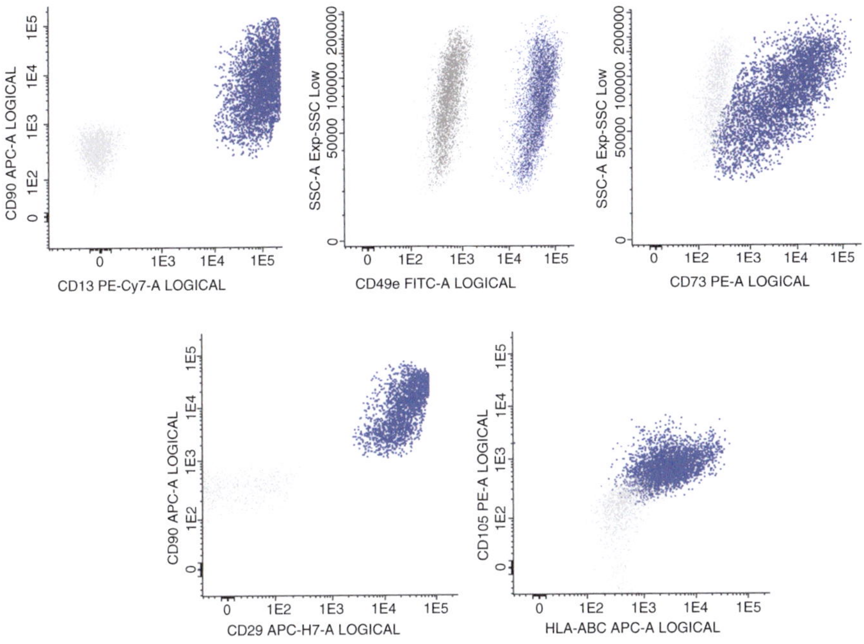

Fig. 12.1 Immunophenotypic characteristics of mesenchymal stem cells from human amniotic membrane (*blue*). Cells represented as *grey* are the hAMSC without staining (negative control)

The combination of cell sorting capability to flow cytometry makes possible the isolation of populations based on their phenotypic heterogeneity in different conditions; after isolation, cell culture and also after in vitro differentiation, through electronic Fluorescence-Activated Cell Sorting (FACS). FACS is an important technology, as heterogeneous cell suspensions can be purified into fractions containing a single cell type based upon virtually unlimited combinations of user-defined parameters. Beyond the exceptionally accurate and analytical nature, sorting function is non-destructive to the cells being processed, with little to no effect on cell viability or function [22]. This methodology also allows to sort, under aseptic conditions, cell populations for expansion in culture. Therefore, the cell sorter becomes a starting point for boundless cellular and molecular investigations [16].

One of the important points to the success of sorting is almost entirely due to the preparation and the state of the samples. It is a prerequisite that cells must be in a suspension, hence the difficulty arises when using cells from adherent cultures or cells from solid tissues. Another point to take into account is the cell sorter adequacy to the samples. Cells should be no more than one fifth the diameter of the nozzle – small cells such as lymphocytes require a 70-μm nozzle -, whereas many cultured adherent cells require a 100-μm nozzle. However, there are several well established methods for preparing samples for flow sorting.

It is recommended that cells before sorting should be washed in cell culture medium or in PBS with 1 % bovine serum albumin and then staining for surface markers using a stain-wash direct immunofluorescence procedure, monoclonal antibodies should be incubating for 15 min at room temperature in the dark. After this incubation period, a wash protocol should be followed: 2 mL of PBS should be added to the cells, the samples should be mixed and then centrifuged. Then, cells should be resuspended in a small volume of medium and aspirated up and down through a pipet several times to help disaggregate clumps. Normally, adherent cells tend to be larger and are recommended to be resuspended at a lower density. It is always better to keep the density high prior to sorting and dilute to an appropriate concentration immediately prior to sorting [16].

Immunocytochemistry Assay

Immunocytochemistry is a procedure that allows the correlation between precise localization of a specific protein or antigen and tissue morphology in fixed cells on slides and tissue sections by binding a specific secondary antibody conjugated with a fluorescent dye and the results could be obtained using a fluorescence or confocal microscope.

Cells should be seeded on glass coverslips coated with an appropriate cell-binding protein or peptide (e.g.: poly-D-Lysine) at 2×10^3 cells/cm^2. For the maintenance of cytoskeleton integrity cells were rinsed at 37 °C with a cytoskeleton buffer, namely PBS and then fixed with 3 % formaldehyde in citrate buffer (CB) solution for 15 min at 37 °C. After that cells should be washed with CB and the membrane is permeabilized with 0.1 % Triton X-100 CB solution for 5 min. Coverslips were blocked with 0.5 % fish skin gelatin in PBS for 20 min at room temperature and incubated overnight at 4 °C with a specific antibody. For example, antibody anti- L-alkaline phosphatase, collagen type I, II, IV and XII, fibronectin, desmin, fibroblast surface protein (FSP), Nanog, Sox-2, Oct-3/4, GFAP, Neuro-D, SSEA-3, SSEA-4, HLA-ABC, CD105, CD29, CD31, CD44, CD54, CD106. Then, the samples were rinsed in PBS and incubated with the secondary antibody (the time and conditions of incubation were related with the manufacturer's instructions). The slides could be assembled with VECTASHIELD® Mounting Media that contains 4′,6-diamidino-2-phenylindole (DAPI) and then analysed in a confocal microscope [9, 12].

The main characteristics of hAMSC are their multipotency and self-renewal capacity. These features seem to be related to the presence of some transcription factors, including Oct-4, SOX-2 and Nanog (with an important role in self-renewal function, however their expression in hAMSC is controversial) [23] and SSEA-4 and L-alkaline phosphatase, expressed in early embryonic development and implicated in the maintenance of the non-differentiation state of these cells. Hence, immunocytochemistry arises as an important additional procedure for the characterization of hAMSC (Fig. 12.2) [24, 25].

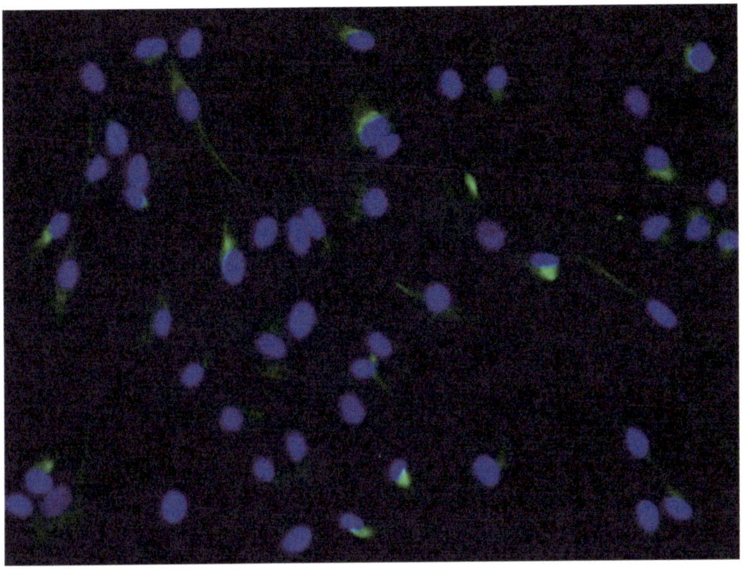

Fig. 12.2 Analysis of Nestin expression by immunocytochemistry of human umbilical cord matrix mesenchymal stem cells. Nestin is stained in *green* and nuclei are counterstained with DAPI (in *blue*)

Molecular Assays

Although some hAMSC features are well-know, their genetic profile is not clear yet. Recent reports referenced that the expression of some specific genes is related to their multipotency and differentiation potential [26]. Therefore, it is important to perform a detailed molecular characterization of these cells that allows the distinction of various subpopulations.

Several procedures may be used to understand the molecular mechanisms of hAMSC. In Table 12.3 are described several markers that can be studied using global methodologies, like transcriptomics, proteomics, real time polymerase chain reaction and others, for molecular characterization of these cells.

The analysis of signature genes of hAMSC revealed that many of them seem to be involved in cellular immune functions, namely tolerance, in biological processes like cell survival and limitation of hAMSC differentiation. The genes that appear to be involved in immune adaptation at the maternoplacental interface are summarized in Table 12.3A. Genes presented in Table 12.3C are associated with the differentiation process. In the case of REX-1, its expression could block MSC differentiation.

The combination between different phenotypes, the successful sorting of heterogeneous populations and the knowledge of the genetic patterns expressed by these cells, can give new insights into the characterization of hAMSC and the understanding of their function. All this information is advantageous for the isolation of the specific cell types, since it is possible to choose which cells are more appropriate for specific therapeutic approaches.

Table 12.3 Description of some important markers identified in mesenchymal stem cells from human amniotic membrane

A – Transcriptomics [27, 28]	
SPON2	Spondin 2, extracellular matrix protein
IFI27	Interferon α-inducible protein 27
BDBKRB1	Bradykinin receptor B1
SCYB5	Small inducible cytokine subfamily B member 5
SCYB6	Small inducible cytokine subfamily B member 6
FOXF1	Forkhead box F1
HAND2	Heart and neural crest derivates expressed 2
TCF21	Transcription factor 21
DPP6	Dipeptidyl aminopeptidase-like protein 6
TDO2	Tryptophan 2,3-dioxygenase
STs	Steroid sulfatase
B – Proteomics [27, 29, 30]	
OGN	Osteoglycin
βIG-H3	Transforming growth factor-beta-induced protein ig-h3
THBS1	Thrombospondin 1
COL	Colagen
TGM2A	Transglutaminase 2 isoform A
ACTB	Beta-actin
FGB	Fibrinogen beta chain
NID2	Nidogen 2
HSPA5	70 kD heat shock protein 5
C – Gene expression [27]	
Oct-3	Octamer-binding protein 3
Oct-4	Octamer-binding protein 4
Rex-1	Reduced expression 1
SCF	Stem cell factor protein
NCAM	Neural cell adhesion molecule
NES	Nestin
BMP-4	Bone morphogenetic protein 4
GATA-4	GATA binding protein 4
HNF-4α	Hepatocyte nuclear factor 4α
FGF5	Fibroblast growth factor 5
PAX6	Paired box protein 6
BMP-2	Bone morphogenetic protein 2
Nanog	Homebox transcription factor Nanog

12.3 hAMSC Differentiation Potential

Recently, experiments with hAMSC cultures were performed in order to evaluate their multilineage differentiation potential, since they seem to have the ability to differentiate into cell type's characteristic of all the three germ layers. However, more studies in vivo need to be made to document the spectrum of their multilineage differentiation [27]. The differentiation process can be confirmed by immunocytochemistry, gene expression and functional analyses [31]. hAMSC from 0 to 4 passages can be used for differentiation assays, after being detached and seeded at a density of $1.5 \times 10^5/cm^2$ in glass chamber slides for all differentiation studies.

12.3.1 *Osteogenic Differentiation*

Osteogenesis was induced by culturing cells in Bulletkit Osteogenic Differentiation Medium or in Bullekit Osteogenic Differentiation Medium (Lonza) or in NH OsteoDiff Medium for up to 3 weeks at 37 °C with 5 % CO_2. The culture medium needs to be changed every 2–3 days. To demonstrate osteogenic differentiation, cultures were fixed and osteoblasts could be identified by immunocytochemical staining of alkaline phosphatase. Calcium deposits, characteristic of osteogenic differentiation, could be visualized by alizarin red staining or von Kossa [7, 10, 32].

12.3.2 *Adipogenic Differentiation*

To evaluate adipogenic differentiation, cells were allowed to become nearly confluent and then cultured for 3 weeks in Bulletkit Adipogenic Differentiation Medium according to the manufacturer's instructions. Cytoplasmic inclusions of neutral lipids were then stained with fresh Oil Red O [7, 32].

12.3.3 *Chondrogenic Differentiation*

hAMSC (5×10^5 cells/cm^2) from passages P0 to P4 were detached using trypsin-EDTA and centrifuged at 300 g for 10 min. The resulting pellet should be cultured in chondrogenic differentiation medium, DMEM with 15 % FBS. The medium was supplemented with 5 mg/mL ascorbic acid (AA), 1/1,000 monotioglycerol and 1 % P/E during the first 2 days to promote the induction of chondrogenesis. The medium should then be replaced by DMEM with 15 % knockout serum, 1 % P/E and supplemented with 1 μl/ml AA, 10 μM dexamethasone, 6 μl/mL transferrin, 1×10^7 M retinoic acid and 1 ng/mL of recombinant human transforming growth factor-b3 (rHuTGF-b3) for 21 days. The medium should be changed every 2–3 days [33]. The presence of metachromatic matrix could be demonstrated by toluidine blue staining [7, 32].

12.3.4 Neural-Like Differentiation

Neural-like differentiation could be induced with the DMEM supplemented with 20 % FBS and 0.05 mg/mL NGF over 21 days, and the cultures were maintained in this medium for 15 days. The neural-like morphology may be evaluated by light microscopy, cells were more elongated and in confluence zones displayed refractile soma and neurite-like processes. Similarly, it is also possible to determine neural-like phenotype by confocal fluorescence microscopy and western blotting to detect GFAP and NEURO-D proteins [12].

References

1. Parolini O, Alviano F, Bagnara GP et al (2008) Concise review – isolation and characterization of cells from human term placenta: outcome of the first international workshop on placenta derived stem cells. Stem Cells 26:300–311
2. Bosio A, Huppert V, Donath S et al (2009) Isolation and enrichment of stem cells. Adv Biochem Eng Biotechnol 114:23–72
3. Mamede AC, Carvalho MJ, Abrantes AM et al (2012) Amniotic membrane: from structure and functions to clinical applications. Cell Tissue Res 349:447–458
4. Dominici M, Le Blanc K, Mueller I et al (2006) Minimal criteria for defining multipotent mesenchymal stromal cells: the international society for cellular therapy position statement. Cytotherapy 8:315–317
5. Schrepfer S, Deuse T, Lange C et al (2007) Simplified protocol to isolate, purify and culture expand mesenchymal stem cells. Stem Cells Dev 16:105–108
6. Harichandan A, Bühring H-J (2011) Prospective isolation of human MSC. Best Pract Res Clin Haematol 24:25–36
7. Soncini M, Vertua E, Gibelli L et al (2007) Isolation and characterization of mesenchymal cells from human fetal membranes. J Tissue Eng Regen Med 1:296–305
8. Barbati A, Mameli MG, Sidoni A, Di Renzo GC (2012) Amniotic membrane: separation of amniotic mesoderm from amniotic epithelium and isolation of their respective mesenchymal stromal and epithelial cells. Curr Protoc Stem Cell Biol 20:1E.8.1–1E.8.15
9. Kim J, Kang HM, Kim H et al (2007) Ex vivo characteristics of human amniotic membrane-derived stem cells. Cloning Stem Cells 9:581–594
10. Alviano F, Fossati V, Marchionni C et al (2007) Term amniotic membrane is a high throughput source for multipotent mesenchymal stem cells with the ability to differentiate into endothelial cells in vitro. BMC Dev Biol 7:1–14
11. Koo BK, Park IY, Kim J et al (2012) Isolation and characterization of chorionic mesenchymal stromal cells from human full term placenta. J Korean Med Sci 27:857–863
12. Leyva-Leyva M, Barrera L, López-Camarillo C et al (2013) Characterization of mesenchymal stem cell subpopulations from human amniotic membrane with dissimilar osteoblastic potential. Stem Cells Dev 22:1275–1287
13. Chamberlain G, Fox J, Ashton B, Middleton J (2007) Concise review: mesenchymal stem cells – their phenotype, differentiation capacity, immunological features and potential for homing. Stem Cells 25:2739–2749
14. Grogan SP, Miyaki S, Asahara H et al (2009) Mesenchymal progenitor cell markers in human articular cartilage: normal distribution and changes in osteoarthritis. Arthritis Res Ther 11:R85
15. Sivasubramaniyan K, Lehnen D, Ghazanfari R et al (2012) Phenotypic and functional heterogeneity of human bone marrow– and amnion-derived MSC subsets. Ann N Y Acad Sci 1266: 94–106

16. Davies D (2007) Cell sorting by flow cytometry. In: Macey M (ed) Flow Cytom. Princ. Appl. Humana Press, Totowa, pp 257–276
17. Ibrahim S, van den Engh G (2007) Flow cytometry and cell sorting. Adv Biochem Eng Biotechnol 106:19–39
18. Van Dongen J, Lhermitte L, Böttcher S et al (2012) EuroFlow antibody panels for standardized n-dimensional flow cytometric immunophenotyping of normal, reactive and malignant leukocytes. Leukemia 26:1908–1975
19. Pedreira CE, Costa ES, Arroyo ME et al (2008) A multidimensional classification approach for the automated analysis of flow cytometry data. IEEE Trans Biomed Eng 55:1155–1162
20. Pedreira CE, Costa ES, Barrena S et al (2008) Generation of flow cytometry data files with a potentially infinite number of dimensions. Cytom A 73A:834–846
21. Laranjeira P, Ribeiro A, Mendes S et al (2011) Immunophenotypic characterization of normal bone marrow stem cells. In: Schmid I (ed) Flow cytom. – recent perspect. InTech, Rijeka, pp 457–478
22. Ibrahim SF, van den Engh G (2003) High-speed cell sorting: fundamentals and recent advances. Curr Opin Biotechnol 14:5–12
23. Ryan JM, Pettit AR, Guillot PV et al (2013) Unravelling the pluripotency paradox in fetal and placental mesenchymal stem cells: oct-4 expression and the case of the emperor's new clothes. Stem Cell Rev Rep 9:408–421
24. Matikainen T, Laine J (2005) Placenta: an alternative source of stem cells. Toxicol Appl Pharmacol 207:544–549
25. Mihu CM, Ciuca DR, Soritau O et al (2009) Isolation and characterization of mesenchymal stem cells from the human amniotic membrane. Rom J Morphol Embryol 50:73–77
26. Allegra A, Altomare R, Curcio P et al (2013) Gene expression of stem cells at different stages of ontological human development. Eur J Obstet Gynecol Reprod Biol 170:381–386
27. Roubelakis MG, Trohatou O, Anagnou NP (2012) Amniotic fluid and amniotic membrane stem cells: marker discovery. Stem Cells Int 107836:1–9
28. Tsai M-S, Hwang S-M, Chen K-D et al (2007) Functional network analysis of the transcriptomes of mesenchymal stem cells derived from amniotic fluid, amniotic membrane, cord blood and bone marrow. Stem Cells 25:2511–2523
29. Baharvand H, Heidari M, Ebrahimi M et al (2007) Proteomic analysis of epithelium-denuded human amniotic membrane as a limbal stem cell nich. Mol Vis 13:1711–1721
30. Hopkinson A, McIntosh RS, Shanmuganathan V et al (2006) Proteomic analysis of amniotic membrane prepared for human transplantation: characterization of proteins and clinical implications. J Proteome Res 5:2226–2235
31. Perin L, Sedrakyan S, Da Sacco S, De Filippo R (2008) Characterization of human amniotic fluid stem cells and their pluripotential capability. Methods Cell Biol 86:85–99
32. Díaz-Prado S, Muiños-López E, Hermida-Gómez T et al (2011) Isolation and characterization of mesenchymal stem cells from human amniotic membrane. Tissue Eng C Methods 17:49–59
33. Díaz-Prado S, Muiños-López E, Hermida-Gómez T et al (2010) Multilineage differentiation potential of cells isolated from the human amniotic membrane. J Cell Biochem 111:846–857

Chapter 13
Preservation of Amniotic Membrane

Mafalda Laranjo

Abstract Human amniotic membrane is an important biomaterial in medicine, it has virtually unlimited availability and low processing cost. Preservation of human amniotic membrane is essential to guarantee a continuous supply in the clinical setting. The establishment of standardized protocols for collecting and preserving human amniotic membrane worldwide is crucial to ensure best quality tissue, according to each application, and safety for the patients. Nowadays preservation of amnion relies on several methodologies, namely, hypothermic storage at 4 °C, usually referred simply as fresh storage, cryopreservation, freeze drying, air-drying and others. Every method used in the processing and preservation of human amniotic membrane affects the properties of the biological material. However the maintenance of some characteristics might be important for each application. Thus, storage conditions should be thought to support the integrity of structures that might influence clinical outcome, like the basement membrane or the extracellular matrix. Within human amniotic membrane preservation methods, each procedure presents advantages and disadvantages. The contraindicated, for safety issues, fresh human amniotic membrane is a ready to use biomaterial. Cryopreservation at −80 °C allows conservation for long periods of time in appropriate cryoprotectant media. Lyophilized human amniotic membrane can be stored at room temperature making its transport simpler. In this chapter, the main issues about human amniotic membrane collection and preservation are reviewed and discussed.

Keywords Human amniotic membrane • Cryopreservation • Glycerol • Lyophilization • Freeze-drying • Deepithelialization • Corneal surface reconstruction • Wound healing

---

M. Laranjo
Biophysics Unit, Faculty of Medicine, Institute for Biomedical Imaging and Life Sciences (IBILI), Center of Investigation in Environment, Genetics and Oncobiology (CIMAGO), University of Coimbra, Coimbra, Portugal
e-mail: mafaldalaranjo@fmed.uc.pt; mafaldalaranjo@gmail.com

A.C. Mamede, M.F. Botelho (eds.), *Amniotic Membrane: Origin, Characterization and Medical Applications*,
DOI 10.1007/978-94-017-9975-1_13

13.1 Introduction

Human amniotic membrane has many beneficial properties that are extensively detailed throughout this book, being the mostly referred, its antimicrobial activity, antiangiogenic properties, inflammation suppression plus no immunogenicity, pain relief, scarring inhibition, wound healing and epithelialization [1–4], and anti-tumorigenic properties [1, 5]. Besides these characteristics, human amniotic membrane acts as an anatomical and vapor barrier [1, 3], being able to preserve a moist atmosphere that promotes healing [6]. It has the ability to maintain the epithelial stem cell niche thus being useful in ocular surface diseases [7]. Furthermore, human amniotic membrane cells are multipotent and may differentiate into several types of cells, namely neurons, hepatocytes, cardiomyocytes, pancreatic cells and chondrocytes [3]. This unique combination of characteristics that is not found in other natural or synthetic biocompatible materials [1], justifies the use of human amniotic membrane in various clinical conditions: in ophthalmic treatment, in partial thickness burn, as a temporary biological dressing, as a cover for microskin grafts, as an epidermal substitute [2, 6], as a substrate for chondrocytes for human articular cartilage repair [1] and as a substrate for the *ex vivo* expansion of limbal epithelial cells used to treat corneal epithelial stem cell deficiency in humans [6, 8, 9].

Moreover, its virtually unlimited availability and low processing cost makes it an important biomaterial in medicine. Therefore, the establishment of standardized protocols for collecting and preserving human amniotic membrane worldwide is crucial to ensure best quality tissue, according to each application, and safety for the patients. In this chapter, the main issues about human amniotic membrane collection and preservation are reviewed and discussed.

13.2 Amniotic Membrane Collection

The standards of quality and safety for donation, procurement, testing, processing, preservation, storage and distribution of human tissues and cells in Europe are under the regulation of Directive 2004/23/EC and the laws of each country for tissue transplantation. In order to comply with these regulations different protocols for preservation, testing and storage might be required [10].

Directive 2004/23/EC demands all safety measures, including in the prevention of the transmission of infectious diseases, are accomplished in donation, procurement, testing, processing, preservation, storage, distribution and use of human tissues and cells, in order to protect public health (Directive 2004/23/EC). To ensure the highest quality and, more important, safety, of tissue, human amniotic membrane collection is recommended after caesarian sections under sterile conditions through approved protocols [2, 11]. However, both C-sections and vaginal delivery human amniotic membrane might have bacterial contamination [2, 12], therefore adequate sterilization procedures are mandatory.

Pregnant women with a social and clinical history compatible with a healthy donor are asked to give their placenta and if informed consent is obtained the screening for diseases is performed. Usually women with history of drugs, alcohol abuse and multiple sexual partners are excluded [2]. The screening for serological diseases must include human immunodeficiency virus (HIV), hepatitis B and C, syphilis, cytomegaly virus and Toxoplasma gondii [2, 11]. The serological tests are carried out in the third trimester, close to the date of caesarean section, and repeated 6 months after delivery [2], in order to avoid false negatives consequent of the "window period" of infection [11]. If both tests are negative and non-reactive, tissue might be used in surgery [2].

There are reports on the use of fresh human amniotic membrane [11, 13], however there is a potential risk of disease transmission associated with this procedure. The death of a recipient of an allograft contaminated with *Clostridium spp.* proved that tissue transplantation can transmit virus or bacteria [10]. Furthermore, if we think that the human amniotic membrane from a single donor has enough biomaterial to be used in several patients the potential danger of spreading a disease is substantial [14]. To prevent situations like this and to comply regulations long-time storage, and appropriate sterilization, of tissue is mandatory; once human amniotic membrane must be quarantined until the results of both serological tests are negative [11].

13.3 Amniotic Membrane Processing

There is no single protocol to the human amniotic membrane processing. The most popular method was suggested Tseng and co-workers, who re-introduced the use of amnion in ophthalmology [15–17]. Usually the placenta is received intact from the maternity hospital in a container appropriate to continue processing under sterile conditions, under a lamellar-flow hood. The placenta is washed from blood clots with saline, usually with antibiotics. Usually an antibiotics combination that covers Gram-negative, Gram-positive bacteria and fungi is included in washing and storage solutions. Tseng *et al.* [16] used 1,000 U/mL penicillin, 20 mg/mL streptomycin, and 2.5 mcg/mL amphotericin B. But, in other procedures other reagents were used, for example: 0.5 % silver nitrate and 0.025 % sodium hypochlorite solution [2].

The human amniotic membrane is stripped from the placenta, commonly by blunt dissection, and flattened, epithelial surface up (mesenchymal side under), into nitrocellulose paper sheets [16, 17]. Human amniotic membrane is cut into pieces of appropriate size the samples are stocked individually in sterile vials with the selected preservative media and stored. As previously stated, the grafts should be quarantined until the results of the serological and bacteriological tests are available.

A matter that recently gained some concern is interdonor variation. Human amniotic membrane is a dynamic tissue that varies during pregnancy to adapt to fetal necessities and to maintain fetal membranes integrity [18]. Furthermore, amnion composition, including growth factors, and characteristics, like thickness and histology, varies in different donors [18, 19]. Concerning biochemical composition, there are variations in the concentration of epidermal growth factor (EGF), basic fibroblast growth factor (bFGF), keratinocyte growth factor (KGF), transforming growth factor (TFG) -α and -β1, nerve growth factor (NGF) and pigment epithelium derived factor (PDEF) levels [18]. It was also found that cytokines like tumor necrosis factor (TNF), interleukin (IL) 6 and 8 vary in amniotic fluid of african americans and caucasians [18]. Additionally, Lopez-Valladares and co-workers [18] verified age influence on human amniotic membrane growth factors, bFGF, KGF, TGF-β1 and hepatocyte growth factor (HGF), content and protein concentrations, being both lower in donors with higher gestational ages and donor ages. Thereby, human amniotic membrane from 38 to 39-week pregnancies and from under 35 years donors has higher growth factor levels [18]. In human amniotic membrane transplantation the biochemical composition of the graft influences therapeutic effects. Therefore the possibility of considering parameters like donor age, gestational age and race should be taken into account in donor selection process or might be registered in collection procedure.

13.4 Amniotic Membrane Preservation Procedures

Preservation of human amniotic membrane is essential to guarantee a continuous supply in the clinical setting [6]. Preservation of amnion relies on several methodologies, namely, hypothermic storage at 4 °C, usually referred simply as fresh storage, freezing, freeze drying, air-drying and others [3, 20]. Every method used in the processing and preservation of human amniotic membrane affects the properties of the biological material [2, 10]. However the maintenance of some characteristics might be important for each application. Thus, storage conditions should be thought to support the integrity of structures that might influence clinical outcome, like the basement membrane or the stromal matrix.

Within human amniotic membrane preservation methods, each procedure presents advantages and disadvantages. The contraindicated, for safety issues, fresh human amniotic membrane is a ready to use biomaterial that could be prepared in hospitals to fulfill its own requirements. However, the risk of widespreading serological diseases makes the preservation methods essential. Cryopreservation at −80 °C allows conservation for long periods of time in appropriate cryoprotectant media. Lyophilized human amniotic membrane can be stored at room temperature making its transport simpler [21]. Other preservation protocols will also be reviewed.

13.4.1 Cryopreservation in Glycerol

Cryopreservation of fetal membranes was first suggested by Tseng and colleagues in the 1990s, when they reintroduced human amniotic membrane in ocular surface reconstruction. The processed samples of human amniotic membrane were preserved at −80 °C in glycerol: Dulbecco's Modified Eagle Medium (DMEM) (1:1) [16, 17]. This procedure was recommended by Food and Drug Administration as a reliable preservation method and continues to be the most used method for human amniotic membrane long-term storage [8, 11, 22–24]. The fact that it has been extensively used in experimental and clinical studies confirms the safety and effectiveness this procedure [8, 25].

The human amniotic membrane can be cryopreserved in glycerol at −80 °C for several months [8] and there are reports that tissue has been stored at −80 °C up to 2 years [2]. Several advantages of the use of glycerol as a cryoprotectant in preservative media have been reported. Storage in glycerol dehydrates the tissues by replacing most of the intracellular water without altering osmotic concentration, preserving cell integrity [26]. Glycerol-preserved human amniotic membrane biological constitution and structure is similar to non-preserved human amniotic membrane [8]. Many works showed that, mostly, human amniotic membrane cells do not survive cryopreservation in glycerol at −80 °C [8]. Cryopreserved human amniotic membrane cells, when removed enzymatically, do not survive in culture [27]. Vital staining and ultrastructural analysis also proved no cell viability after cryopreservation in glycerol [27]. However maintenance of human amniotic membrane properties while epithelial cells loose viability is considered one of the advantages of cryopreservation in glycerol, thus human amniotic membrane becomes non immunogenic and safe for clinical practice [11, 17].

Glycerol has antiviral and antibacterial properties, dependent on concentration, time and temperature [2, 26, 28]. It was shown that HIV-1 positive cadaver skin preserved in glycerol 85 % at 4 °C resulted in virus inactivation after 5 days [28]. However preservation in glycerol *per se* is not a guarantee of sterility even after several months [8].

The limitations of this preservation procedure is the need of a deep-freeze facility that is expensive and unavailable particularly in underdeveloped countries and the consequent difficulties for transportation in appropriate temperature [8].

In the literature, one can find several variations to Tseng and colleagues [15, 17] procedure. Although the majority of protocols uses 50 % glycerol in cell culture media like DMEM [16, 17] or Minimum Essential Medium (MEM) [14], other formulations were reported. Adds and colleagues, frozen the tissue in a 50 % glycerol-Hanks's balanced salts solution at −80 °C [12]. Niknejad *et al.* [21] cryopreserved the samples in 40 % DMEM/F12, 10 % FBS and 50 % glycerol.

The company Bio-Tissue, Inc, founded by Scheffer Tseng commercializes several products based on glycerol cryopreserved human amniotic membrane to be used in ocular surface applications. AmnioGraft®, an allograft 50–100 μm thick available in various sizes to fit any ocular defect; AmnioGuardTM, a ticker allograft (300–400 μm) appropriate for glaucoma; and PROKERA®, a class II medical device composed of human amniotic membrane thermoplastic ring which is a treatment option for ocular surface and corneal wound healing. All these products are based on human amniotic membrane obtained, under informed consent, from donors who underwent elective caesarian section delivery and performed serological testing. The products are preserved at −80 °C in DMEM:glycerol (1:1) with antibiotics. Shipping is made in NanoCoolTM systems that maintain temperature under −21 °C until 72 h [29].

13.4.2 Cryopreservation in Other Media

Also in the 1990, Tsubota's group popularized a different cryopreservation procedure that has been used alternatively to glycerol:DMEM preservation [2]. In this procedure rinsing of the human amniotic membrane with antibiotics in saline is replaced by increasing concentrations of dimethyl sulphoxide (DMSO). According to Tsubota's procedure, pieces of amnion, cut into squares of 10 cm, are rinsed during 5 min in solutions of DMSO in phosphate buffered saline, PBS, 0.5 M (4 %w/v), 1.0 M (8 %w/v). 1.5 M (12 % w/v). Then, the samples are placed in appropriate sterile containers and preserved at −80 °C until use. Preoperatively, the vessel is thawed at room temperature. The amnion is rinsed, saline containing 100 mg of dibekacin sulphate was suggested, and the amniotic membrane is separated from the underlying chorion with forceps [30]. The samples can be maintained, with concentrations of DMSO about 10 % or 0.15 M, at −80 °C during several months [8].

Díaz-Prado *et al.* [1] also used medium with 10 % DMSO as cryoprotectant and a strict protocol of temperature decreasing until −150 °C. Cryopreserved human amniotic membrane was successfully used as a system to proliferate human chondrocytes for cell therapy in human articular cartilage repair [1]. Hennerbichler and colleagues proposed a glycerol-free preservation whose main feature of the procedure is the retention of 13–18 % of cell viability [3].

There are a few reports on the clinical use of DMSO cryopreserved human amniotic membrane for ocular surface reconstruction [8, 31, 32]. Azurara-Blanco used human amniotic membrane preserved in medium with 10 % DMSO and the treatment was effective in promoting corneal healing [31]. Rama and colleagues also preserved human amniotic membrane in 10 % DMSO in RPMI, at −80 °C and then at −140 °C, for transplantation in ocular surface diseases [33]. The antiinflammatory effects of human amniotic membrane transplantation in ocular surface disorders, that showed tissue affinity for antiinflammatory cells, was also performed in grafts preserved at −80 °C in DMSO solutions until use [34].

13.4.3 Lyophilization

Lyophilization or freeze-drying is the removal of all the water from the tissue by sublimation [8, 21]. Concerning structure, physical and biological properties, irreversible alterations occur. Nevertheless, it results in inhibition of destructive chemical reactions that lead to tissue alteration; therefore, the samples can be long-term stored at room temperature without deterioration. Transportation is simple, on the contrary of cryopreserved human amniotic membrane [8, 19]. Indeed, freeze-drying represents a cheaper method for preserving amniotic membranes once the costs of storage and shipment are reduced [35].

Nakamura and colleagues [7, 20] reported human amniotic membrane freeze-drying. Human amniotic membrane is washed in saline with antibiotics (penicillin, 10,000 U/mL; streptomycin, 10,000 μg/mL; and amphotericin B, 25 μg/mL) and cut into 4×4 cm pieces. The samples are treated with 0.02 % ethylene diamine tetraacetic acid (EDTA) at 37 °C for 2 h with the purpose of removing epithelial cells. At this point the samples are freeze-dried and packed in vacuum to be radiation sterilized with 25 Gy [7, 20].

In the case of lyophilization there are also several procedures described. Niknejad *et al.* [21] prefrozen the human amniotic membrane and lyophilized, in a freeze dryer, at −55 °C for 24 h. The samples were rehydrated in PBS during 30 min prior use [21].

Freeze-dried and air-dryed human amniotic membrane are already available commercially from different companies, however there is few data describing its clinical outcome [25].

13.4.4 Fresh Amniotic Membrane

Despite the already mentioned risk of transmission of HIV, Hepatitis C or other, there are some reports on the use of fresh human amniotic membrane for clinical practice [2, 13, 36, 37]. Besides the safety issue, Adds and colleagues [12] recalled practical disadvantages to the use of fresh human amniotic membrane compared with frozen material. The most important is the risk of disease transmission, but also the viability of finding a suitable donor, allowing processing and testing, at the moment of surgery, and also the fact that the major part of the human amniotic membrane is discarded in this condition [12].

Some authors debate about some theoretical advantages of the use of fresh human amniotic membrane, for instance, the intact epithelial cell layer which is the source of important cytokines and growth factors that are pointed as responsible for the characteristics of even, the mechanisms of action, of this tissue. However many studies have been showing that fresh human amniotic membrane does not carry advantages towards preserved human amniotic membrane [12].

Short-term preservation for research or clinical use and is mainly performed under 0 °C [3, 13, 36]. Fresh amnion can be maintain in viable conditions up to 6 weeks if properly stored at 4 °C in silver nitrate solution, in 20 % glycerin solution, or in sterile saline after passage through one rinse of 0.025 % sodium hypochlorite solution [26]. Samples preserved fresh, at 4 °C, in 85 % glycerol remain intact for over 1 year [2].

13.4.5 Trehalose

Trehalose, $C_{12}H_{22}O_{11}$, a natural occurring disaccharide, confers resistance to desiccation in several biologic systems, being found in organism capable of tolerate high levels of dehydration. Furthermore, it is being currently used in an eye drop solution for dry-eye syndrome [8, 38]. Once lyophilization affects the physical and biological structure of human amniotic membrane, mainly due to water loss, and trehalose can replace some water content in the cells, it might have a positive effect in terms of stabilization of proteins and other components. In order to improve the quality of sterilized freeze-dried human amniotic membrane, pre-treatment with trehalose was proposed. The procedure starts with washing the human amniotic membrane in saline with antibiotics and antimycotics and cut into pieces. The samples were decellularized and incubated in 10 % trehalose during 2 h at 37 °C. Then the procedure of freeze-drying and sterilization by irradiation was performed. Trehalose treated lyophilized human amniotic membrane is superior to trehalose non treated human amniotic membrane, having collagen types I, III, IV, V and VII, fibronectin, and laminin-5 in its structure. Moreover, trehalose proved effective in retaining the characteristics of native human amniotic membrane in terms of adaptability, ultrastructure, physical and biological properties [38].

Although this procedure has not been used in the clinics, *in vivo* pre-clinical studies are promising; corneal surface reconstruction of rabbit eye showed improved clarity, no neovascularization and superior epithelialization rate [8].

13.4.6 Other Preservation Procedures

Other procedures for processing and storing human amniotic membrane were used in ophthalmology and for other purposes. These include air drying, hyper-drying, and other methodologies based on chemical solutions [2, 11].

Air drying preservation technique, considering its lower cost and simpler storage, may be set for applications like wound dressing [6]. Sigh *et al.* applied air-dried gamma irradiation sterilized human amniotic membrane to wounds of 22 patients and had excellent results with complete reepithelialization of the wound in 15–25 days [39]. In alternative, dehydration over silica gel with further sterilization of human amniotic membrane was reported. The authors referred the advantage of drying the human amniotic membrane without being fixed to nitrocellulose paper and considered the results comparable to fresh human amniotic membrane in the treatment of ocular severe chemical and thermal burns [40].

Hyper-drying of human amniotic membrane under far-infrared rays and microwaves was suggested to be used instead of freeze-drying. The human amniotic membrane is washed and dried under far-infrared rays and microwaves at temperatures inferior to 60 °C by using a hyper-drying device. Then, sterilization is performed 25 kGy gamma radiation [11].

Human amniotic membrane crosslinking by chemical means, by treatment with glutaraldehyde, or by radiation, with gamma-ray and electron beam was also described. Comparison of both methods showed that radiation cross-linked human amniotic membrane degraded faster than chemically cross-linked ones. Glutaraldehyde cross-linked human amniotic membrane showed interesting physical properties, with no alteration of tensile strength, water content and permeation properties [41]. Superior resistance to enzymatic digestion associated to better transparency and less wrinkling was also reported, with preservation during 90 days in opposition to fresh or cryopreserved human amniotic membrane that are dissolved within 7 days [42]. Human amniotic membrane cross-linking with glutaraldehyde was used in therapy for corneal perforations. In this case, cryopreserved human amniotic membrane was used, cross-linked and then hyper-dried. This biomaterial was used in three patients and the corneal perforations were repaired within 28 days [43].

Other reagents were also proposed for cross-linking. It is the case of carbodiimide hydrochloride/N- hydroxysuccinimide. This biomaterial was thought to overcome human amniotic membrane limited biomechanical strength and rapid biodegradation and proved to be able to maintain limbal epithelial cell culture *in vitro* and *in vivo* [44]. In the same direction, aluminium sulfate [$Al_2(SO_4)_3$] was used and superior mechanical properties, like 125 % increase of tensile strength and capacity to support limbal epithelial cell cultures, were reported [45]. In a study to evaluate human amniotic membrane transplantation in the management of bullous keropathy, Georgiadis *et al.* referred storage in absolute alcohol at 4 °C, as an alternative protocol used in the local eye bank [23].

13.5 Amniotic Membrane Sterilization Procedures

13.5.1 Microbiological Quality Control

A good practice in human amniotic membrane processing and storage is to perform microbiological quality control. When human amniotic membrane is stored in preservative media simple procedures can be performed.

Qureshi and colleagues [14] described the microbiological quality control performed at the Al-Shifa Trust Eye Hospital (Rawalpindi, Pakistan). After human amniotic membrane processing for cryopreservation in glycerol, a random bottle from the batch is left for 1 h at room temperature. Then 5 mL of cryopreservation media is inoculated in 100 mL of brain heart infusion medium and 100 mL of thioglycolate broth medium. The media are incubated for 21 days, and if there is no growth of bacteria or fungi, the batch is considered free of microbiological contamination. Despite this control, the human amniotic membrane batch is used only after 3–4 months when the second serological control was performed to the donor [14].

13.5.2 Gamma Irradiation

Ionizing radiation might be used to sterilize preserved membrane attaining a level of biological contamination safe for clinical usage [6]. Ionizing radiation is used worldwide for sterilization of medical products. Its penetrating nature and negligible temperature rise make it suitable for pre-packaged material [46]. Radiation eliminates bacteria, viruses and fungy. Sterility assurance level, equivalent to 10^{-6} is achieved with 25 kGy [6, 8]. Gama radiation was considered the most reliable and effective method for sterilization of tissue allografts [39] with many advantages over processes as heating and chemicals. However, radiation also has some effects in biological properties and tissue integrity [8].

Sterilization by gamma irradiation is usually performed after lyophilization. This procedure was introduced by the International Atomic Energy Agency with the purpose to establish a system with quality guidelines to retrieve, process human amniotic membrane and distribute donor tissue [11]. The procedure was described as follows: the human amniotic membrane is washed, pasteurized at 60 °C, treated with 70 % ethanol, washed thoroughly and freeze-dried. Packed and sealed samples of human amniotic membrane are sterilized by 25 kGy gamma radiation of cobalt 60 in an appropriate ISO certified facility. The tissue can be stored at room temperature, protected from light until 6 months [47].

Djefal and colleagues maked efforts to validate a dose for sterilizing lyophilized human amniotic membrane and also substantiated 25 kGy [46]. Nakamura *et al.* also irradiated lyophilized human amniotic membrane with 25 kGy to sterilize it [7, 20].

The use of freeze-dried and gamma sterilized human amniotic membrane for the treatment of burns was reported in a few studies that proved the safety of the procedure [11].

Gamma irradiation influence in glycerol preserved human amniotic membrane was evaluated. The tissue was submitted to increasing concentrations of glycerol until 87 % and placed in an appropriate bag to be irradiated. Glycerol protects tissue from indirect effects of ionizing radiation, limiting free radical formation. Structure of glycerol preserved human amniotic membrane was similar to fresh human amniotic membrane until 25 kGy [6].

13.5.3 Chemical Solutions

Paracetic acid is a standard sterilizing agent highly effective against bacteria, viruses, and spores due to its high oxidizing potential. It is appropriate to be used in biomaterials because it breaks into two non-toxic residues: acetic acid and oxygen peroxide [8]. The structure of paracetic acid treated human amniotic membrane is satisfactorily preserved [10] with retention of basement membrane components like collagen type IV, fibronectin and laminin. Also, collagen types I and III are more abundant than in gamma sterilized human amniotic membrane. Paracetic acid treatment does not induce significant reduction in human amniotic membrane tensile strength and elasticity [8].

Reports of clinical application of paracetic acid sterilized human amniotic membrane are not known. However, *ex vivo* expansion of limbal epithelial cells under paracetic acid sterilized decellularized human amniotic membrane was performed and compared with non-sterilized decellularized human amniotic membrane. Cell proliferation rate and morphology are similar, as well as, the number of cells expressing protein 63 isoform delta-Np63-alpha (ΔNp63α) and protein ATP-binding cassette sub-family G member 2 (ABCG2), putative markers of limbal stem cells [4]. Therefore, once paracetic acid destroys all contaminants, in opposition to disinfectants, like antibiotics, it might be a good option.

Paracetic acid/ethanol sterilization was also used by Pruss *et al.* in a strenuous procedure that must be performed under negative pressure and permanent stirring to avoid air bubbles. Paracetic acid/ethanol sterilization inactivates bacteria, fungi and viruses. However it may contribute to changes in basement membrane and denaturation of the soluble growth factors/cytokines, as a consequence of protein denaturation [25]. Ethylene oxide was proposed to sterilize human amniotic membrane after lyophilization [11].

13.6 Amniotic Membrane Deepithelization

As previously presented in this book, human amniotic membrane is composed of a single layer of columnar epithelial cells that rest under a basement membrane composed mainly of collagen IV and VII, laminin and fibronectin. The underlying matrix includes collagen type I, III and V, interspersed with fibroblast mesenchymal cells. Epithelial cells are adhered by desmosomes and hemi-desmosomes. These cells are responsible by producing important human amniotic membrane cytokines and growth factors that contribute to properties like proper growth of ocular surface cells or wound healing.

Several methodologies and reagents were described to remove the layer of epithelial cells from human amniotic membrane and mostly are performed to use it as a substrate for cell growth. As a matter of fact, there is no consensus if this procedure is beneficial or not. There are arguments that intact human amniotic membrane is beneficial because of higher concentration of citokines and growth factors [8] and that cultivated stem cells retain their function longer with less differentiation [48]. However, deepithelialized human amniotic membrane promotes better cell proliferation and differentiation and more uniform cell growth while intact human amniotic membrane retards migration and differentiation of cultivated cells [8, 49]. In fact, for corneal epithelial cell culture denuded human amniotic membrane seems to be the preferred choice [8, 9]. The methodology used to denude human amniotic membrane should maintain its structural integrity as well as its biological function [8, 48] which is, indeed, a significant limitation of this procedure independently of using enzymatic means, dispase, trypsin, thermolysin, or chemical reagents, EDTA, urea, ethanol.

Dispase is an enzyme, crystallized from cultures of *Bacillus polymyxa*, that has proteolytic activity under basement membrane components like laminin, collagens and fibronectin [8, 49]. Therefore, there are irreversible physical and biological damage results of its action. Prolonged incubation results in significant alterations in ultrastructure of human amniotic membrane, particularly basement membrane. Treatment with dispase digests proteins like collagen IV, fibronectin and laminin, and growth factors, namely TGF-α, TGF-β1, transforming growth factor β2 receptor (TGF-β2R), platelet-derived growth factor subunit A (PDGF-A), vascular endothelial growth factor (VEGF), and epidermal growth factor receptor (EGFR) [49, 50]. There are reports of the use of 1.2 U/mL Dispase II, at room temperature, from 5 min to 2 h, to remove the epithelial cells from human amniotic membrane [8]. This procedure was successfully used for corneal surface reconstruction [8, 51]. However Hopkinson *et al.* stated it was not possible to identify the basement membrane components after only after 10 min of incubation.

Trypsin, mainly used combined with EDTA, is commonly used in cell culture to detach cells from growing substrates, commonly flasks or plates. It has been widely used do denude human amniotic membrane in concentrations from 0.1 to 0.25 % at 37 °C for 30 min [9, 52]. It was reported that this procedure maintains basement membrane components however extracellular matrix, including growth factors, is compromised [8, 50].

Thermolysin is a heat-stable metalloproteinase, isolated from *Bacillus stearothermophilus* [49], that was also used to obtain denude human amniotic membrane resulting in intact basal lamina. Thermolysin seems to act specifically at hemidesmosome complex, being a more specific reagent. Treatment consisted of 125 μg/mL for 9 min [8, 51].

Ethylene diamine tetraacetic acid (EDTA) is a chelating agent that contributes for the disruption of calcium dependent adhesions. EDTA was used from 0.02 to 0.25 %, 10 min to 2 h at 37 °C, followed by scrapping. 90–100 % of the cells were removed by EDTA resulting in the disaggregation and some destruction of the basal membrane structure and molecular composition [7, 8, 20, 51]. Zhang and colleagues showed that EDTA denuded human amniotic membrane had intact basal lamina and smooth basement membrane surface shown under transmission and scanning electron microscopy [50].

Urea, a denaturant protein frequently used in labs to solubilize proteins was also reported as a reagent to deepithelize human amniotic membrane. It is advantageous because only 5 min 5M ice-cold urea for and gentle scraping are sufficient to remove cells, it is a easily available reagent and basement membrane components, collagens I, II, IV, VI and VII, laminin-5, fibronectin, elastin and thrombospondin, as well as growth factors, TGF-α, -β1 and – β2 receptor, EGFR, KGF, bFGF, VEGF, and PDGF remained present [8, 50]. Plus, human limbal epithelial cells showed high adherence and proliferation [50].

Ethanol 20–30 s, followed by scrapping was also used. It is also a quick method with an easily available reagent that maintains basement membrane and extracellular matrix components [8]. However epithelial cell remains are observed after the procedure [50].

ACELAGRAFT™ is a deepithelized dehydrated human amniotic membrane, whose cells and associated growth factors were removed, produced for commercial use. ACELAGRAFT™ can be stored at room temperature; therefore, transport and shipment are simple. Studies showed the potential of this product in wound healing and other applications are under investigation. For preparation of this product, human amniotic membrane previously removed from placentas and properly washed, is denuded. Epithelial cell removal is performed by incubation with deoxycholic acid. Then tissue is dried using a gel dryer [22]. This product is commercialized in the USA, with Food and Drug administration approval, from the company Celgene Corporation.

13.7 Features of Preservation Methods

Different preservatives and storage temperatures have an impact on the membrane structure. As discussed, the beneficial effects of human amniotic membrane were hypothesized, usually taking into account the presence of some components [2], however, the mechanism of action may vary according to the preservation procedure used.

13.7.1 Epithelial Cells Viability

In the majority of the protocols performed, for example, glycerol-preserving, lyophilizing or air-drying the viability of human amniotic membrane epithelial cells is lost [3, 15, 17, 20, 35, 39]. However, some procedures, like tissue culturing, refrigerating or cryopreserving with DMSO, may, in certain conditions, maintain cell viability variable between 40 and 90 % [2, 3, 35]. It was demonstrated that cryopreservation significantly reduced cell viability and the properties of human amniotic membrane are provided by itself as a matrix or scaffold, not by proliferative cells [2, 53]. Adds *et al.* [12] confirmed this and showed that epithelial cells are non-viable in cryopreserved samples (50 % glycerol in Hanks's balanced salts solution at −80 °C) and also in fresh hypothermically stored (CPTES – corneal-potassium tris-EDTA solution, containing 2.5 % chondroitin sulphate at 4 °C) human amniotic membrane.

Kubo and colleagues shown that 50 % of the amniotic cells are viable and proliferative after 2 months of cryopreservation in media with DMSO, however viability was not appreciated after 18 months [54].

In an attempt to clarify how the temperature and preservative media influences viability of amnion cells Hennerbichler and colleagues, in 2007, evaluated several human amniotic membrane storage conditions. They confirmed that cell viability decreases during storage. Samples kept above 0 °C last 28 days have a remaining viability of 15–35 %. However storage in glycerol at 4 °C induced immediate cell death. Moreover, freezing reduces viability to 13–18 % with no influence of preservation media [3].

Up to this time, human amniotic membrane is mainly used as a biomaterial devoid of epithelial cells as consequence of processing and preservation, like cryopreserved in glycerol, or as a decellularized matrix. However, as discussed, some studies demonstrated that human amniotic membrane cell viability can be retained under certain conditions. Epithelial human amniotic membrane cells, as we know, bear some interesting characteristics like stem cell characteristics, expression of growth factors, increased immunogenicity [53]. Accordingly, preservation methods, which partly retain cell viability in human amniotic membrane might be beneficial in some applications, like for research of the improvement of amnion as a viable biomaterial in wound healing, or in future *in vivo* skin models [3]. Human amniotic membrane has important characteristics, namely good adherence to wound, no immunological reaction, beneficial bacteriostatic properties, acts as a barrier that prevents water and protein loss, that make it an appropriate skin substitute [6, 19].

13.7.2 Histological and Biophysical Properties

As we know, human amniotic membrane favors the healing processes but also diminishes the levels of pain and discomfort. Once it was previously proved the application of amniotic fluid to augment corneal sensitivity and nerve regeneration, Dua and colleagues stated this might be due to a mechanical or physical effect [2]. Hence, maintenance of structural, biophysical and histological properties in human amniotic membrane preservation is very important. Microscopy studies revealed that the protocols used for processing and preserving human amniotic membrane influences its histological and biophysical properties [6, 10, 39]. Preservation processes like cryopreservation, lyophilization or air-drying cause protein loss and damage therefore affect human amniotic membrane structure and morphology.

In ophthalmology, although cell viability is not a requirement, human amniotic membrane structure is particularly important [2, 6], maintenance of basement membrane and stromal matrix integrity seems crucial to promote rapid reepithelialization [12], therefore, storage conditions are thought to keep these structures intact.

Scanning electron microscopy studies of human amniotic membrane preserved by air-drying and cryopreserved in glycerol showed air dried human amniotic membrane presents big alterations in histology and biophysical properties, with condensation of microvilli and of intercellular channels, while glycerol preserved human amniotic membrane structure remains similar to the fresh. Glycerol prevents loss of fluid, proteins and electrolytes. It is also considered radioprotective being able to minimize the indirect effects of radiation once it removes water, the target of radiolysis [6]. However, the use of doses inferior to 25 kGy is being considered in tissue banking in order to minimize damage. Ab Hamid and colleagues [6] confirmed that radiation damage might be prevented by the use of glycerol as a preservative until 35 kGy while cell morphology of air dried amnion changed at 25 kGy. Radiation at 15 kGy did not affect the surface morphology in air dried and the glycerol preserved amnions [6].

Concerning adherence, folding and structural maintenance glycerol preserved is comparable to fresh human amniotic membrane in case of both non-irradiated and irradiated glycerol amnions [6]. For some applications, like vascular tissue engineering, it is important that human amniotic membrane is still suturable after preservation. It was reported that even lyophilized human amniotic membrane became flexible after rehydration being possible to suture. However neither lyophilized nor cryopreserved human amniotic membrane maintained the same suture retention strength than fresh amnion [21].

Maral and co-workers compared human amniotic membrane long-term storage in 85 % glycerol, 4 °C, with fresh human amniotic membrane and with glycerol preserved skin has a wound dressing. The authors concluded that glycerol-preserved human amniotic membrane efficiency, in terms of adherence, structural maintenance and duration of action was comparable to fresh amnion on rat wounds. Therefore, human amniotic membrane undergone conservation procedures retains its structural properties and effectiveness as a biologic dressing [26]. Storage time could influence the histological and physical properties of human amniotic membrane, however, the basement membranes kept the same distribution of collagen IV, collagen VII, laminin, laminin 5, and fibronectin up to 2 years when cryopreserved in 50 % glycerol [55].

Glycerol:MEM (1:1) preserved human amniotic membrane was compared to DMSO preserved in terms of histology. Glycerol:MEM human amniotic membrane presented similar appearance to fresh, with superior integrity of basement membrane but epithelial cells were higher. DMSO preserved human amniotic membrane had reduced intercellular junctions and detachment of epithelial cells [28].

Nakamura and colleagues elaborated a series of papers on the improvement of human amniotic membrane preparation and storage [7, 20, 38]. The use of human amniotic membrane still concerns a few problems, namely, cryopreservation procedure, dependent on several washing with antibiotics in saline, do not guaranties complete sterilization, and requires expensive and space consuming −80 °C freezers that limits its transport and storage particularly in developing countries [20]. The possibility of storing human amniotic membrane at room temperature led the authors to evaluate the use of sterilized freeze-dried human amniotic membrane as substrate for cultivating autologous corneal epithelial cells for ocular surface reconstruction. The sterilized, freeze-dried human amniotic membrane retains the characteristics of the cryopreserved, namely physical, biological, and morphologic, being appropriate for ocular surface reconstruction [20]. It can be stored at room temperature, being easily transported and stored, is easily re-hydrated before use and is contamination free [7]. The sterilized, freeze-dried amniotic membrane transplantation adapted well to the sclera of patients submitted to pterygium surgery and complete epithelialization was achieved within 2 weeks, similarly to use of cryopreserved human amniotic membrane. Once the freeze-drying process affects in some extent the physical and biological characteristics of human amniotic membrane further improvements in the procedure are needed to adapt the amnion for various applications [38].

Macroscopically lyophilized human amniotic membrane is thinner and weakest than cryopreserved [8]. Light and electron microscopy showed both freeze-dried and cryopreserved human amniotic membrane basement membrane, which is essential for ocular reconstruction procedures, remains intact [19]. Freeze-dried human amniotic membrane shows tissue typical structure with amniotic epithelium, basement membrane, compact and fibroblast layer. However in some areas structure was disrupted and epithelial cell layer was loss. Cryopreserved human amniotic membrane do not maintain epithelial cell viability, however epithelial layer remained intact [21]. Gamma radiation sterilization does not alter significantly the structure of lyophilized human amniotic membrane [8]. Lyophilized human amniotic membrane basement membrane, as well as, extracellular matrix, comprising collagen types I, III, IV, V and VII, fibronectin and laminin-5 do not show significant differences comparing 25 kGy irradiated lyophilized human amniotic membrane versus cryopreserved [8]. Lyophilized human amniotic membrane is also associated with epithelial vacuolar degeneration and flattening of the epithelial surface, likewise cryopreservation induces some vacuolar degeneration and stromal oedema [18]. Although, lyophilization diminishes the total amount of protein and specific growth factors, it maintains the histological structure of human amniotic membrane. Hence, lyophilized human amniotic membrane, considered not such a good method as cryopreserved, also maintains the most important characteristics of the tissue possibly it could be considered when deep-freeze facilities are not available for storage or transportation.

Procedures like acid pretreatment and subsequent air drying, on the contrary of cryopreservation, leads to complete loss of biological properties, thus theoretically efficacy in clinical application is limited. Air-dried human amniotic membrane was already used as dressing in burn wound care and its biophysical characteristics were evaluated. Fluid handling capacity (3.79–4.2 g/10 cm^2/24 h) is of particular importance to deal with the exudates production of a wound and is in accordance with the water loss reported from burns (5 g/10 cm^2/24 h). In terms of infrared spectra there are no significant changes even after 2 or 5 years of storage, neither in terms of impermeability to bacteria [39]. Collagen IV, Collagen VII, laminins including laminin 5 and fibronectin form a continuous layer below the epithelium in cryopreserved human amniotic membrane. In air-dried human amniotic membrane collagen VII and laminin 5, are not detectable, therefore structure is highly altered as a result of preservation [25].

The properties of decellularized and dehydrated human amniotic membrane were compared with cryopreserved human amniotic membrane. Histology and ultrastructure showed significant differences like disruption of the trilaminar structure of basement membrane, loss of components like collagen IV and VII, laminin, and fibronectin and lower levels of growth factors in decellularized and dehydrated human amniotic membrane. However, a case report showed that clinical transplant of decellularized and dehydrated human amniotic membrane was successful with complete corneal epithelialization 3–4 weeks after procedure. Therefore, the authors emphasized the practical advantages of the decellularized and dehydrated human amniotic membrane; it is simple to preserve once it can be maintained at room temperature and rehydrated before use, sterility is assured. Therefore it can be an appropriate option for smaller centers, where the cost of purchasing and maintaining cryopreservation might be an obstacle, or even for tertiary centers in situations of donor shortage [24].

13.7.3 Scaffold for Cell Proliferation

Human amniotic membrane is a biomaterial used in several clinical applications. The possibility of using it as a chondrocyte substrate/carrier for cartilage therapy and transplantation was evaluated. Therefore air dried and freeze dried human amniotic membrane were used in comparison to chondrocytes grown in plastic dishes. Both procedures of human amniotic membrane preservation provided chondrocyte proliferation, glycosaminoglycan expression and attachment. Cells implanted on human amniotic membrane revealed superior capacity to produce extracellular matrix independently of the number of cells present. There is also a superior expression of glycosaminoglycan per cell production in cells grown on human amniotic membrane, what could be explained by the alteration of conformation consequent of drying. Hence, human amniotic membrane could potentially be used has a chondrocyte carrier, increasing cell proliferation and extracellular matrix production [56].

Niknejad and colleagues aimed to verify if preservation methods influence human amniotic membrane performance as a substrate for *ex vivo* endothelial cell expansion. An issue on the preservation procedures of human amniotic membrane is the alterations of the extracellular matrix that might influence the attachment of cells when human amniotic membrane is used as expansion substrate [21]. Cryopreserved and lyophilized human amniotic membrane was compared to fresh human amniotic membrane and all groups had similar extracellular matrix components. Still, lyophilization induced alterations in terms of histology, thickness, strength, elongation at break and suture retention. Lyophilized human amniotic membrane becomes thinner, probably due to loss of water, while cryopreserved becomes thicker, probably due to the incorporation of glycerol in the tissue that promotes swelling. Despite this, cells grown on lyophilized human amniotic membrane showed superior attachment to the basement membrane and retained endothelial characteristics like cytoplasm with Weibel–Palade bodies, possibly due to the direct contact with basement membrane exposed during the process of lyophilization [21]. Therefore, lyophilization seems to be a superior technique when the purpose is culturing endothelial cells [21].

Human limbal epithelial cells expanded *in vitro* under human amniotic membrane achieved appreciable success rate in ocular surface reconstruction and visual outcome [9], therefore, it is one of the most popular uses of human amniotic membrane as a scaffold for cell proliferation. Although the majority of the reports employ 50 % glycerol preserved denuded human amniotic membrane several authors verified the influence of human amniotic membrane preservation methods on its performance as a substrate for *ex vivo* human limbal epithelial cells expansion. Thomasen and colleagues evaluated air-dried human amniotic membrane for the expansion of limbal epithelial cells and verified that cell proliferation is significantly higher in cryopreserved than air-dried human amniotic membrane. Possibly alterations in basement membrane composition, loss of collagen VII and laminin 5, plus loss of soluble factors might explain the limited growth of limbal epithelial cells [25]. Shortt *et al.* [4]

performed an interesting study that brought some relevant questions. Removal of human amniotic membrane epithelial cells prior seeding limbal epithelial cells resulted in a higher percentage of confluence and lower cell density than intact human amniotic membrane; therefore the authors proposed possibly denuded human amniotic membrane not increase proliferation, but rather facilitates cell migration resulting in larger cells. Although, small cell size is considered an indicator of stem cells. However, human amniotic membrane epithelial cell removal did not affect the percentage of cells expressing the putative limbal stem cells markers, ΔNp63α, ≤4 % and ABCG2, ≤3 %. The most surprising result from this work was that glycerol cryopreservation of human amniotic membrane (glycerol:HBSS, 1:1) resulted in poor morphology and a low proportion of cells expressing ΔNp63α (≤6 %) and ABCG2 (≤8 %) compared to human amniotic membrane frozen at −80 °C in HBSS, where excellent morphology and high levels of ΔNp63α (≤68 %) and ABCG2 (≤68 %) expression were observed ($p<0.001$). This study suggests that intact (non-decellularized) human amniotic membrane frozen in HBSS alone is the best expansion of limbal epithelial cells [4].

As discussed, human amniotic membrane can be used epithelized or denuded and the type of preparation appropriate depends on the type of cells to be expanded. It is known that the presence of epithelial human amniotic membrane cells may interfere with cells attachment. Basement membrane reinforces cell attachment, facilitates migration and may promote differentiation of epithelial cells [14]. There are various methods to denude human amniotic membrane with different structural effects. It would be important to develop methodologies that maintain basement membrane architecture.

13.7.4 Molecular Features

The need of storing human amniotic membrane during at least 6 months in order to comply with serologic control brings issues regarding the maintenance of molecules important on amnion properties. There are several molecules showed to be kept on human amniotic membrane preserved at −80 °C [2]. Is it the case of EGF, TGF- α, KGF, HGF, bFGF, TGF-β1, and – β2, TGF- β3, keratinocyte growth factor receptor (KGFR) and hepatocyte growth factor receptor (HGFR) [14, 57]. Both glycerol:MEM (1:1) and DMSO cryopreserved human amniotic membrane present the most of the cytokines and growth factors present in fresh amnions. Differences were registered only for fibroblast growth factor 4 (FGF-4), bFGF, prostaglandin E2 (PGE2) e KGF [28].

Total protein of tissue-suspensions obtained from fresh, cryopreserved and lyophilized human amniotic membrane was quantified. Cryopreserved human amniotic membrane show higher liberation of proteins and specific growth to the suspension correlated with freeze-dried [57]. Intact human amniotic membrane presents higher level of growth factors than deepithelialized suggesting its epithelial origin [58]. Although lyophilization leads to loss of protein content, composition in growth factors is similar to cryopreserved human amniotic membrane [19].

Sterilization of human amniotic membrane; with gamma radiation, was reported not to alter significantly the content of growth factors [39].

Human amniotic membrane release soluble proteins, growth factors and cytokines, with beneficial effects in wound healing, inflammation and anti-neovascularization however, this condition might be altered according to preservation methodology. Koh and colleagues [59] investigated thc alteration of the tissue inhibitor of metalloproteinase (TIMP) in cryopreserved and freeze-dried human amniotic membrane. TIMPs have a role in the modulation of matrix metalloproteinases relevant in the transplantation of inflammatory cornea. TIMP was expressed in cryopreserved human amniotic membrane until 12 months in comparative amount of the fresh human amniotic membrane and TIMP-1 expression tend to decrease in freeze-dried [59]. Storage time might influence human amniotic membrane characteristics but proteins TIMP-1 and interleukin 1 receptor antagonist (IL-1ra) were present in the samples of human amniotic membrane stored until 2 years [25]. TIMP-1, IL-1ra and TGF-β1 were detected in conditioned media of cryopreserved human amniotic membrane, while only IL-1ra was detected in air-dried human amniotic membrane conditioned media [25].

Trefoil factor family (TFF) peptide 3 is a protein involved in ocular surface restitution after corneal injury and wound healing. This peptide is produced by conjunctival goblet cells and is present in human amniotic membrane. Cryopreservation has no influence in its secretion rate [60].

Concerning inflammation and complement Fust and colleagues investigated the expression of CD59 and demonstrated its presence on the surface of fresh and −80 cryopreserved non-viable cells. Therefore, cryopreserved human amniotic membrane cells keep complement inhibiting ability during storage [27].

Concerning angiogenic factor, release of relevant proteins is dependent largely on the storage procedure and is related with maintenance of cell viability. Fresh human amniotic membrane secretes high levels of angiogenin, growth related oncogene (GRO), IL-6, IL-8, TIMP-1, TIMP-2 and monocyte chemoattractant protein-1 (MCP-1) and low levels of EGF, interferon-γ, insulin like growth factor-I (IGF-I), leptin, TGF-β1 and thrombopoietin. Although DMSO cryopreserved human amniotic membrane levels of GRO, TIMP-1 and TIMP-2 are intermediate, levels of angiogenin, interferon-γ, IL-6 and MCP-1 are low and EGF, IGF-1, leptin, TGFβ1 and thrombopoietin are not detectable. Glycerol preserved at 4 °C human amniotic membrane releases only TIMP-1 and -2. Therefore, these data should be considered for selection of an amniotic membrane product for a specific clinical application [61].

13.8 Conclusion

To our knowledge the best method of preservation of human amniotic membrane is not selected yet and the way storage influences the mechanisms of action of this tissue is far from understood. Probably there is no gold standard for preservation media, temperature, sterilization procedures whose choice might depend on each application. Knowledge of the human amniotic membrane mechanisms of action, which is incompletely explored, could also be useful in the selection of preservation procedures. By now, there is good knowledge of the physical, histological and biochemical consequences on human amniotic membrane of the commonly used preservation methods. In this regard it is important to stimulate clinicians and researchers to continue investigation on human amniotic membrane pathways and applications. Moreover, it is important to take into account that human amniotic membrane preparation in a laboratory requires expertise in preparation of media and solutions as well as good laboratory practice of sterilization and preservation of biological materials. Establishment and commercialization of products like AmnioGraft®, ACELAGRAFT™ or PROKERA® confirm the importance of human amniotic membrane therapies in clinical practice and contributes to make this biomaterial accessible. Due to the lack of uniformity in human amniotic membrane procurement, another important feature of these products is standardization in tissue processing and preservation.

References

1. Díaz-Prado S, Rendal-Vázquez ME, Muiños-López E et al (2010) Potential use of the human amniotic membrane as a scaffold in human articular cartilage repair. Cell Tissue Bank 11:183–195
2. Dua HS, Gomes JA, King AJ, Maharajan VS (2004) The amniotic membrane in ophthalmology. Surv Ophthalmol 49:51–77
3. Hennerbichler S, Reichl B, Pleiner D et al (2007) The influence of various storage conditions on cell viability in amniotic membrane. Cell Tissue Bank 8:1–8. doi:10.1007/s10561-006-9002-3
4. Shortt AJ, Secker GA, Lomas RJ et al (2009) The effect of amniotic membrane preparation method on its ability to serve as a substrate for the ex-vivo expansion of limbal epithelial cells. Biomaterials 30:1056–1065
5. Mamede AC, Carvalho MJ, Abrantes AM et al (2012) Amniotic membrane: from structure and functions to clinical applications. Cell Tissue Res 349:447–458
6. Ab Hamid SS, Zahari NK, Yusof N, Hassan A (2014) Scanning electron microscopic assessment on surface morphology of preserved human amniotic membrane after gamma sterilisation. Cell Tissue Bank 15:15–24
7. Nakamura T, Inatomi T, Sekiyama E et al (2006) Novel clinical application of sterilized, freeze-dried amniotic membrane to treat patients with pterygium. Acta Ophthalmol Scand 84:401–405
8. Riau AK, Beuerman RW, Lim LS, Mehta JS (2010) Preservation, sterilization and de-epithelialization of human amniotic membrane for use in ocular surface reconstruction. Biomaterials 31:216–225
9. Mariappan I, Maddileti S, Savy S et al (2010) In vitro culture and expansion of human limbal epithelial cells. Nat Protoc 5:1470–1479
10. Von Versen-Hoeynck F, Steinfeld AP, Becker J et al (2008) Sterilization and preservation influence the biophysical properties of human amnion grafts. Biologicals 36:248–255
11. Kesting MR, Wolff K-D, Nobis CP, Rohleder NH (2012) Amniotic membrane in oral and maxillofacial surgery. Oral Maxillofac Surg 18:153–164
12. Adds PJ, Hunt CJ, Dart JK (2001) Amniotic membrane grafts, "fresh" or frozen? A clinical and in vitro comparison. Br J Ophthalmol 85:905–907
13. Ganatra M, Durrani K (1996) Method of obtaining and preparation of fresh human amniotic membrane for clinical use. J Pak Med Assoc 46:126–128

14. Qureshi IZ, Fareeha A, Khan WA (2010) Technique for processing and preservation of human amniotic membrane for ocular surface reconstruction. World Acad Sci Eng Technol 45:757–760
15. Kim J, Tseng S (1995) Transplantation of preserved human amniotic membrane for surface reconstruction in severely damaged rabbit corneas. Cornea 14:473–484
16. Kim JC, Tseng SC (1995) The effects on inhibition of corneal neovascularization after human amniotic membrane transplantation in severely damaged rabbit corneas. Korean J Ophthalmol 9:32–46
17. Lee S, Tseng S (1997) Amniotic membrane transplantation for persistent epithelial defects with ulceration. Am J Ophthalmol 123:303–312
18. López-Valladares MJ, Rodríguez-Ares MT, Touriño R et al (2010) Donor age and gestational age influence on growth factor levels in human amniotic membrane. Acta Ophthalmol 88:e211–e216
19. Rodríguez-Ares MT, López-Valladares MJ, Touriño R et al (2009) Effects of lyophilization on human amniotic membrane. Acta Ophthalmol 87:396–403
20. Nakamura T, Yoshitani M, Rigby H et al (2004) Sterilized, freeze-dried amniotic membrane: a useful substrate for ocular surface reconstruction. Invest Ophthalmol Vis Sci 45:93–99
21. Niknejad H, Deihim T, Solati-Hashjin M, Peirovi H (2011) The effects of preservation procedures on amniotic membrane's ability to serve as a substrate for cultivation of endothelial cells. Cryobiology 63:145–151
22. Bhatia M, Pereira M, Rana H et al (2007) The mechanism of cell interaction and response on decellularized human amniotic membrane: implications in wound healing. Wounds 19:207–217
23. Georgiadis NS, Ziakas NG, Boboridis KG et al (2008) Cryopreserved amniotic membrane transplantation for the management of symptomatic bullous keratopathy. Clin Experiment Ophthalmol 36:130–135
24. Lim LS, Poh RW, Riau AK et al (2010) Biological and ultrastructural properties of acelagraft, a freeze-dried γ-irradiated human amniotic membrane. Arch Ophthalmol 128:1303–1310
25. Thomasen H, Pauklin M, Steuhl K-P, Meller D (2009) Comparison of cryopreserved and air-dried human amniotic membrane for ophthalmologic applications. Graefes Arch Clin Exp Ophthalmol 247:1691–1700
26. Maral T, Borman H, Arslan H et al (1999) Effectiveness of human amnion preserved long-term in glycerol as a temporary biological dressing. Burns 25:625–635
27. Füst A, Pállinger É, Stündl A et al (2012) Both freshly prepared and frozen-stored amniotic membrane cells express the complement inhibitor CD59. Sci World J 2012:815615
28. Deolinda de Oliveira Pena J, de Melo GB, Gomes JÁP et al (2007) Ultrastructural and growth factor analysis of amniotic membrane preserved by different methods for ocular surgery. Arq Bras Oftalmol 70:756–762
29. Bio-Tissue (2013) Biotissue. www.biotissue.com
30. Shimazaki J, Shinozaki N, Tsubota K (1998) Transplantation of amniotic membrane and limbal autograft for patients with recurrent pterygium associated with symblepharon. Br J Ophthalmol 82:235–240
31. Azuara-Blanco A, Pillai C, Dua HS (1999) Amniotic membrane transplantation for ocular surface reconstruction. Br J Ophthalmol 83:399–402
32. Shimazaki J, Aiba M, Goto E et al (2002) Transplantation of human limbal epithelium cultivated on amniotic membrane for the treatment of severe ocular surface disorders. Ophthalmology 109:1285–1290
33. Rama P, Giannini R, Bruni A et al (2001) Further evaluation of amniotic membrane banking for transplantation in ocular surface diseases. Cell Tissue Bank 2:155–163
34. Shimmura S, Shimazaki J, Ohashi Y, Tsubota K (2001) Antiinflammatory effects of amniotic membrane transplantation in ocular surface disorders. Cornea 20:408–413
35. Ricci E, Vanosi G, Lindenmair A et al (2013) Anti-fibrotic effects of fresh and cryopreserved human amniotic membrane in a rat liver fibrosis model. Cell Tissue Bank 14:475–488
36. Mejía L, Acosta C, Santamaría J (2000) Use of nonpreserved human amniotic membrane for the reconstruction of the ocular surface. Cornea 19:288–291

37. Panda A (1999) Amniotic membrane transplantation in ophthalmology (fresh versus preserved tissue). Br J Ophthalmol 83:1409–1414
38. Nakamura T, Sekiyama E, Takaoka M et al (2008) The use of trehalose-treated freeze-dried amniotic membrane for ocular surface reconstruction. Biomaterials 29:3729–3737
39. Singh R, Chacharkar M (2011) Dried gamma-irradiated amniotic membrane as dressing in burn wound care. J Tissue Viability 20:49–54
40. Miljudin E, Zolotaryov A, Volova L, Ahmerova U (2004) Silica gel dessication of amniotic membrane with related epithelium cells for ocular surface reconstruction. Cell Tissue Bank 5:271–275
41. Fujisato T, Tomihata K, Tabata Y et al (1999) Cross-linking of amniotic membranes. J Biomater Sci Polym Ed 10:1171–1181
42. Spoerl E, Wollensak G, Reber F, Pillunat L (2004) Cross-linking of human amniotic membrane by glutaraldehyde. Ophthalmic Res 36:71–77
43. Kitagawa K, Okabe M, Yanagisawa S et al (2011) Use of a hyperdried cross-linked amniotic membrane as initial therapy for corneal perforations. Jpn J Ophthalmol 55:16–21
44. Ma DH-K, Lai J-Y, Cheng H-Y et al (2010) Carbodiimide cross-linked amniotic membranes for cultivation of limbal epithelial cells. Biomaterials 31:6647–6658
45. Sekar S, Sasirekha K, Krishnakumar S, Sastry TP (2013) A novel cross-linked human amniotic membrane for corneal implantations. Proc Inst Mech Eng H J Eng Med 227:221–228
46. Djefal A, Tahtat D, Khodja AN et al (2007) Validation and substantiation of 25 kGy as sterilization dose for lyophilized human amnion membrane. Cell Tissue Bank 8:9–12
47. Gajiwala K, Gajiwala AL (2004) Evaluation of lyophilised, gamma-irradiated amnion as a biological dressing. Cell Tissue Bank 5:73–80
48. de Melo GB, Gomes JÁP, da Glória MA et al (2007) Morphological assessment of different amniotic membrane epithelial denuding techniques. Arq Bras Oftalmol 70:407–411
49. Lim LS, Riau A, Poh R et al (2009) Effect of dispase denudation on amniotic membrane. Mol Vis 15:1962–1970
50. Zhang T, Yam GH-F, Riau AK et al (2013) The effect of amniotic membrane de-epithelialization method on its biological properties and ability to promote limbal epithelial cell culture. Invest Ophthalmol Vis Sci 54:3072–3081
51. Hopkinson A, Shanmuganathan VA, Gray T et al (2008) Optimization of amniotic membrane (AM) denuding for tissue engineering. Tissue Eng C 14:371–381
52. Covre JL, Loureiro RR, Cristivam PC et al (2011) Comparison between amniotic membrane with and without epithelium as a support for human limbal epithelial cells cultured ex vivo. Arq Bras Oftalmol 74:114–117
53. Kruse FE, Joussen AM, Rohrschneider K et al (2000) Cryopreserved human amniotic membrane for ocular surface reconstruction. Graefes Arch Clin Exp Ophthalmol 238:68–75
54. Kubo M, Sonoda Y, Muramatsu R, Usui M (2001) Immunogenicity of human amniotic membrane in experimental xenotransplantation. Invest Ophthalmol Vis Sci 42:1539–1546
55. Thomasen H, Pauklin M, Noelle B et al (2011) The effect of long-term storage on the biological and histological properties of cryopreserved amniotic membrane. Curr Eye Res 36:247–255
56. Krishnamurithy G, Shilpa P, Ahmad R et al (2011) Human amniotic membrane as a chondrocyte carrier vehicle/substrate: in vitro study. J Biomed Mater Res A 99A:500–506
57. Russo A, Bonci P, Bonci P (2012) The effects of different preservation processes on the total protein and growth factor content in a new biological product developed from human amniotic membrane. Cell Tissue Bank 13:353–361
58. Koizumi N, Fullwood NJ, Bairaktaris G et al (2000) Cultivation of corneal epithelial cells on intact and denuded human amniotic membrane. Invest Ophthalmol Vis Sci 41:2506–2513
59. Koh JW, Shin YJ, Oh JY et al (2007) The expression of TIMPs in cryo-preserved and freeze-dried amniotic membrane. Curr Eye Res 32:611–616
60. Schulze U, Hampel U, Sel S et al (2012) Fresh and cryopreserved amniotic membrane secrete the trefoil factor family peptide 3 that is well known to promote wound healing. Histochem Cell Biol 138:243–250
61. Wolbank S, Hildner F, Redl H et al (2009) Impact of human amniotic membrane preparation on release of angiogenic factors. J Tissue Eng Regen Med 3:651–654

Chapter 14
Ethical and Legal Concerns of the Use of Amniotic Membrane

Guilherme de Oliveira and André Pereira

Abstract In this chapter the legal issues concerning the use of amniotic membrane are analyzed, in particular the problems of informed consent, data protection, transplantation of tissues, biomedical research, biobanks and the prohibition of financial gain; the European Directives concerning technical quality and safety of human tissues and cells will also be briefly presented.

Keywords Legal issues • Protection of human tissues • Informed consent • Biobank • Data protection

14.1 Amniotic Membrane as Human Tissue: The Protection of the Law

Since the beginning of the twentieth century, amniotic membrane has been used in several medical areas: from skin transplantation to surgery, including biological dressing for burned skin, skin wounds, and chronic ulcers of the leg or as an adjunctive tissue in surgical reconstruction of artificial vagina, and for repairing omphaloceles. Since the 1940s its use in treating a variety of ocular surface disorders has been pointed out. Nowadays, this tissue is relevant since it was reintroduced to ophthalmologists. "Several studies have addressed this subject and the scope of the application of amniotic membrane transplantation (AMT) in the management of ocular surface disorders is ever increasing" [1–3].

Amniotic membrane is human tissue; therefore ethical and legal issues concerning the protection and respect for human tissue must be taken into consideration, as the principle of respect for the dignity of the human being indirectly applies to all

G. de Oliveira (✉) • A. Pereira
Centre for Biomedical Law, University of Coimbra, Coimbra, Portugal
e-mail: guilhermedeoliveira@me.com

A.C. Mamede, M.F. Botelho (eds.), *Amniotic Membrane: Origin, Characterization and Medical Applications*,
DOI 10.1007/978-94-017-9975-1_14

human material. That is, amniotic membrane is not a normal thing, a res. It is a res extra commercium, a thing that is outside the normal legal trade.

This special protection has two sources: a subjective or individual one, and an objective or collective one. Firstly, human tissue deserves legal protection since it is or it was part of a human body and the individual has the right to determine what shall be done with his body and parts of it and has the right to control the destination of her personal health information and genetic information. Secondly, human tissue shall be respected and protected as an indirect protection to the dignity of the human life and special respect for the human species. These principles are to be found, at least to a certain extent, in Article 1 of the Portuguese Constitution and Article 1 of the Convention on Human Rights and Biomedicine.

Human tissue has a limited protection, since it has been separated from the human body and neither human life nor physical integrity are being violated when it is used in research or transplantation. On the other hand, the respect for the will of the person and the respect for the dignity of the human body and parts of it require the respect for some material and procedural conditions. The scope of this chapter is to point out the relevant sources of law concerning the protection of tissues of human origin.

Biomedical use of human tissue has been subject to legal regulation at international level: the World Health Organization (WHO) enacted the WHO Guiding Principles on Human Cell, Tissue and Organ Transplantation (WHO: 21 May 2010 – WHA63.22); the Council of Europe approved the Convention on Human Rights and Biomedicine, the Additional Protocol to the Convention on Human Rights and Biomedicine concerning Transplantation of Organs and Tissues of Human Origin[1] and the Additional Protocol to the Convention on Human Rights and Biomedicine, concerning Biomedical Research. The European Union enacted three directives of great importance in this respect: Directive 2004/23/EC on setting the standards of quality and safety for the donation, procurement, testing, processing, preservation, storage and distribution of human tissues and cells "intended for human applications, in order to ensure a high level of protection of human health"; Commission Directive 2006/17/EC of 8 February 2006 implementing Directive 2004/23/EC of the European Parliament and of the Council as regards certain technical requirements for the donation, procurement and testing of human tissues and cells intended for human applications; Commission Directive 2006/86/EC of 24 October 2006 implementing Directive 2004/23/EC of the European Parliament and of the Council as regards traceability requirements, notification of serious adverse reactions and events and certain technical requirements for the coding, processing, preservation, storage and distribution of human tissues and cells intended for human applications. These three directives, the first European Union Tissue and Cells Directive and its two technical directives are concerned only with the quality and safety issues surrounding the medical use of

[1] Additional Protocol to the Convention on Human Rights and Biomedicine concerning Transplantation of Organs and Tissues of Human Origin, Strasbourg, 24.I.2002. This Protocol applies to the transplantation of organs and tissues of human origin carried out for therapeutic purposes. This protocol has already 12 ratifications/accessions and 9 other States have signed it but not yet ratified (like Portugal). Entered into force in 1.5.2006.

human tissues and cells, not the ethical issues. The European Union Tissue and Cells Directives (EUTCD) set out to establish a harmonized approach to the regulation of tissues and cells across Europe for human applications. The Directives set a benchmark for the standards that must be met when carrying out any activity involving tissues and cells for human application (patient treatment). The Directives also require that systems be put in place to ensure that all tissues and cells used in human application are traceable from donor to recipient.

At European level, Portugal is among the first countries concerning cornea, amniotic membrane and cardiac tissue procurement. In 2009 were performed, per million people, 5.6 amniotic membrane, higher than the EU average rate (4.73 procurements) [4].

Amniotic membranes are tissues of human origin, therefore Directive 2004/23/EC of the European Parliament and of the Council of 31 March 2004 on setting standards of quality and safety for the donation, procurement, testing, processing, preservation, storage and distribution of human tissues and cells is applicable. The technical requirements are listed in Directive 2004/23/EC but described only in Directives 2006/17/EC and 2006/86/EC.[2] These Directives have been transposed into Portuguese law by Law no. 12/2009, of 26 March.[3]

14.2 Informed Consent

First of all, it must be stressed that whatever scope there is for the amniotic membranes, research, transplantation or other, the well-being of the woman and the foetus or the baby are the paramount concern of the medical and nursing team during delivery.[4]

Secondly, information about research projects should also be separate from information about routine clinical care and the information for collection and its scope should be given before delivery, so that the woman has time to reflect about the consent.[5]

---

[2] See also Commission Decision of 3 August 2010 establishing guidelines concerning the conditions of inspections and control measures, and on the training and qualification of officials, in the field of human tissues and cells provided for in Directive 2004/23/EC of the European Parliament and of the Council.

[3] Moreover, although not directly applicable to human tissues, one shall mention the Directive 2010/53/EU of the European Parliament and of the Council of 7 July 2010 on standards of quality and safety of human organs intended for transplantation, implemented in Portugal by Law no. 36/2013, 12 June.

[4] See Recommendation 4.1.1. of the National Statement on Ethical Conduct in Human Research, Australia, 2007 (Updated December 2013).

[5] See Recommendation 4.1.6. of the National Statement on Ethical Conduct in Human Research, Australia, 2007 (Updated December 2013).

14.2.1 *Informed Consent to Remove the Amniotic Membrane for a Therapeutic Purpose, i.e., Transplantation*

Wherever tissues are removed from human beings, and possibly transplanted into other human beings, the activities involved in the collection and use of such tissues are subject to ethical requirements intended to safeguard respect for human beings, their dignity and autonomy. As the issue of safety is vital, the protection must extend to tissue donors and recipients, and to all health care professionals – whose work involves collecting, manipulating and using human tissues. The removal of amniotic membrane can be considered as a removal of tissue from a living person.

This tissue has special characteristics since: (a) the "removal" is a natural phenomenon that occurs during delivery; (b) this tissue is only to be found in women and (c), in general, it shall be collected only in living women.

Therefore, the requirements for the "removal" may be less demanding than any parts of the human body and the consent to remove and transplant post mortem do not have to be considered in this text.

Finally, it is the woman who shall be given adequate information in a comprehensible form and that information shall be documented. It shall be the woman (and not the couple) who has the right to be informed and to give consent, to refuse or to withdraw consent, since the tissue was part of the woman's body. In fact, even in case of foetal disputes, European Human Rights Law recognizes the woman shall have the right to decide.[6]

The Additional Protocol on Transplantation states: "removal of organs or tissue from a living person may be carried out solely for the therapeutic benefit of the recipient and where there is no suitable organ or tissue available from a deceased person and no other alternative therapeutic method of comparable effectiveness" (Article 9). Amniotic membrane is to be found in living and pregnant women and as long as there is no alternative method of comparable effectiveness, this tissue can be used for transplantation. Article 11 obliges the medical team to evaluate the risks for the donor and "the removal may not be carried out if there is a serious risk to the life or health of the donor." In the case of amniotic membrane, generally there is no risk for the woman, since it is not an invasive procedure. However, the woman must be informed and give her consent not for the removal as such, but about the scope of the removal, be it for transplantation or for research.[7]

In case the woman gives consent to use the amniotic membrane for research, but later doctors want to use if for a therapeutic procedure, Article 20 of the Additional Protocol on Transplantation states that "it may only be implanted if the consequences and possible risks have been explained to that person and his/her informed

[6] See European Commission of Human Rights, Petition no. 8416/78 (W.P. vs. UK); in USA, see Planned Parenthood of SE Pennsylvania v Casey [1992] 112 S Ct 2791 (US Sup Ct).

[7] That is stated in Articles 12 and 13 of the Additional Protocol: "The donor (…) shall beforehand be given appropriate information as to the purpose and nature of the removal as well as on its consequences and risks. (…)." "An organ or tissue may be removed from a living donor only after the person concerned has given free, informed and specific consent to it either in written form or before an official body. The person concerned may freely withdraw consent at any time."

consent, or appropriate authorization in the case of a person not able to consent, has been obtained".[8]

Moreover, the requisites of a valid consent are:

1. The consent of the donor and recipient must be free, enlightened, informed and unequivocal and the donor can identify the beneficiary.
2. The consent of the donor and the recipient is provided before:
   (a) A physician designated by the medical director of the establishment where the procurement takes place in the case organ, tissue and regenerative cell transplantation;
   (b) A physician designated by the medical director of the establishment where the procurement takes place and not belonging the transplant team, in the case of transplantation organs, tissues or cells not regenerable.[9]

These norms protect the personality right of the individuals to determine what shall be done with parts of their body, even after separation from the body and has its source in the dignity of the human life, the right to moral integrity of the person as well as the right to self-determination of personal information.[10]

14.2.2 Informed Consent and Ethical Assessment in Biomedical Research with Amniotic Membranes

Research with human beings, including human tissues and cells is regulated by the Convention on Human Rights and Biomedicine, Chapter V, namely articles 15–17.[11] Moreover the Additional Protocol on Biomedical Research[12] also applies in research with human tissue. Its scope includes the full range of research activities

---

[8] A similar norm is to be found in Article 22 of the Biomedicine Convention: "When in the course of an intervention any part of a human body is removed, it may be stored and used for a purpose other than that for which it was removed, only if this is done in conformity with appropriate information and consent procedures." See also Article 18 (5) of the Portuguese Law 12/2005: "The biological samples collected for a particular purpose should not be used for other health care or biomedical research purposes unless the person from whom it was collected gives consent to it (…)."

[9] See Article 24 of Portuguese Law 12/2009, of 26 March and Article 8 of Law 12/93, of 22 April, changed by Law 22/2007, of 29 June. The problem of donation by minors or adults with diminished capacity due to metal illness is regulated in the following paragraphs. It allows donation of renewable tissues (that could the case of amniotic membranes) in the case of minors after parental authorization, in case of mental ill patients after court authorization (see also Annex IV of Law 12/2009). This Law Establishes the legal regime concerning the quality and safety the donation, procurement, testing, processing, preservation, storage, distribution and application of tissues and cells of human origin, transposing into national law Directive 2004/23/EC of the European Parliament and of the Council of 31 March, Directive 2006/17/EC, the Commission of 8 February and Directive 2006/86/EC, the Commission of 24 October.

[10] See Articles 1, 26 and 35 of the Portuguese Constitution.

[11] Article 18 of the Oviedo Convention is not important in this case, since the amniotic membrane tissue does not contain an embryo anymore.

[12] Additional Protocol to the Convention on Human Rights and Biomedicine, concerning Biomedical Research, Strasbourg, 25.I.2005. There were so far (28.12.2013) 9 ratifications/acces-

in the health field involving interventions on human beings; the term "intervention" includes: (i) a physical intervention, and (ii) any other intervention in so far as it involves a risk to the psychological health of the person concerned." Thus, the use of amniotic membranes for research my cause a risk to the psychological health of the woman concerned, therefore the conditions for research included in the Additional Protocol shall be respected.

Some documents of soft law are also very important to take into account, notably the Helsinki Declaration of the World Medical Association,[13] the UNESCO Universal Declaration on Bioethics and Human Rights (2005); the CIOMS/WHO (Council for International Organizations of Medical Sciences/World Health Organization) International Ethical Guidelines for Biomedical Research Involving Human Subjects (2002) as well as the abovementioned WHO Guiding Principles on Human Cell, Tissue and Organ Transplantation (WHO: 21 May 2010 – WHA63.22).

One of the main requisites for research with human tissues is Ethical assessment. "Medical research involving human subjects may only be conducted if the importance of the objective outweighs the risks and burdens to the research subjects" (Article 16 of the Helsinki Declaration). Before any research of clinical application, the ethical control and the positive opinion of the ethics committee is mandatory.

In Portugal, the Institutional Review Boards are regulated by Decree-Law n. 97/95. The local Ethics Committee shall give a binding opinion concerning research protocols with human tissues.

In case there is a clinical trial on medicinal products for human use, Law 21/2014, of 16 April, applies. The Clinical Trials Act implemented the Directive 2001/20/EC of the European Parliament and of the Council of 4 April 2001 on the approximation of the laws, regulations and administrative provisions of the Member States relating to the implementation of good clinical practice in the conduct of clinical trials on medicinal products for human use. In this case the CEIC (Ethics Committee for Clinical Research): an independent body composed by health care professionals and others, with the aim of assuring the protection of rights, safety and well-being of participants in clinical trials and the public in general, and who shall, as a rule, give the single opinion. CEIC has the power to give the ethical and scientific Opinion, which is mandatory for the realization of clinical trials [5, 6].

In case we are doing research with a medical device, Law 21/2014, of 16 April applies and requires the ethical approval of an independent Committee.

Ethical assessment together with quality assurance might be as important or even more important than informed consent, namely when we are dealing with detachable parts of detachable human body.

---

sions and 13 signatures not followed by ratification (the case of Portugal). It entered into force in 1.9.2007.

[13] Last revision: 64th WMA General Assembly, Fortaleza, Brazil, October 2013.

The Additional Protocol on Biomedical Research reinforces that "Research may only be undertaken if the research project has been approved by the competent body after independent examination of its scientific merit, including assessment of the importance of the aim of research, and multidisciplinary review of its ethical acceptability (Article 7)."

Ethical assessment is probably more important or even more important than informed consent. However, the later shall be carefully considered as well. After the removal of the amniotic membrane, there might be a legitimate interest to use that tissue for biomedical research. In that case, Chapter IV of the Additional Protocol about Biomedical Research provides detailed rules about information and consent.

In Portugal, Article 18 (4) of Law 12/2005 requires, as a rule, a specific consent for each research, as it states "the biological samples collected for a particular purpose should not be used for other health care or biomedical research purposes unless the person from whom it was collected gives consent to it".[14]

However, it is possible to do research without informed consent for the specific purpose in the following cases: (1) after irreversible anonymisation[15]; (2) in case of retrospective use of the samples, or in special situations in which it is not possible to obtain consent from the persons involved – due to the amount of data or of subjects, to their age or other similar reason – the material and the data can only be processed for the purposes of scientific research or for collecting epidemiological or statistical data (Article 19 (6) of Law 12/2005) [7].

14.2.3 Informed Consent to Store the Amniotic Membrane in a Biobank: The Portuguese Example (Law 12/2005, 26 January)

The collection biological products (v.g., amniotic membranes) is subject to separate informed consents for health care purposes and for biomedical research purposes.[16] Moreover, the consent must include the purpose of the collection and the duration of storage of the samples and its by-products. The person to whom the biological

[14] In the international literature, some Authors defend the concept of broad consent, or an open consent for research with samples stored in a Biobank – see [8]. The model termed 'broad consent' has been adapted by many current Biobank projects, like UK Biobank, CARTaGENE (Montreal, QC, Canada) and the Norwegian HUNT study.

[15] Anonymisation is the process of turning data into a form, which does not identify individuals and where identification is not likely to take place. This allows for a much wider use of the information.

[16] Zahid et al. [9] emphasize the main ethical aspects of Biobanks of human tissue as being confidentiality, informed consent, organizational policies, and respect for scientific standards.

material belongs can withdraw, at any time, his/her consent or, in case of death or incapacity of the person in question, his/her relatives can withdraw it; if that is the case, the stored biological samples and its by-products must be destructed for good.

According to Article 19 (1) of the Portuguese Law 12/2005, "biological material databanks" are defined as any collection of biological samples or its by-products, with or without a storage time limit, previously accumulated or prospectively performed, obtained through routine health care provision, whether in screening programs or for research purposes, and that includes samples that are identified, identifiable, anonymised or anonym."

There are several conditions to create and operate a Biobank, among which, according to Portuguese law:

(a) The need to obtain authorization from an entity accredited by the health authorities and by the National Data Protection Commission, if the Biobank is associated to personal information;
(b) Biological material databanks should only be created for the purposes of providing health care services, including the diagnostic and the prevention of diseases, or for the purposes of basic or health related research;
(c) Written informed consent is necessary to obtain and use material for a biological material databank; the consent form should include information about the purposes of the databank, the person responsible for it, the types of research it performs, its potential risks and benefits, the conditions and duration of storage, the measures taken to guarantee the privacy and confidentiality of the persons involved, as well as the prevision regarding the possible disclosure or not of the results obtained from the materials in question;
(d) The privacy and confidentiality must always be ensured – the storage of identified material should be avoided, the access to the collections of biological material should be controlled, the number of persons authorized to access it should be restricted and its safety should be guaranteed, particularly in terms of losses, changes and destruction.[17]

14.3 Data Protection

14.3.1 In Case of Transplant

In case of transplant of human tissue, e.g., amniotic membrane, following Directive 2004/23/EC, Member States must take all necessary measures to ensure that all data collected and to which third parties have access are rendered anonymous. The

[17] In Portugal, the collection and storage of biological samples from the Biobank-IMM (Faculty of Medicine – University of Lisbon) are authorized by the Ethics Committee of the North Lisbon Hospital Centre – Hospital de Santa Maria. The collection and processing of clinical data associated to samples are authorized by the National Commission for Data Protection.

recipient's identity must never be divulged to the donor or his or her family, or vice versa. To this end, measures must be adopted to ensure data security and prevent unauthorized modifications to files and records. As there is the demand for traceability, the rules concerning data protection must be respected.

Under EU law, Directive 95/46/EC of the European Parliament and of the Council of 24 October 1995 on the protection of individuals with regard to the processing of personal data and on the free movement of such data[18], personal data can only be gathered legally under strict conditions, for a legitimate purpose. Furthermore, persons or organizations, which collect and manage your personal information must protect it from misuse and must respect certain rights of the data owners, which are guaranteed by EU law.

Common EU rules have been established to ensure that personal data enjoy a high standard of protection everywhere in the EU and the person has the right to complain and obtain redress if data is misused anywhere within the EU.

The EU's Data Protection Directive also foresees specific rules for the transfer of personal data outside the EU to ensure the best possible protection of your data when it is exported abroad.

The respect for Data Protection Law means: respect for the safety rules, respect for the right to information, right to access, right to object and prohibition of automated decisions (Article 10–13 Data protection Act).[19]

Three situations shall be distinguished:

A. If we are not using the amniotic membranes for transplantation, than the requirements of Art 7 (4) of Data Protection Act must be respected. That is, health information data is subject to special protection; therefore, processing of such data is only legal if (Article 7 of the Law 67/98):

    (i) It is required for the purposes of preventive medicine, medical diagnosis, the provision of care or treatment or the management of health-care services, and
    (ii) Where those data are processed by a health professional subject under national law or rules established by national competent bodies to the obligation of professional secrecy or by another person also subject to an equivalent obligation of secrecy;
    (iii) The data processing center shall be notified to the Portuguese Data Protection Authority and
    (iv) To ensure the technical safety of its processing.

    Thus, it is possible to process personal data in connection with transplant of amniotic membranes, since it aims the provision of care or treatment.

B. However, if the amniotic membranes will be transplanted to another person, than Article 23 of Law 12/2009 applies, which means there is an express legal

[18] Implemented in Portuguese Law by Law 67/98, of 26 October.

[19] See also Articles 12, 14 and 15 of Directive 95/46/EC of the European Parliament and of the Council of 24 October 1995 on the protection of individuals with regard to the processing of personal data and on the free movement of such data.

authorization to process personal data, as long as the requirements of Data protection Act are respected.

C. If the amniotic membranes are stored in biobanks, than Article 19 of Law 12/2005 applies and requires the respect for the rules of Data Protection Act. Accordingly, Portuguese law, in case of biomedical research with biobanks of human tissue or cells, demands anonymisation as a rule. Article 19 of Law 12/2005 states: "9. Only anonym or irreversibly anonymised samples can be used". Nevertheless, it accepts exceptions: "the use of identified or identifiable samples should be limited to studies that cannot be conducted in any other way."[20] However (11) "If there is an absolute need to use identified or identifiable samples, these should be coded and the identifying codes must be kept separately, but always in a public institution".[21] It shall be stressed that these norms refer to samples and not only to personal data.

14.3.2 In Case of Research

The processing of personal data in connection with research is only possible, if there is consent of the person involved and safeguarding all other legal and technical conditions are met.

In the field of research and the use of personal data, it is useful to mention the taxonomy of: (1) Identified: The tissue source is known and the individual's identity is tied to the sample; (2) Identifiable: The tissue source is tied to the specimen through the use of a link (e.g., a code number), but the identity of the source is not directly known without tracing the link; (3) Anonymized: The tissue source's identity is irrevocably unlinked from the specimen, so that the individual's identity cannot be discerned (i.e., the tissue is not identifiable); (4) Anonymous: The tissue source's identity is never known, since the specimen is collected with no identifiers at all (i.e., the sample is unidentified).

Therefore, in general, in case of identified and identifiable tissue, informed consent of the patient is mandatory and as Law 12/2009 does not expressly allows it, we advocate that the authorization of the National Data Protection Authority shall be obtained in order to build a database with information about the tissues identified or identifiable.

It shall be mentioned that in the near future, there might be some relevant changes. In fact, in 2012, the Commission proposed a major reform of the EU legal framework on the protection of personal data. The new proposals will strengthen individual rights and tackle the challenges of globalization and new technologies.

---

[20] The Helsinki Declaration states: "32. For medical research using identifiable human material or data, such as research on material or data contained in Biobanks or similar repositories, physicians must seek informed consent for its collection, storage and/or reuse. There may be exceptional situations where consent would be impossible or impracticable to obtain for such research. In such situations the research may be done only after consideration and approval of a research ethics committee."

[21] Lowrance [10] argues for a change in European Data protection law, so that health research is fostered and, by the same token, confidentiality of data is respected.

14.4 Prohibition of Financial Gain

The WHO Guiding Principles – Commentary to Principle 5 state: "Payment for cells, tissues and organs is likely to take unfair advantage of the poorest and most vulnerable groups, undermines altruistic donation, and leads to profiteering and human trafficking. Such payment conveys the idea that some persons lack dignity, that they are mere objects to be used by others."

U.S. federal law prohibits the buying and selling of human organs, it does allow fees for the recovering, processing and transporting of human tissue.[22]

By the same path, prohibition of financial gain is a paramount principle in European Bioethics,[23] consecrated in Article 3 of the European Charter on Fundamental Rights, in Article 21 of the Biomedicine Convention. The Additional Protocol on Transplant of organs and tissues states: Article 21 (1): "The human body and its parts shall not, as such, give rise to financial gain or comparable advantage." However, it adds: "The aforementioned provision shall not prevent payments which do not constitute a financial gain or a comparable advantage, in particular: – compensation of living donors for loss of earnings and any other justifiable expenses caused by the removal or by the related medical examinations; – payment of a justifiable fee for legitimate medical or related technical services rendered in connection with transplantation; – compensation in case of undue damage resulting from the removal of organs or tissues from living persons.

On the other hand, "advertising the need for, or availability of, organs or tissues, with a view to offering or seeking financial gain or comparable advantage, shall be prohibited".

Following Directive 2004/23/EC, Member States must encourage voluntary and unpaid donations of tissues and cells. However, donors may receive compensation strictly limited to making good the expenses and inconveniences related to the donation (e.g. travel expenses). No promotion and publicity activities are allowed in support of the donation of human tissues and cells with a view to offering or seeking financial gain or comparable advantage. The general rule is that Member States must endeavor to ensure that the procurement of tissues and cells is carried out on a non-profit basis. By the same path, article 22 of the Additional Protocol prohibits organ and tissue trafficking.

In Portugal it is forbidden to give financial compensation or remuneration of the donor. However, living donors may receive compensation strictly limited to reim-

---

[22] National Organ Transplant Act: 42 U.S. CODE § 274E – Prohibition of Organ Purchases.

[23] Article 12 of the Directive 2004/23/CE is related to the principles governing tissues and cells donation. In its paragraph one establishes that "1. Member States shall endeavor to ensure voluntary and unpaid donation of tissues and cells. Donors may receive compensation, which is strictly limited to making good the expenses and inconveniences related to the donation. In that case, Member States define the conditions under which compensation may be granted. Member States shall report to the Commission on these measures before 7 April 2006 and thereafter every three years. On the basis of these reports the Commission shall inform the European Parliament and the Council of any necessary further measures it intends to take at Community level".

bursement of expenses incurred or losses immediately resultant from the donation (Article 22 (1) and (3) of Law 12/2009, of 26 March). The law specifies that such compensation involves the right of the donor to have medical care until complete recovery and the right to be compensated for damages experienced in the process of organ donation and procurement (Article 9 of Law 12/93, of 22 April, amended by Law 22/2007, of 29 June, and Article 22 (3) of Law 12/2009).

Another relevant aspect is that the woman shall be informed about the possible commercial benefits of the use of the amniotic membranes in research or in transplantation, but that she shall not expect any financial profit from it. Article 13 of the Biomedical Research Protocol indicates several issues that must be informed, including "vii. any foreseen potential further uses, including commercial uses, of the research results, data or biological materials." However, the financial benefit of the investigation is not to be shared with the subjects who consented to research.[24]

On the other hand, Article 18 of Law 12/2005 states: "8. The commercial use, the patent registration or any type of financial gains derived from biological samples, as such, is strictly forbidden."[25]

14.5 Technical Quality and Safety of Human Tissues and Cells

In Portugal, the Authority for Blood Services and Transplantation must authorize banks of tissues and cells, with regard to the activities of procurement, testing, processing, storage and distribution. Concerning the requirements for the procurement of tissue and cells of human origin, apart from all the requirements of the European Directives, Law n. 12/2009 added the following demand: "the encoded data should be entered in to a national register, held by the Blood and Transplantations Services Authority."

[24] See Article 21 of the Biomedicine Convention and Article 3 Charter of Fundamental Rights of the European Union. In the USA, see the Moore v. Regents of the University of California (51 Cal. 3d 120; 271 Cal. Rptr. 146; 793 P.2d 479) – this was a landmark Supreme Court of California decision filed on July 9, 1990, which dealt with the issue of property rights in one's own body parts. John Moore underwent treatment for hairy cell leukemia at the UCLA Medical Center. Moore's cancer was later developed into a cell line that was commercialized. The California Supreme Court ruled that Moore had no right to any share of the profits realized from the commercialization of anything developed from his discarded body parts. See also Recommendation 4.1.20 (d) and (e) of the National Statement on Ethical Conduct in Human Research, Australia, 2007 (Updated December 2013) – "the woman shall be informed, "(d) whether there is potential for commercial application of outcomes of the research, including the development of cell lines; (e) that she will not be entitled to a share in the profits of any commercial applications."

[25] See about the problem of patenting of human genes, the decision of the European Court of Justice, Case C-34/10 Oliver Brüstle v. Greenpeace e.V.; See the analysis of Pereira [11].

14.5.1 Traceability

Article 8 of the Directive 2004/23/EC establishes that: "Member States shall ensure that all tissues and cells procured, processed, stored or distributed on their territory can be traced from the donor to the recipient and vice versa. This traceability shall also apply to all relevant data relating to products and materials coming into contact with these tissues and cells". Article 9 of the Directive 2006/86/CE establishes in this regard that: "1. Tissue establishments shall have effective and accurate systems to uniquely identify and label cells/tissues received and distributed. 2. Tissue establishments and organizations responsible for human application shall retain the data set out in Annex VI for at least 30 years, in an appropriate and readable storage medium". Moreover, Member States shall establish a system for the identification of human tissues and cells, in order to ensure the traceability of all human tissues and cells pursuant to Article 8 and the Commission, in cooperation with the Member States, shall design a single European coding system to provide information on the main characteristics and properties of tissues and cells (Article 25 Directive 2004/23/CE). In Portugal, Article 8 of Law 12/2009 implement these norms.

14.6 Import/Export Tissues Between EU Member States and Third Countries

According to Article 9 of the Directive 2004/23/EC, all imports of tissues and cells from third countries are undertaken by tissue establishments accredited, designated, authorized or licensed for the purpose of those activities, and that imported tissues and cells can be traced from the donor to the recipient and vice versa. Member States and tissues establishments that receive such imports from third countries shall ensure that they meet standards of quality and safety equivalent to the ones laid down in this Directive".

Concerning the exports of tissues and cells to third countries, the Directive states that Member States shall take all necessary measures to ensure that all exports of tissues and cells to third countries are undertaken by tissue establishments accredited, designated, authorized or licensed for the purpose of those activities. Those Member States that send such exports to third countries shall ensure that the exports comply with the requirements of this Directive".

In Portugal, tissues can only be imported from third countries where they came from banks of tissues and cells allowed for these activities and meeting the standards of quality and traceability provided by Portuguese law, ensuring a system for the reporting of adverse reactions incidents equivalent the one provided by Portuguese law (Article 9 of Law 12/2009).

Imports/exports need to be authorized by the ASST. The export must be made through banks authorized for such activities. Applications for export need to be authorized by ASST and only where there is sufficient availability of tissues and cells in tissues national banks or for justified compatibility reasons. Requests for import shall refer to the original institution and are allowed only by Authority for Blood Services and Transplantation when: there is proven benefit in the use of tissues and cells, the purpose of tissues and cells is human use, there is no availability in tissue or cells national banks and compatibility grounds are justified by a doctor. In emergency situations, imports/exports are authorized directly by the Authority for Blood Services and Transplantation, provided that the supplier has the accreditation, designation or authorization according to Portuguese law or complies with standards of quality and safety equivalent to the Portuguese ones.

Concerning import/export and trade of human tissues, not only the rules of law 12/2009 must be respected, but also Article 19 of Law 12/2005 shall be taken into consideration, as it states: (1) "Commercial entities cannot store or use human biological material that has not been anonymised" and (2) "The transfer of a large number of samples or of biological material collections to other national or foreign entities must always respect the purposes for which the bank was originally created and for which consent was obtained, and must also be approved by the responsible ethic commissions".

References

1. Dua HS, Azuara-Blanco A (1999) Amniotic membrane transplantation. Br J Ophthalmol 83:748–752
2. Niknejad H, Peirovi H, Jorjani M et al (2008) Properties of the amniotic membrane for potential use in tissue engineering. Eur Cell Mater 15:88–99
3. Mamede AC, Carvalho MJ, Abrantes AM et al (2012) Amniotic membrane: from structure and functions to clinical applications. Cell Tissue Res 349:447–458
4. ASST (2009) Statistical report – organs and tissues donation and transplantation. http://www.asst.min-saude.pt/SiteCollectionDocuments/RelatorioEstatistico2009.pdf
5. Loureiro JC, Pereira AD (2004) Portuguese report. Clin Trials Med Eur Rules Trial 259–293
6. Pereira AD (2011) Country report Portugal. In: Deutsch E, Duttge G, Schreiber H-L et al (eds) Die Implementierung der GCP-Richtlininen und ihre Austrahlungswirkungen. Springer, Berlin, pp 255–270
7. de Oliveira G (2013) Consentimento informado para biobancos. Livro homenagem a António Barbosa Melo. Almedina, Coimbra, pp 257–283
8. Steinsbekk K, Myskja B, Solberg B (2013) Broad consent versus dynamic consent in biobank research: is passive participation an ethical problem? Eur J Hum Genet 21:897–902
9. Zahid HM, Chakma K, Miah M, Nessa A (2010) Ethical perception of tissue banking in Bangladesh. Bangladesh J Bioeth 1:11–19
10. Lowrance WW (2013) Privacy, confidentiality and health research. Cambridge University Press, Cambridge 204
11. Pereira AD (2012) Human genes as corporate assets? Nature, morals and the role of intellectual property in healthcare innovation. Rev Port direito da saúde – Lex Med 17:127–146

Index

A
Abbe-McIndoe operation, 171
Adipogenic differentiation, 68, 205
AECs. *See* Amniotic epithelial cells (AECs)
AF. *See* Amniotic fluid (AF)
AIS. *See* Amnion infection syndrome (AIS)
Akt-FKHR, 144
Alveolar bone (AB), 154, 155
Alveolar mucosa, 154
AM. *See* Amniotic membrane (AM)
Amnion
  after 7–10 weeks, gestation, 10
  and amniotic cavity, formation, 10
  biochemical properties (*see* Biochemical properties, amniotic membrane)
  for burn (*see* Burn wound dressing)
  connective tissue, 15
  cytochemical and histochemical studies, 15
  epithelium, 15
  fecundation, 10
  fetal membranes, 13, 14
  fusion with chorion, 11–12
  as graft overlay, 129
  layering, chorionic plate, 17
  and mesodermal derivatives, 11
  nonvillous chorion, 16
  preparation and preservation methods, 125
  spongy layer, 16
  thickness after separation, 13–14
  ultrastructural studies, 15–16
  23 weeks, pregnancy, 12
  zone of fusion, 15, 16
Amnion infection syndrome (AIS), 93
Amniotic epithelial cells (AECs)
  and AMSCs, 26
  hAECs (*see* Human amniotic epithelial cells (hAECs))
Amniotic fluid (AF)
  amnionitis, 93
  amniotic epithelium, 21, 25
  contamination, 96
  fetal lungs, 16
  hAMSC, 199
  intra-amniotic pressures, 50
  meconium-stained, 93
  mesenchymal-epithelial cells, 14
  oligohydramnios, 42
  polyhydramnios, 42
  prostaglandins (PG), 88
  uterine tenderness, 94
  volume anomalies, 96–97
Amniotic membrane (AM)
  in cancer (*see* Cancer)
  drug permeability, 178–180
  in gynaecology (*see* Gynaecology)
  MSC (*see* Mesenchymal stem cells (MSC))
  in oral medicine, 154–161
Amniotic membrane transplantation (AMT)
  conjunctiva, 182
  cornea healing, 181
  drug administration, 177–178
  limbal stem cell deficiency, 109
  microstructure, 184–185
  ocular burns, acute stage, 109
  ocular surface diseases, 180–181, 231
  pharmacological factors, 185–187
  postoperative period, 181
  Stevens–Johnson syndrome, 109
  surgical factor
    carrier molecules, 183

A.C. Mamede, M.F. Botelho (eds.), *Amniotic Membrane: Origin, Characterization and Medical Applications*,
DOI 10.1007/978-94-017-9975-1

Amniotic membrane transplantation (AMT) (*Cont.*)
cyclodextrin (CD), 183–184
de-epithelialized amniotic membrane, 182
drug release capacity and barrier function, 181
drug reservoir function, 183
epithelium acts, 182
Franz-diffusion cell system, 182–183
grafting/patch technique procedure, 181–182
hydrophilic molecules, 183
localization and etiology, 181
multilayer graft transplantation, 182
ocular surface, 182
ofloxacin, 183
pharmacological effect, 182
sandwich technique, 182
VEGF-blockade, 184
tear film alterations, 187
Amniotic mesenchymal stem cells (AMSCs), 26
AMSCs. *See* Amniotic mesenchymal stem cells (AMSCs)
AMT. *See* Amniotic membrane transplantation (AMT)
Anatomic barrier, 168
Angiogenesis, 20, 28, 35, 80, 133, 142, 181
Anterior segment optical coherence tomography (AS-OCT), 115–116
Anti-angiogenic activity
concept, 143
endothelial cell growth, 143
human amnion, 143
interleukin (IL)-1 antagonist receptor, 143
matrix proteins, 143
properties, 20, 142, 143, 210
vascular endothelial cells, 144
Anti-cancer therapies
amniotic membrane/amniotic membrane-derived cells, 148–149
hAECs, 147
Apexogenesis, 160–161
Aquaporins (AQPs), 25, 45
ART. *See* Assisted reproduction technology (ART)
AS. *See* Asherman syndrome (AS)
Asherman syndrome (AS)
AM application
anatomic barrier, 168
endogenous stem cells, 167–168
fibrin matrix, 168
hAC, 167
hysteroscopic adhesiolysis, 167–170
immunological benefits, 168
lyophilized membrane, 168
pregnancy rate, 170
dysfunctional endometrium, 166
hysteroscopy, 166, 167
hysterotomy approach, 166
iatrogenic trauma, 166
IUD, 166
medical treatment, 167
permanent adhesions, 166
post-operative evaluation, 167
pregnancy complication, 166
recurrence rate, 166
risk factors, 166, 167
stenosis of cervical ostium, 165
AS-OCT. *See* Anterior segment optical coherence tomography (AS-OCT)
Assisted reproduction technology (ART), 82

B
Band keratopathy, 110–111
BAX, 142
BCL-2 family, 142
Biobank
anonymisation, 240
biological samples, 238
Data Protection Act, 240
Portuguese law, creation and operation, 238
Biochemical properties, amniotic membrane
amniotic epithelium, 21–22
AQPs, 25
avascular stroma, 22
basement membrane, 22
biological structure, 20–21
characteristics, 20
collagen, 23
ECM materials, 23
elastin, 24
fetal membrane, 19
fibronectins, 24
hyaluronic acid (HA), 25
laminins, 24
metabolic functions, 24–25
physiologic processes, 35
prostaglandins, 25
proteoglycans, 23–24
proteomics, 26, 28–36
tissue inhibitors, 25
transcriptomics, 26–28

Biophysics
mechanical properties, 47–55
nutrients transportation, 42–47
pregnancy, 41–42
BM-MSCs. *See* Bone marrow mesenchymal stem cells (BM-MSCs)
Bone marrow mesenchymal stem cells (BM-MSCs), 147
Bovine-derived collagen, 154
Bullous keratopathy, 108
Burn wound dressing
adherence, 130
amnion, 124–125
anti-bacterial properties, 133–134
biological functions, 126
deep (dermal) partial-thickness, 127–128
disadvantages, 134–135
dried amnion, 125
feto-maternal relationship, 134
fluid/water loss, 131
full-thickness burns, 124, 127–128
gamma-irradiation, 125
glycerolization, 125
glycoprotein, 134
graft overlay, 129
hand and facial burns, 128–129
light, thin and transparent, 131
neovascularization, 132
non-fatal burns injuries, 124
pain reduction, 130
radiation-sterilized, 131–132
scar tissue, 134
split-skin graft donor site, 130
superficial (dermal) partial-thickness burns, 124, 126–127
wound healing, 132

C

CA. *See* Chorioamnionitis (CA)
Cancer therapy
angiogenesis (anti-angiogenesis), 142
anti-angiogenic activity, 143–144
anti-carcinogenic properties, 142
BM-MSCs, 147
capacity, 141
carcinogenesis, 140
cause of death in developed countries, 140
cell physiology, 141
characterization, 140–141
DNA-damaging agents, 141
genomic instability, 141
hAECs, 147–148
hallmarks of, 141
hAMCs, 146–148
and immune system, 142
immunoregulatory activity, 144
inflammatory process, 141
intracellular events, 140
pro-apoptotic activity, 144
resistance to apoptosis, 142
resistance to cell death and apoptosis, 142
somatic evolution, 140
stem characteristics, 144–145
Carcinogenesis, 140
CDs. *See* Cyclodextrins (CDs)
Cell sorting, 9, 201–202
Cerebral white matter injury (CWMI), 95, 98
Chitin membranes, 154
Chondrogenic differentiation, 65, 68, 205
Chorioamnionitis (CA)
AIS, 93
amnionitis, 93
FIRS (*see* Fetal inflammatory response syndrome (FIRS))
foetal membranes, 93
funisitis, 94
IL-1 and TNF-α, 94
pro-inflammatory cytokines, 94
risk factors, 93
symptoms, 94
Chorioamniotic membranes
as bag of waters, 4
CA, 93, 94
characteristics, 20
intact placenta and fetus, 12
layers, 19–20
Chorion
and amnion, 8, 13 (*see also* Amnion)
CA (*see* Chorioamnionitis (CA))
cavity, 11
chorion frondosum, 7, 12
fetal membranes, 13, 14
fusion with amnion, 11–12
growth, chorionic villi, 8
layering, chorionic plate, 17
mesoderm, 16
nonvillous chorion, 16
and normal pregnancy (*see* Pregnancy)
smooth chorion/chorion laeve, 8
stromal cells, 17
thickness after separation, 13–14
trophoblastic layer, 17
23 weeks, pregnancy, 12
Chorion frondosum, placenta, 7, 12, 14
Citrate buffer (CB) solution, 202

Conjunctiva
  bulbar, 179
  and corneal ulcer, 107
  production of lipids, 179
  protection of ocular surface, 179
  pterygium, 111
  reconstruction, 105, 112
  squamous nonkeratinized cells, 179
  symblepharon, 111–112
  tarsal, 179
  transport process, 179
Conjunctival applications
  conjunctival reconstruction, 112
  pterygium, 111
  symblepharon, 111–112
Cornea
  bloodandlymphaticvessels, 179
  Bowman'slayer, 179
  stroma, 180
  zonulae occludentes, 179–180
Corneal applications
  band keratopathy, 110–111
  corneal ulcers and perforations, 107–108
  limbal stem cell deficiency, 109–110
  painful bullous keratopathy, 108
Corneal collegenous shell, 180
Corneal stroma derived cells (CSDC), 116
Corneal surface reconstruction, 216, 220
Cryopreservation
  DMEM preservation, 214
  in glycerol, 212–213
  human chondrocytes, 214
  ocular surface reconstruction, 214
CSDC. *See* Corneal stroma derived cells (CSDC)
CWMI. *See* Cerebral white matter injury (CWMI)
Cyclodextrins (CDs), 188

D
Darwin's theory, 140
Data protection
  personal data research, 240
  transplantation
    amniotic membranes, 239–240
    biobanks, 240
    EU law, 239
    health information data, 239
    human tissue, 238
Data Protection Act, 239, 240
Deepithelialization
  ACELAGRAFT™, 220–221
  Bacillus polymyxa, 219
  cell growth, 219
  columnar epithelial cells, 219
  EDTA, 220
  thermolysin, 220
  trypsin, 220
  urea, 220
Differentiation potential, hAMSC
  adipogenic, 205
  cell type's characteristics, 204
  chondrogenic, 205
  neural-like, 206
  osteogenic, 205
DMEM. *See* Dulbecco's Modified Eagle's Medium (DMEM)
Drug administration
  after AMT, 177–178
Drug loaded nanoparticles, 188
Drug penetration
  conjunctiva, 179
  diclofenac, 183
  graft and patch, 182
  lymphatic vessels, 181
  ocular surface, 181
  sclera, 180
  transcorneal and transconjunctival, 185
  ultrastructure, AM, 185
Drug permeability in normal eye
  after AMT (*see* Amniotic membrane transplantation (AMT))
  and bioavailability, 190
  conjunctiva, 179
  cornea, 179–180
  DDS, 187–189
  sclera, 180
  tear film, 178–179
Dulbecco's Modified Eagle's Medium (DMEM), 60, 61, 198, 205, 206, 213, 214
Dysfunctional endometrium, 166

E
Early placenta
  blastocyst, 5
  cell fate decisions, 5
  chorion frondosum, 7
  chorionic villous, 7, 8
  cytotrophoblast, 6
  embryo-fetus and mother, 4–5
  fecundation, 5, 6
  first and third placental villi, 7
  growth, chorionic villi, 8
  molecular mechanisms, 5
  second trimester placenta, 6

smooth chorion/chorion laeve, 8
syncytiotrophoblast, 6–7
trophoectoderm, 5
vasculosyncytial membranes, 7
ECM. *See* Extracellular matrix (ECM)
EDTA. *See* Ethylene diamine tetraacetic acid (EDTA)
Endodontic retreatment surgery, 154
Ethylene diamine tetraacetic acid (EDTA), 205, 215, 219–221
Extracellular matrix (ECM)
cellular migration and proliferation, 20, 225
EDTA, 220
fibronectins, 24
glycoproteins, 22
IL-1 and TNF-α, 94
mimecan, 28
MMP, 80
proteoglycans, 23–24

F
FACS. *See* Fluorescence-activated cell sorting (FACS)
FBS. *See* Fetal bovine serum (FBS)
Fecundation, 5, 6, 10, 11
Female Sexual Function Index (FSFI), 172–173
Fetal bovine serum (FBS), 60, 61, 197, 213
Fetal inflammatory response syndrome (FIRS)
AF contamination, 96
CWMI, 92
intrauterine inflammation, 94
neurological and respiratory outcome, 95–96
Financial gain prohibition, 241–242
FIRS. *See* Fetal inflammatory response syndrome (FIRS)
First-line therapy, cancer, 141
Fluorescence-activated cell sorting (FACS), 201
Franck's dilator method, 171
Franz-diffusion cellsystem, 182–183
Freeze-drying technique, 214–216, 223
FSFI. *See* Female Sexual Function Index (FSFI)
Full-thickness burns, 124, 127–128

G
Gene expression, 22, 70, 81, 199, 203
Glaucoma surgery, 112
Glycerol
air-drying, 222
amniotic membrane, 125, 130
cell viability, 221
cryopreservation, 212–213
DMEM preservation, 214
gamma irradiation, 218
TIMP, 227
tissue swelling, 225
Graft *versus* host disease (GVHD), 196
Guided tissue regeneration (GTR), 157, 158
Gynaecology
AS, 165–170
vagina, 170–173

H
hAC. *See* Human amnion cells (hAC)
hAMSCs. *See* Human amniotic mesenchymal stromal/stem cells (hAMSCs)
Hand and facial burns, 128–129
Heat shock protein 90 (HSP90), 148
Henderson-Hasselbach equation, 186
Hertwig's epithelial root sheath (HERS), 160
HLA-G. *See* Human leukocyte antigen (HLA)-G
HSP90. *See* Heat shock protein 90 (HSP90)
Human amnion cells (hAC), 167
Human amniotic epithelial cells (hAECs)
antibodies, 63
bleomycin-induced lung injury, 71
epidermal growth factor (EGF), 60
hepatic cells, 67–68
in vitro culture, 67
immunofluorescence staining, 59
iPS, 70
isolation protocols, 60
lineage-associated markers, 63
mesenchymal and hematopoietic markers, 63
myogenic lineage, 68
neuronal, glial and oligodendrocyte markers, 67
pancreatic differentiation conditions, 68
phenotype/morphology, 61
self-renewal/clonogenic capacity, 66
spheroids, 61
subpopulations, 64–65
surface markers and transcription factors, 61–62
teratoma formation, 66
Human amniotic membrane
air drying preservation technique, 216
anatomical and vapor barrier, 210
antibiotics combination, 211

Human amniotic membrane (*Cont.*)
  biochemical composition, 211–212
  collection, 210–211
  disease transmission, 215
  epithelial cell layer, 215
  glutaraldehyde, 216–217
  hydrochloride/N-hydroxysuccinimide, 217
  hyper-drying, 216
  placenta, 211
  preservation methods
    cell proliferation, 225–226
    epithelial cells viability, 221–222
    histological and biophysical properties, 222–224
    molecular features, 226–227
  sterilization (*see* Sterilization procedures)
Human amniotic membrane derived-mesenchymal cells (hAMCs), 146–148
Human amniotic mesenchymal stromal/stem cells (hAMSCs)
  angiogenic potential, 69
  cardiomyocytes, 71
  characterization
    immunocytochemistry assay, 202–203
    immunophenotyping, 199–202
    markers, 197
    molecular assays, 203–204
  differentiation, 204–206
  ectodermal lineage, 69
  hepatocytes, 69
  iPS, 69, 70
  isolation and culture, 197, 198
  isolation protocols, 61
  lineage-associated markers, 63–64
  myogenic differentiation, 68
  osteogenic, adipogenic and chondrogenic cell lineages, 68
  phenotypic characterization, 63
  pluripotency marker expression, 62
  self-renewal/clonogenic capacity, 66
  subpopulations, 64–65
  tissue fragments, 60
Human leukocyte antigen (HLA)-G, 83, 144
Human skin allografts, 154
Human tissues, Protection Law
  AMT, 231
  directives, 232–233
  financial gain prohibition, 241–242
  human rights and biomedicine, 232
  indirect protection, 232
  legal protection, 232
  skin transplantation, 231
  technical quality and safety, 242–243
  tissues import/export, 243–244
  WHO guiding principles, 232
Hydroxyprostaglandin dehydrogenase (PGDH), 88

I

Immunocytochemistry assay, 62, 202–203
Immunophenotype
  cell sorting, 201–202
  dead cells and cell debris, 199
  EuroFlow methodology, 199
  flow cytometry, 199
  gene expression, 199
  hAMSC characterization, 199, 201
  Infinicyt Software, 199, 201
  International Society of Cellular Therapy, 199
  monoclonal antibodies, 199, 200
  monocyte-macrophage-like immunophenotype, 64
  MSC characterization, 199, 200
Immunoregulatory activity
  concept, 144
  Fas ligand, 144
  HLA-G, 144
  properties, 142, 143
Induced pluripotent stem (iPS) cells, 69–70
Informed consent, amniotic membranes
  biobank, 237–238
  ethical assessment, research
    Clinical Trials Act, 236
    data collecting, 237
    human tissues and cells, 235
    interventions, 236
  transplantation
    organs/tissue removal, 234
    therapeutic benefit, 234
    tissue characteristics, 234
International Society of Cellular Therapy, 199
Intrauterine adhesions (IUA). *See* Asherman syndrome (AS)
Intrauterine device (IUD), 166
Intra-uterine infection
  CA, 93–94
  chorion-amniotic cavity, 92
  foetal involvement, 94–96
  genital infection, 91
  inflammatory process, 92–93
  perinatal morbidity and mortality, 92
  vaginal and cervical flora, 92
iPS cells. *See* Induced pluripotent stem (iPS) cells
IUD. *See* Intrauterine device (IUD)

L

Limbal stem cell deficiency, 109–110
Lining mucosa, 154
Lyophilization
  amniotic membrane, 214–215
  freeze-dried and air-dryed, 215
  transportation, 214

M

Masticatory mucosa, 154
Matrix metalloproteinases (MMP)
  foetal membranes, 93
  and IL-8, 81
  IL-1 and TNF-α, 94
  production, 20
  regulation, 80
  tissue inhibitors, 25
Mayer-Rokitansky-Küster-Hauser syndrome (MRKH)
  characterization, 170–171
  definition, 170
  genetic defects, 170
  sporadic disease, 170
Mechanical properties
  amniotic fluid, 50
  burst testing, 52–53
  cesarean delivery, 55
  chorion, 50–51
  collagen production, 54
  elastic deformation, 47–48
  gestational age, 55
  Hooke's law, 48
  intrinsicmaterialquantities, 51
  membrane thickness, 50
  non-deformable materials, 48
  physical properties, membranes, 51
  plastic materials, 49
  PPROM, 54
  puncture tests, 53–54
  rheological properties, 48
  soft tissues, 50
  tensilestrength, chorioamnion, 50, 52, 54
  uniaxial tension test, 52
  viscoelastic materials, 49–50
  Young's modulus, 48, 49, 51
Merckel's cartilage, 154
Mesenchymal stem cell growth medium (MSCGM), 197
Mesenchymal stem cells (MSC)
  growth medium, 197
  GVHD, 196
  hAMSC (*see* Human amniotic mesenchymal stromal/stem cells (hAMSCs))
  immunomodulatory properties, 196
  immunophenotypic characterization, 196
  isolation and characterization, 196
  plasticity, 196
Micelles, 189
MMP. *See* Matrix metalloproteinases (MMP)
Molecular assays, 203–204
Mouth, 154, 155
MRKH. *See* Mayer-Rokitansky-Küster-Hauser syndrome (MRKH)
MSC. *See* Mesenchymal stem cells (MSC)
MSCGM. *See* Mesenchymal stem cell growth medium (MSCGM)
Mucosal and skin autografts, 154
Myometrial scoring, 166

N

Neovascularization, 132
Nestin expression, 202–203
Neural-like differentiation, 206
Nitric oxide (NO), 144
Nuclear factor kappa-light-chain-enhancer of activated B cells (NFkB), 144
Nutrients transportation, diffusion process
  amniotic fluid volume, 42
  cell cultures, 47
  difficulties, 47
  diffusion force, 43
  Fick's First Law of diffusion, 44–45
  glucose, 47
  hydrostatic pressure, 45
  lipophilic molecules, 42
  mean velocity, 43
  molecular mobility, 43
  osmotic pressure and water flow, 45–46
  perfusion models, 47
  polyhydramnios, 42
  potential energy, 43
  solute characteristics, 45
  syncytiotrophoblast, 42
  tissue preparations, 47
  water flux, 45

O

Ocular drug delivery systems (DDS)
  bioadhesive preparations, 188
  bioavailability, 187, 188
  improved corneal permeability, 187–188
  improving residence time, eye surface, 187
  liposomes, 189
  micelles, 189
  mucoadhesives, 189

Ocular drug delivery systems (DDS) (*Cont.*)
  nanocapsules, 189
  nanoparticles, 189
  pharmaceutical developments, 189
  polymerics, 189
  residence time, 189
  silver-impregnated amniotic membranes, 189
  site specificity, 187
  sol-gel transition, 188
  viscosity enhancing agents (polymers), 188
  viscous formulations (ointments), 188
Ocular surface diseases, 180–181, 210, 214
Oestrogen-progestin treatment, 167
Ophthalmic pharmacotherapy, 177
Ophthalmology
  AS-OCT, 115–116
  clinical conditions, 107
  conjunctival applications, 111–112
  corneal applications, 107–111
  corneal confocal microscopy, 115–116
  glaucoma surgery, 112
  graft and patch techniques
    blanket-fold, 113
    crossed sutures, limbus to limbus, 114
    cryopreservation, 114
    epithelium, 113, 114
    fibrin glue, 115
    fluffed-up, 113
    Prokera®, 115
  limitations, 116–117
  mechanism of action, 106
  orbit and oculoplastic surgery, 113
  scleral melt, 113
  strabismus, 113
Oral medicine
  apexogenesis, 160–161
  bone surface, 154
  human amniotic membrane, 154
  mouth, embryogenesis of, 154, 155
  mucosal and skin autografts, 154
  oroantral fistula, 159–160
  periapical surgery, 154–159
  periostium, 154
  surgical procedures in oral cavity, 153
Oroantral fistula, 159–160
Osteogenic differentiation, 65, 68, 205

P

PBS. *See* Phosphate-buffered saline (PBS)
PEDF. *See* Pigment epithelium derived factor (PEDF)
Peptide mass fingerprinting (PMF), 28
Periapical surgery
  advantage of cells isolation, 158
  apical margin lesions, 157
  cavity in root apex, 155
  classification, 157
  cystic lesion, 156
  endodontic retreatment, 154
  endodontic surgery, 156
  growth factors, 158–159
  GTR, 157, 158
  histological evaluation, 157
  histomorphometry, 157
  human amniotic membrane, 158
  large apical cystic lesions, 157
  lesions in bone tunnel, 157
  microsurgery, 155–156
  procedures, 156
  protocols, 157
  PRP, 157–158
  tissue regeneration of apical surgery, 156–157
  wound healing, 156
Permeability
  drug (*see* Drug permeability in normal eye)
  membrane, 44, 45, 47
PGDH. *See* Hydroxyprostaglandin dehydrogenase (PGDH)
PGHS. *See* Prostaglandin synthase (PGHS)
Pharmacology, AMT
  application of ointments, 187
  contact time, 185
  formulation of eyedrop, 185
  Henderson-Hasselbach equation, 186
  hydrophilic molecules, 186
  ionisable drugs, 186
  lipophilic molecules, 186
  meibomian gland, 187
  ophthalmic solution, 185
  physico-chemical characteristics, 185
  preservatives, 187
  protein binding and metabolism, 185
  tear film, 187
  transcorneal and transconjunctival drug penetration, 185
  viscosity, 186
Phosphate-buffered saline (PBS), 197, 214, 215
Pigment epithelium derived factor (PEDF), 143
Placental membranes
  amnion (*see* Amnion)
  as chorioamniotic membranes, 4
  chorion (*see* Chorion)
  early placenta, 4–8

nutrient transport functions (*see* Nutrients transportation, diffusion process)
Platelet-rich plasma (PRP), 157–158
PMF. *See* Peptide mass fingerprinting (PMF)
PPROM. *See* Prevent the preterm premature rupture of membranes (PPROM)
Pregnancy
amniochorion, 80
amnion and chorion layers, 78–79
decidual leukocytes, populations, 79–80
foetal membranes, 78
immune regulation (*see* Reproductive immunology)
inflammatory process, 84–85
labour, inflammatory mechanisms
arachidonic acid, 87–88
chorion PGDH, 88
cytokines and chemokines, 87
glucocorticoids, 88
parturition, physiological process, 87
PGHS activity, 88
progesterone, 86
uterine activation, 86
membrane formation and degradation
apoptosis, 81
hormonal factors, 81
infection, 81
mechanical factors, 82
membrane apposition, defects, 81–82
MMP, 80, 81
oxidative stress, 82
mesenchymal cells, 78
perinatal infection and prematurity, 78
spiral arteries' remodelling, 80
Preterm birth (PTB)
bacteria, 90
CA, 92, 93
causes, 89
inflammatory response, 91
pathogenic agents, 89
pathways, 89, 90
perinatal mortality, 89
polymorphisms, 90
relaxin, 81
TLRs, 90–91
Prevent the preterm premature rupture of membranes (PPROM)
chorion-decidual interface, 81
decidual relaxin, 81
gestational membrane, 54
hydramnios, 82
mechanisms, 81
PTB, 89, 93
Pro-apoptotic activity
concept, 144
interferon-g, 144
properties, 142, 143
Prostaglandin synthase (PGHS), 88
Proteome
cytoplasmic proteins, 26
epithelium-denuded, 28
membrane proteins, 29, 36
PMF, 28
soluble proteins, 29–35
THBS1, mimecan, βIG-H3 and integrin α6, 26, 28
PRP. *See* Platelet-rich plasma (PRP)
PTB. *See* Preterm birth (PTB)
Pterygium, 111
Pulp cavity, 154, 155
Pulp regeneration, 160

R
Reproductive immunology
cytokines, 84
decidualization, 84
development, 82
embryo, 84
HLA-G expression, 83
interaction zones, 84
maternal-foetal tolerance, 83
oocyte donation, ART, 82
trophoblast functions, 83–84
Roswell Park Memorial Institute medium (RPMI), 197, 214
RPMI. *See* Roswell Park Memorial Institute medium (RPMI)

S
Sclera, 112, 113, 180, 223
Scleral melt, 113
Sigmoidal colpoplasty, 171
Specialized mucosa, 154
Stem cells
amniotic epithelium, 59–60
anti-cancer effect, 145
cytotrophoblast, 6
definition, 58–59
fetal development, 58
fibroblast/stromal layer, 60
gastrulation, 58
GTR, 157
hAECs (*see* Human amniotic epithelial cells (hAECs))
hAMSCs (*see* Human amniotic mesenchymal stromal/stemcells (hAMSCs))

Stem cells (*Cont.*)
 limbal, 109–110, 185
 maternal womb, 59
 pluripotency markers, 61–62
 tumor-homing, 145
Stem cells tumor-homing, 145
Stemness, cell population
 in vitro differentiation, 67–69
 in vivo pluripotency, 66–67
 in vivo studies, 70–72
 reprogramming, 69–70
 self-renewal/clonogeniccapacity, 65–66
 teratoma formation, 66–67
Sterilization procedures
 chemical solutions, 218–219
 gamma irradiation, 217–218
 microbiological quality control, 217
Stevens-Johnson syndrome, 109, 111, 181
Strabismus, 113
Superficial (dermal) partial-thickness burns, 124, 126–127
Symblepharon, 111–112

T

Tear film, 106, 178–179
Thrombospondin (THBS1), 26, 28
Tissue inhibitors of metalloproteinase polypeptides (TIMPs), 20, 25, 80, 158, 227
Toll-like receptors (TLRs), 90–91, 94
TP53 tumor suppressor gene, 142
Transforming growth factor beta (TGF-β), 20, 106, 147, 158
Trehalose, 215–216
Tumor necrosis factor alpha (TNF-α), 87, 91, 94, 144, 147
Tumor necrosis factor beta (TNF-β), 147

V

Vagina
 Abbe-McIndoe operation, 171
 AM application
  FSFI, 172–173
  gestations of healthy women, 172
  mean operating time, 172
  neovagina length and vaginal cavity, 172
  squamous epithelium, 171
  transverse incision, 172
  urinary tract infection, 172
 differential diagnosis, 171
 Franck's dilator method, 171
 imaging technique, 171
 management, 171
 MRKH (*see* Mayer-Rokitansky-Küster-Hauser syndrome (MRKH))
 septum and imperforate hymen, 171
 sigmoidal colpoplasty, 171
 Veccchietti operation, 171
Vaginoplasty, 172–173. *See also* Vagina
Veccchietti operation, 171

Z

Zonulae occludentes, 178–179

MIX
Papier aus verantwortungsvollen Quellen
Paper from responsible sources
FSC® C105338
FSC
www.fsc.org

If you have any concerns about our products,
you can contact us on
ProductSafety@springernature.com

In case Publisher is established outside the EU,
the EU authorized representative is:
Springer Nature Customer Service Center GmbH
Europaplatz 3, 69115 Heidelberg, Germany

Printed by Libri Plureos GmbH
in Hamburg, Germany